D1600334

ENDANGERED AND THREATENED FISHES IN THE KLAMATH RIVER BASIN

CAUSES OF DECLINE AND STRATEGIES FOR RECOVERY

Committee on Endangered and Threatened Fishes
in the Klamath River Basin

Board on Environmental Studies and Toxicology

Division on Earth and Life Studies

NATIONAL RESEARCH COUNCIL
OF THE NATIONAL ACADEMIES

THE NATIONAL ACADEMIES PRESS
Washington, DC
www.nap.edu

THE NATIONAL ACADEMIES PRESS 500 Fifth Street, N.W. Washington, DC 20001

NOTICE: The project that is the subject of this report was approved by the Governing Board of the National Research Council, whose members are drawn from the councils of the National Academy of Sciences, the National Academy of Engineering, and the Institute of Medicine. The members of the committee responsible for the report were chosen for their special competences and with regard for appropriate balance.

This project was supported by Grant 98210-1-G092 between the National Academy of Sciences and the U.S. Department of the Interior and the U.S. Department of Commerce. Any opinions, findings, conclusions, or recommendations expressed in this publication are those of the authors and do not necessarily reflect the view of the organizations or agencies that provided support for this project.

Library of Congress Cataloging-in-Publication Data

Endangered and threatened fishes in the Klamath River Basin : causes of decline and strategies for recovery.
p. cm.
Includes bibliographical references (p.).
ISBN 0-309-09097-0 (hardcover) — ISBN 0-309-52808-9 (pdf)
1. Rare fishes—Klamath River Watershed (Or. and Calif.) 2. Fishes—Conservation—Klamath River Watershed (Or. and Calif.) I. National Academies Press (U.S.).
QL617.73.U6E53 2004
333.95′68′09795—dc22
2004001241

Additional copies of this report are available from:

The National Academies Press
500 Fifth Street, NW
Box 285
Washington, DC 20055

800-624-6242
202-334-3313 (in the Washington metropolitan area)
http://www.nap.edu

Printed in the United States of America.

THE NATIONAL ACADEMIES

Advisers to the Nation on Science, Engineering, and Medicine

The **National Academy of Sciences** is a private, nonprofit, self-perpetuating society of distinguished scholars engaged in scientific and engineering research, dedicated to the furtherance of science and technology and to their use for the general welfare. Upon the authority of the charter granted to it by the Congress in 1863, the Academy has a mandate that requires it to advise the federal government on scientific and technical matters. Dr. Bruce M. Alberts is president of the National Academy of Sciences.

The **National Academy of Engineering** was established in 1964, under the charter of the National Academy of Sciences, as a parallel organization of outstanding engineers. It is autonomous in its administration and in the selection of its members, sharing with the National Academy of Sciences the responsibility for advising the federal government. The National Academy of Engineering also sponsors engineering programs aimed at meeting national needs, encourages education and research, and recognizes the superior achievements of engineers. Dr. Wm. A. Wulf is president of the National Academy of Engineering.

The **Institute of Medicine** was established in 1970 by the National Academy of Sciences to secure the services of eminent members of appropriate professions in the examination of policy matters pertaining to the health of the public. The Institute acts under the responsibility given to the National Academy of Sciences by its congressional charter to be an adviser to the federal government and, upon its own initiative, to identify issues of medical care, research, and education. Dr. Harvey V. Fineberg is president of the Institute of Medicine.

The **National Research Council** was organized by the National Academy of Sciences in 1916 to associate the broad community of science and technology with the Academy's purposes of furthering knowledge and advising the federal government. Functioning in accordance with general policies determined by the Academy, the Council has become the principal operating agency of both the National Academy of Sciences and the National Academy of Engineering in providing services to the government, the public, and the scientific and engineering communities. The Council is administered jointly by both Academies and the Institute of Medicine. Dr. Bruce M. Alberts and Dr. Wm. A. Wulf are chair and vice chair, respectively, of the National Research Council.

www.national-academies.org

COMMITTEE ON ENDANGERED AND THREATENED FISHES IN THE KLAMATH RIVER BASIN

Members

WILLIAM M. LEWIS, JR. (*Chair*), University of Colorado, Boulder
RICHARD M. ADAMS, Oregon State University, Corvallis
ELLIS B. COWLING, North Carolina State University, Raleigh
EUGENE S. HELFMAN, University of Georgia, Athens
CHARLES D. D. HOWARD, Consulting Engineer, Victoria, British Columbia, Canada
ROBERT J. HUGGETT, Michigan State University, East Lansing
NANCY E. LANGSTON, University of Wisconsin, Madison
JEFFREY F. MOUNT, University of California, Davis
PETER B. MOYLE, University of California, Davis
TAMMY J. NEWCOMB, Michigan Department of Natural Resources, Lansing
MICHAEL L. PACE, Institute of Ecosystem Studies, Millbrook, NY
J.B. RUHL, Florida State University, Tallahassee

Staff

SUZANNE VAN DRUNICK, Project Director
DAVID J. POLICANSKY, Associate Director and Senior Program Director for Applied Ecology
NORMAN GROSSBLATT, Senior Editor
KELLY CLARK, Assistant Editor
MIRSADA KARALIC-LONCAREVIC, Research Assistant
BRYAN P. SHIPLEY, Research Assistant

Sponsors

U.S. DEPARTMENT OF THE INTERIOR
U.S. DEPARTMENT OF COMMERCE

BOARD ON ENVIRONMENTAL STUDIES AND TOXICOLOGY[1]

[1]This study was planned, overseen, and supported by the Board on Environmental Studies and Toxicology.

OTHER REPORTS OF THE BOARD ON ENVIRONMENTAL STUDIES AND TOXICOLOGY

Cumulative Environmental Effects of Alaska North Slope Oil and Gas Development (2003)
Estimating the Public Health Benefits of Proposed Air Pollution Regulations (2002)
Biosolids Applied to Land: Advancing Standards and Practices (2002)
Ecological Dynamics on Yellowstone's Northern Range (2002)
The Airliner Cabin Environment and Health of Passengers and Crew (2002)
Arsenic in Drinking Water: 2001 Update (2001)
Evaluating Vehicle Emissions Inspection and Maintenance Programs (2001)
Compensating for Wetland Losses Under the Clean Water Act (2001)
A Risk-Management Strategy for PCB-Contaminated Sediments (2001)
Acute Exposure Guideline Levels for Selected Airborne Chemicals (3 volumes; 2000–2003)
Toxicological Effects of Methylmercury (2000)
Strengthening Science at the U.S. Environmental Protection Agency (2000)
Scientific Frontiers in Developmental Toxicology and Risk Assessment (2000)
Ecological Indicators for the Nation (2000)
Modeling Mobile-Source Emissions (2000)
Waste Incineration and Public Health (1999)
Hormonally Active Agents in the Environment (1999)
Research Priorities for Airborne Particulate Matter (4 volumes, 1998–2003)
Ozone-Forming Potential of Reformulated Gasoline (1999)
Arsenic in Drinking Water (1999)
The National Research Council's Committee on Toxicology: The First 50 Years (1997)
Carcinogens and Anticarcinogens in the Human Diet (1996)
Upstream: Salmon and Society in the Pacific Northwest (1996)
Science and the Endangered Species Act (1995)
Wetlands: Characteristics and Boundaries (1995)
Biologic Markers (5 volumes, 1989–1995)
Review of EPA's Environmental Monitoring and Assessment Program (3 volumes, 1994–1995)
Science and Judgment in Risk Assessment (1994)
Pesticides in the Diets of Infants and Children (1993)
Dolphins and the Tuna Industry (1992)

Science and the National Parks (1992)
Human Exposure Assessment for Airborne Pollutants (1991)
Rethinking the Ozone Problem in Urban and Regional Air Pollution (1991)
Decline of the Sea Turtles (1990)

Copies of these reports may be ordered from the National Academies Press
(800) 624-6242 or (202) 334-3313
www.nap.edu

Acknowledgments

This project was supported by the U.S. Bureau of Reclamation (USBR), the U.S. Fish and Wildlife Service (USFWS), and the National Marine Fisheries Service (NMFS).

Many people assisted the committee and National Research Council staff in creating this report. We are grateful for the support provided by the following:

Pablo Arroyave, U.S. Bureau of Reclamation
Edward Bartell, Water for Life Foundation
John Bartholow, U.S. Geological Survey
Michael Belchik, Yurok Tribe
Antonio Bentivoglio, U.S. Fish and Wildlife Service
Gary Black, Siskiyou Resource Conservation District
Randy Brown, U.S. Fish and Wildlife Service
Mark Buettner, U.S. Fish and Wildlife Service
Donald Buth, University of California, Los Angeles
James Carpenter, Carpenter Design Inc.
William Chesney, California Department of Fish and Game
Paul Cleary, Oregon Department of Water Resources
David Cottingham, Marine Mammal Commission
John Crawford, Tule Lake Irrigation District
Earl Danosky, Tule Lake Irrigation District
Michael Deas, Watercourse Engineering Inc.
Thomas Dowling, Arizona State University
Larry Dunsmoor, Klamath Tribes Natural Resources

John Fay, U.S. Fish and Wildlife Service
Mary Freeman, U.S. Geological Survey
Thomas Hardy, Utah Water Research Laboratory
William Hogarth, National Marine Fisheries Service
Becky Hyde, Yainex Ranch Owner
Cecil Jennings, U.S. Geological Survey
Marshall Jones, U.S. Fish and Wildlife Service
John Keys, III, U.S. Bureau of Reclamation
Jacob Kann, Aquatic Ecosystem Sciences LLC
Steve Kirk, Oregon Department of Environmental Quality
Don Knowles, National Marine Fisheries Service
Ron Larson, U.S. Fish and Wildlife Service
James Lecky, National Marine Fisheries Service
Steven Lewis, U.S. Fish and Wildlife Service
Loren Little, Modoc Irrigation District
Douglas Markle, Oregon State University
Graham Matthews, Graham Matthews & Associates
Martin Miller, U.S. Fish and Wildlife Service
Ken Maurer, Scott Valley resident
David Mauser, U.S. Fish and Wildlife Service
Frank McCormick, U.S. Forest Service
Chris Mobley, National Marine Fisheries Service
Curt Mullis, U.S. Fish and Wildlife Service
Joseph Nelson, University of Alberta
Roger Nicholson, Fort Klamath Rancher
Todd Olson, PacifiCorp
Felice Pace, Klamath Forest Alliance
Ronnie Pierce, Karuk Tribe
Richard Raymond, E&S Environmental Chemistry Inc.
Donald Reck, National Marine Fisheries Service
Michael Rode, California Department of Fish and Game
Kimball Rushton, Iron Gate Hatchery
Michael Ryan, U.S. Bureau of Reclamation
Ken Rykbost, Oregon State University Klamath Experiment Station
David Sabo, U.S. Bureau of Reclamation
Gary Scoppettone, U.S. Geological Survey
Tom Shaw, U.S. Fish and Wildlife Service
Rip Shively, U.S. Geological Survey
Daniel Snyder, U.S. Geological Survey
David Solem, Klamath Irrigation District
Sari Sommarstrom, Consultant and President, California Watershed
Management Council

Glen Spain, Pacific Coast Federation of Fisherman's Associations, Institute for Fisheries Resources
Marshall Staunton, Upper Klamath Basin Working Group
Mark Stern, The Nature Conservancy
Ronald Sutton, U.S. Bureau of Reclamation
Doug Tedrick, U.S. Bureau of Indian Affairs
Larry Todd, U.S. Bureau of Reclamation
Greg Tranah, Harvard School of Public Health
Manuel Ulibarri, Dexter National Fish Hatchery and Technology Center
Carl Ullman, Klamath Tribes
David Vogel, Natural Resource Scientists, Inc.
Nancy Vucinich, Pyramid Lake Fisheries
David Walters, U.S. Environmental Protection Agency
Wedge Watkins, U.S. Bureau of Land Management
David Webb, Shasta Coordinated Resources Management and Planning
Faye Weekley, U.S. Fish and Wildlife Service
Thomas Weimer, U.S. Department of the Interior
Sam Williamson, U.S. Geological Survey
Sue Ellen Wooldridge, U.S. Department of the Interior

The committee's work also benefited from written and oral testimony submitted by the public, whose participation is much appreciated.

Acknowledgment of Review Participants

This report has been reviewed in draft form by persons chosen for their diverse perspectives and technical expertise in accordance with procedures approved by the National Research Council Report Review Committee. The purpose of this independent review is to provide candid and critical comments that will assist the institution in making its published report as sound as possible and to ensure that the report meets institutional standards of objectivity, evidence, and responsiveness to the study charge. The review comments and draft manuscript remain confidential to protect the integrity of the deliberative process. The committee and the NRC thank the following for their review of this report:

Don Chapman, consultant, McCall, Idaho
Jeff Curtis, Trout Unlimited
Kurt Fausch, Colorado State University
Wilford Gardner, University of California, Berkeley
Stanley Gregory, Oregon State University
Roger Kasperson, Stockholm Environment Institute
James Kitchell, University of Wisconsin
Mark Stern, The Nature Conservancy
David Vogel, Natural Resource Scientists, Inc.
Robert Wetzel, University of North Carolina, Chapel Hill
M. Gordon Wolman, Johns Hopkins University

Although the reviewers listed above have provided many constructive comments and suggestions, they were not asked to endorse the conclusions or recommendations, nor did they see the final draft of the report before its release. The review of this report was overseen by Paul Risser, Oregon State University, and Stephen Berry, University of Chicago. Appointed by the National Research Council, they were responsible for making certain that an independent examination of this report was carried out in accordance with institutional procedures and that all review comments were carefully considered. Responsibility for the final content of this report rests entirely with the committee and the National Research Council.

Preface

The federal Endangered Species Act (ESA) of the United States has the admirable goal of minimizing extinction rates through regulations and actions that are intended to produce recovery of species that are in critical decline. For any given species listed under the act, agencies implementing the ESA must choose from an immense array of possibilities the ones most likely to lead to recovery, and in doing so they must forego the luxury of an extended interval of monitoring or experimentation.

Remedies for the recovery of species often have harmful or at least frustrating effects on people and institutions. In such instances, the affected parties often are especially dissatisfied with the implementation of remedies that are not absolutely secure scientifically. But the ESA does not allow delay, which would defeat its purpose. Thus, some of the remedies prescribed by agencies ultimately will prove ineffective and may cause economic or social disruption without any tangible benefit to listed species.

The National Research Council's Committee on Endangered and Threatened Fishes in the Klamath River Basin deals in its final report with three Klamath basin fish species listed under the federal ESA. The committee's work is broad in that it encompasses the entire actual or potential range of those species in the Klamath basin, regardless of the boundaries set by ownership or management, and with all the potential environmental changes that could suppress or promote the welfare of the species. The committee, in response to its charge, has given particular attention to evaluation of the certainty underlying specific kinds of remedies that might lead to the recovery of species. The issues that the committee has dealt with are specific to the basin, but the Klamath basin presents in microcosm most of the problems that are generally identified with implementation of the ESA. Especially prominent in the Klamath basin is controversy over the extent to which remedies that have uncertain outcomes should be pursued even though they are economically or socially painful.

One issue especially well highlighted by the Klamath basin is the relative weight that should be given to professional judgment as opposed to direct empirical evidence that appears to be contradictory to that judgment. Whereas professional judgment is essential for successful ESA implementations where site-specific information is absent, its use is more problematic when initial judgments fail empirical tests. Reversal of an initial judgment may seem to be an abandonment of duty or principle, but it is unrealistic to expect that all initial judgments will be proved scientifically sound. By raising this issue in specific terms in its interim report, the committee has generated considerable controversy in the Klamath basin. The committee believes, however, that a rational and consistent resolution of the issue works toward the long-term stability and effectiveness of the ESA. The committee's final report gives a more detailed view of the committee's approach.

The committee owes a great debt of gratitude to the National Research Council staff members who have guided it through the production of the final report. Suzanne van Drunick, project director, has been especially critical to the success of the committee; David Policansky, James Reisa, and Bryan Shipley also helped the committee in numerous ways; Norman Grossblatt, Mirsada Karalic-Loncarevic, and Kelly Clark helped with the many details that made the report ready for publication. The committee is also appreciative of James MacMahon and other board members for their oversight of this study. The committee is grateful to Leslie Northcott of the University of Colorado for helping to produce the manuscript of the report and to Marylee Murphy and Rebecca Anthony of the University of Colorado for their work on figures and tables.

The committee benefited immensely from the help and advice of scientists and administrators who have dealt with environmental issues in the Klamath basin and to contributions from the citizens, organizations, and tribes working and living in the basin. The committee's highest hope is that its work will be a contribution to the long-term general welfare of everyone who resides in, visits, or cares about the Klamath basin.

The National Research Council process for producing the report involves extensive reliance on external reviewers. The committee thanks the reviewers of its final report for their thoughtful contributions.

William M. Lewis, Jr., *Chair*
Committee on Endangered and Threatened
Fishes in the Klamath River Basin

Contents

Box, Figures, and Tables

BOX

FIGURES

TABLES

Summary

Two endemic fishes of the upper Klamath basin (Figure S-1), the shortnose sucker (*Chasmistes brevirostris*) and the Lost River sucker (*Deltistes luxatus*), were listed as endangered under the federal Endangered Species Act (ESA) in 1988 by the U.S. Fish and Wildlife Service (USFWS). USFWS cited overfishing, water management, habitat alteration, nonnative species, poor water quality, and several other factors as likely contributors to the decline of the fishes, which once were very abundant. In 1997, the Southern Oregon Northern California Coast "evolutionarily significant unit" of coho salmon (*Oncorhynchus kisutch*), which is native to the Klamath basin and several adjacent drainages, was listed by the National Marine Fisheries Service (NMFS) as threatened under the ESA. NMFS cited water management, water quality, loss of habitat, overfishing, and several other potential causes of decline for the coho salmon.

In 2001, in response to biological assessments prepared by the U.S. Bureau of Reclamation (USBR), the two listing agencies issued biological opinions that required USBR to take numerous actions, including maintenance of higher water levels in Upper Klamath Lake and two reservoirs on the Lost River and higher flow of the Klamath River below Iron Gate Dam. Release of the two biological opinions coincided with a severe drought. Because of the new biological opinions and the drought, USBR was prohibited from releasing large amounts of water to farmers served by its Klamath Project, which diverts waters from Upper Klamath Lake and the upper Lost River for use in irrigation through USBR's Klamath Project. The unexpected restrictions on water supply, which severely impaired or eliminated

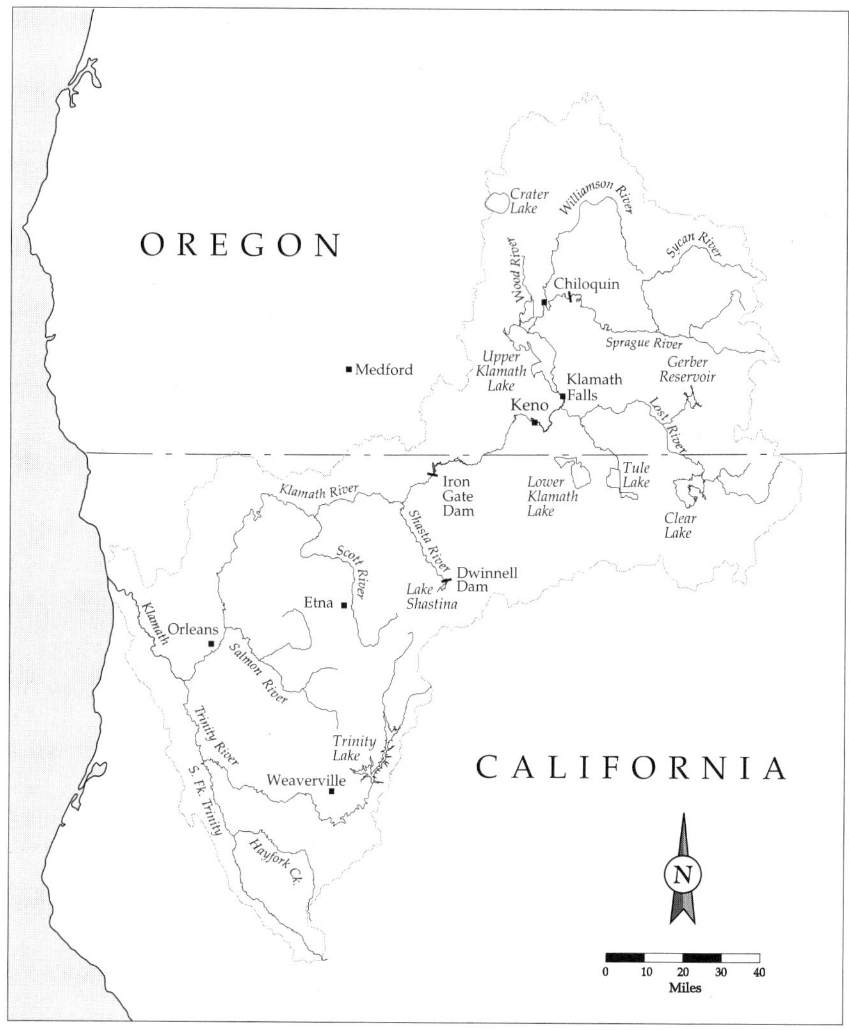

FIGURE S-1 Map of the Klamath River basin showing surface waters and landmarks. Source: Modified from USFWS.

agricultural production on the 220,000 acres irrigated by the Klamath Project, caused agricultural water users and others to question the basis for water restrictions, while other parties, fearing adverse effects of the Klamath Project on the endangered and threatened fishes, supported the restrictions.

In late 2001, the Department of the Interior and the Department of Commerce asked the National Academies to form a committee (the Committee on Endangered and Threatened Fishes in the Klamath River Basin) to evaluate the strength of scientific support for the biological assessments and biological opinions on the three listed species, and to identify requirements for recovery of the species. The committee was charged to complete an interim report in early 2002, focusing on effects of the Klamath Project, and to complete a final report in 2003 that would take a broad view of the scientific aspects of the continued survival of the listed species (Box S-1). This is the committee's final report.

In its interim report of February 2002, the committee found substantial scientific support for all recommendations made by the two listing agencies for the benefit of the endangered and threatened species, except for recommendations requiring more stringent controls over water levels in Upper Klamath Lake and flows at Iron Gate Dam. The committee also noted, however, that USBR had not provided any substantial scientific support for its own proposal of revised operating procedures, which might have led to lower mean water levels or lower minimum flows.

In 2002, USBR issued a new biological assessment that dealt with the two endangered sucker species and the threatened coho salmon. In response, USFWS prepared another biological opinion on the suckers, and NMFS prepared another biological opinion on the coho salmon. These documents reflect a closer interaction between the agencies than in previous years. USBR moved toward more restrictive operational practices than it had previously proposed and toward development of reserve water supplies; USFWS and NMFS were more cautious in requiring actions whose basis would be contradicted by site-specific studies, and they acknowledged the need to consult with parties in addition to USBR. The biological assessment and the two biological opinions for 2002 cover a 10-yr interval (2002–2012), during which time the listing agencies may require additional consultation and may revise their biological opinions.

PRINCIPAL FINDINGS OF THE COMMITTEE

Lost River and Shortnose Suckers

Upper Klamath Lake

Although suckers of all age classes are present in Upper Klamath Lake, population densities of suckers are low, and there are no signs that the populations are returning to their previously high abundance.

Suckers spawn in tributaries to Upper Klamath Lake, but they are blocked from much potentially suitable spawning habitat by Chiloquin Dam on the Sprague River (Figure S-1). Numerous smaller blockages and diversions also are present but are poorly documented. Expansion of spawning on the Sprague River could increase the abundance of fry descending to

Box S-1. Statement of Task

The committee will review the government's biological opinions regarding the effects of Klamath Project operations on species in the Klamath River Basin listed under the Endangered Species Act, including coho salmon and shortnose and Lost River suckers. The committee will assess whether the biological opinions are consistent with the available scientific information. It will consider hydrologic and other environmental parameters (including water quality and habitat availability) affecting those species at critical times in their life cycles, the probable consequences to them of not realizing those environmental parameters, and the interrelationship of these environmental conditions necessary to recover and sustain the listed species.

To complete its charge, the committee will:

1. Review and evaluate the science underlying the Biological Assessments (USBR 2001a,b) and Biological Opinions (USFWS 2001; NMFS 2001).
2. Review and evaluate environmental parameters critical to the survival and recovery of listed species.
3. Identify scientific information relevant to evaluating the effects of project operations that has become available since USFWS and NMFS prepared the biological opinions.
4. Identify gaps in the knowledge and scientific information that are needed to develop comprehensive strategies for recovering listed species and provide an estimate of the time and funding it would require.

A brief interim report will be provided by January 31, 2002. The interim report will focus on the February 2001 biological assessments of the Bureau of Reclamation and the April 2001 biological opinions of the U.S. Fish and Wildlife Service and National Marine Fisheries Service regarding the effects of operations of the Bureau of Reclamation's Klamath Project on listed species. The committee will provide a preliminary assessment of the scientific information used by the Bureau of Reclamation, the Fish and Wildlife Service, and the National Marine Fisheries Service, as cited in those documents, and will consider to what degree the analysis of effects in the biological opinions of the Fish and Wildlife Service and National Marine Fisheries Service is consistent with that scientific information. The committee will identify any relevant scientific information it is aware of that has become available since the Fish and Wildlife Service and National Marine Fisheries Service prepared the biological opinions. The committee will also consider any other relevant scientific information of which it is aware.

The final report will thoroughly address the scientific aspects related to the continued survival of coho salmon and shortnose and Lost River suckers in the Klamath River Basin. The committee will identify gaps in the knowledge and scientific information that are needed and provide approximate estimates of the time and funding needed to fill those gaps, if such estimates are possible. The committee will also provide an assessment of scientific considerations relevant to strategies for promoting the recovery of listed species in the Klamath Basin.

Upper Klamath Lake and would beneficially extend the interval over which they arrive at the lake.

The water quality of the tributaries to Upper Klamath Lake is poor for some native fishes but is probably adequate for the listed suckers. The tributaries do, however, show loss of riparian vegetation and wetland (largely due to agricultural practices), which could adversely influence the survival of fry. The physical condition of channels in general and spawning areas in particular is degraded, but the nature and extent of degradation is poorly documented for the tributaries.

Endangered suckers also spawn near springs that emerge at the margin of Upper Klamath Lake. Some apparently suitable spawning sites are no longer used, probably because entire groups of fish that used the sites were eliminated during the era of fishing, which ended in 1987. Lakeside spawning behaviors are associated with a specific range of depth. During dry years, the amount of appropriate spawning substrate with appropriate water depth is reduced by drawdown of the lake. Data on year-class strength show no indication of a relationship between year-class strength and water level, which might be expected if drawdown were strongly suppressing production of fry.

Fry are strongly dispersed from their points of origin by currents and ultimately are found in shallow water in or near emergent vegetation at the margins of the lake. Loss of such vegetation, especially near the tributary mouths, could be disadvantageous to the fry. The area around the lake associated with preferred depths and presence of emergent vegetation varies with water level; drawdown, especially in dry years, reduces this area. Standardized sampling of fry and studies of year-class strength for large fish do not, however, indicate associations between water level and abundance of larvae.

Juveniles seek somewhat deeper water than larvae. There is substantial juvenile mortality, but current information is insufficient to show whether it is extraordinary in comparison with mortality in other lakes that have more favorable living conditions.

Subadult and adult fish seek deeper water than younger fish and congregate in specific areas of Upper Klamath Lake. In contrast to the tributaries, poor water quality in the lake itself appears to be their greatest vulnerability. Direct evidence of harm to large fish by poor water quality includes physical indications of stress and mass mortality of large fish ("fish kills") at times of exceptionally poor water quality.

Mass mortality of large fish occurs during the second half of the growing season, but not in all years. Upper Klamath Lake is hypertrophic (extremely productive) because its rich supplies of phosphorus lead to extreme abundance of phytoplankton dominated by *Aphanizomenon flos-aquae*, a nitrogen-fixing bluegreen (cyanobacterial) alga. High abundances of

Aphanizomenon induce high pH through high rates of photosynthesis. Although strong algal blooms of this type occur each year, conditions for mass mortality are associated with a specific sequence of weather events involving calm weather succeeded by windy weather.

Low concentrations of dissolved oxygen probably are the immediate cause of death of endangered suckers during episodes of mass mortality, but other water-quality factors may contribute to stress. Mass mortality of large fish in Upper Klamath Lake has occurred for many decades, but anthropogenic factors, especially those leading to strong dominance of *Aphanizomenon*, probably have increased its severity and frequency. Poor water quality may also challenge the sucker populations in other ways. High pH, for example, could be harmful to young fish even if they are not subject to the mass mortality of larger fish.

Because hypertrophic conditions indicate very high supplies of phosphorus, much attention has been given to the possibility of reducing the phosphorus load passing from the watershed to Upper Klamath Lake. The prospects for suppressing algal blooms by this means in Upper Klamath Lake seem poor, however, because about 60% of the external phosphorus load is derived from natural sources. In addition, the anthropogenic component of load is accounted for by dispersed sources, which are difficult to control, and the internal load (phosphorus released from lake sediments) is about double the external load.

The key change over the last 50 yr in Upper Klamath Lake probably was the rise of *Aphanizomenon*, which replaced diatoms as the dominant type of algae. Diatoms probably were limited by nitrogen depletion and thus were unable to use fully the rich phosphorus supplies of the lake, whereas *Aphanizomenon* is able to fix nitrogen and thus can fully exploit the high availability of phosphorus, which causes it to reach very high abundances. Various anthropogenic factors could have contributed to the rise of *Aphanizomenon*; one example is increased transparency of the lake caused by disconnection of its associated wetlands, which were sources of dark humic compounds. Reestablishment of these sources would seem advisable but may be impractical because the organic deposits in the wetlands oxidized extensively after the wetlands were drained.

There is no evidence of a causal connection between water level and water quality or fish mortality over the broad operating range in the 1990s, the period for which the most complete data are available for Upper Klamath Lake. Neither mass mortality of fish nor extremes of poor water quality shows any detectable relationship to water level. Thus, despite theoretical speculations, there is no basis in evidence for optimism that manipulation of water levels has the potential to moderate mass mortality of suckers in Upper Klamath Lake. Planning must anticipate that poor water quality will continue to affect the sucker populations of Upper Klamath Lake.

Suckers in Upper Klamath Lake also are affected by entrainment from the Link River near the outflow of the lake. Screens installed at the main irrigation-water withdrawal point probably will be beneficial, but loss of small fish still can be expected. The Link River Dam intakes still are not screened.

Nonnative fishes, which are diverse and abundant in Upper Klamath Lake, may be suppressing the populations of endangered suckers there, but no practical mechanisms for reducing their abundance are known.

Other Locations in the Klamath Basin

Below Upper Klamath Lake, waters of the upper basin collect through the Lost River system, which is regulated by the Klamath Project (Figure S-1). The headwaters include tributaries to Clear Lake and Gerber Reservoir. These tributaries support recurrently successful spawning of endangered suckers, as shown by the apparently stable populations of suckers in the two reservoirs. Unprecedented drawdown of both reservoirs in the drought year of 1992 coincided with deteriorating body condition and increased incidence of parasitism in the suckers. Thus, the conditions of 1992 have been used by USFWS in setting thresholds of water level for these lakes.

On the Lost River below Gerber Reservoir and Clear Lake (Figure S-1), all waters are strongly affected by the Klamath Project and are unsuitable for suckers, although they still offer some opportunities for restoration, especially through increases in water depth for Tule Lake Sumps and Lower Klamath Lake.

Reservoirs of the main stem Klamath have created new habitat capable of holding endangered suckers, but recruitment of young fish has not been observed. Reservoirs have low potential to support self-sustaining populations.

Coho Salmon

The peak migration of adult coho salmon in the Klamath basin occurs between late October and mid-November; the fish spawn primarily in tributaries. Fry reach peak abundance in March and April, and can disperse as far as the tributary mouths, but most appear to stay close to the areas where they originate. Coho develop through the juvenile stage in the tributaries over about 1 yr. They may occupy the main stem at times but are nearly absent from it by late summer, when the water is warmest. Winter habitat in the tributaries is critical for the juvenile coho but has not been well studied.

Juveniles smoltify and migrate downstream in spring, with a peak in April. Short transit times facilitated by high flow could be favorable to the migrating smolts, although this has not been demonstrated for the Klamath River. Smolts spend approximately 1 mo in the estuary and then enter the ocean, where they spend about 1.5 yr before returning to the Klamath River. Ocean conditions such as productivity affect the strength of year classes.

The most important cause of impairment of coho salmon probably is excessively high summer temperatures in tributary waters. Coho salmon, unlike Chinook salmon, remain in freshwater for an entire year, during which they mainly occupy tributaries, where summer water temperatures can be dangerously high. Causes of extreme temperatures include diversion of cold flows for use in agriculture, flow depletion that leads to warming of cool water, and destruction of riparian vegetation that leads to loss of shading. Temperatures also are excessively high in the main stem, but at present high temperatures there probably are more relevant to other species that are more likely than coho to use the main stem for rearing. Decrease in main-stem temperatures by augmentation of main-stem flows is problematic because augmentation water must be derived from the surface layer of Iron Gate Reservoir, which is very warm in summer. Projections of benefit to be expected from possible thermal manipulations may not have taken into account the exceptional importance of nocturnal thermal minimums in determining the energetic balance of coho exposed to high temperatures; nocturnal minimums can be as important as daily maximums in determining the survival of juvenile coho salmon.

Barriers to passage caused by dams and diversion structures are important to coho salmon. The main-stem dams on the Klamath River block spawning movements, as do Dwinnell Dam on the upper Shasta River and the Trinity River Diversion project on the Trinity River. Numerous small dams used by individual irrigators or ditch companies also block movement of coho in tributaries. Dams also have contributed to habitat degradation.

Coho habitat has been seriously degraded in the tributaries. Lack of cover and impairment of substrate through deposition of sediments are common. Woody debris, which is critical as cover for young fish, has largely been lost as a result of human activity. Excessive depletion of flow may separate fish from adequate habitat in the last half of summer.

Competition between hatchery coho and the smaller wild coho during migration to the estuary may be severe. Probably even more important are competition and predation from large numbers of Chinook salmon and steelhead that are released from hatcheries to the main stem when smoltification of the coho is in progress.

The Klamath River Fish Kill of 2002

During the second half of September 2002, numerous fish died in the lowermost 40 mi of the Klamath River main stem, 150 mi below Iron Gate

Dam (Figure S-1). Most of the dead fish were adult Chinook salmon that had just entered the lower Klamath River. At least 33,000 Chinook, of a total estimated spawning run of about 130,000, died. The immediate cause of death was massive infection by two types of pathogens that are widely distributed and generally become harmful to fish under stress, particularly if crowding occurs. The fish kill, although important for Chinook salmon, did not involve many coho salmon (about 1% of the total dead fish) because coho enter the river later than Chinook, and thus were mostly absent when conditions leading to mass mortality occurred.

The California Department of Fish and Game (CDFG), through an analysis of environmental conditions over 5 yr of low flow within the last 15 yr, showed that neither the flows nor the temperatures that occurred in the second half of September 2002 were unprecedented. A study by the U.S. Geological Survey (USGS) supports this conclusion. Thus, no obvious explanation of the fish kill based on unique flow or temperature conditions is possible.

CDFG has proposed that the shape of the channel in the lowermost reaches of the Klamath main stem changed in 1997–1998 under the influence of high flows, which caused fish entering the river to be unable to proceed upstream under low-flow conditions. An alternate hypothesis is that an unusual combination of temperature, flow, and migration conditions occurred in 2002, possibly in association with weather that prevented the river from showing nocturnal cooling to an extent that would usually be expected.

The two hypotheses—or others that may be proposed—are difficult to test because the conditions coinciding with the fish kill were unexpected and therefore largely unmonitored. If a lasting change in channel configuration was responsible, recurrence of the episode can be expected with similar low flows in the future. If other factors were responsible, recurrence may be much less likely. It is unclear what the effect of specific amounts of additional flow drawn from controllable upstream sources (waters from reservoirs on the Trinity River or Iron Gate Reservoir) would have been. Flows from the Trinity River could be most effective in lowering temperature.

Legal, Regulatory, and Administrative Context of Recovery Actions

Adaptive management is accepted in principle by the listing agencies but has not been implemented in the Klamath basin for the benefit of the listed species, except as part of the Trinity River Restoration Project. Information collected through monitoring and research has been valuable, but the absence of an integrated, evolving management plan connected to monitoring, research, review, and periodic readjustment of management actions will hamper progress in the future. Although agencies must meet the requirements of the ESA, many actions that could benefit the listed species can also be justified from the viewpoint of ecosystem management favorable to numerous other species, some of which are perilously close to listing, and to ecosystem functions that have great practical value.

Specifically with reference to ESA Section 4(f), USFWS and NMFS recovery planning for the three listed species has stalled and needs to be revived. Jeopardy consultations, which have focused on operation of the Klamath Project, must be broadened geographically because critical environmental resources of the listed species are found not only in but also beyond the Klamath Project. Furthermore, USFWS and NMFS appear to have overlooked take (mortality and impairment) of the listed species that is incidental to agricultural practice, private water management, and other activities beyond the control of USBR, and thus have not taken full advantage of their authorities under ESA Section 9.

The listing agencies have been criticized for using pseudoscientific reasoning ("junk science") in justifying their requirements for the protection of species in the upper Klamath basin. The committee disagrees with this criticism. The ESA allows the agencies to use a wide array of information sources in protecting listed species. The agencies can be expected, when information is scarce, to extend their recommendations beyond rigorously tested hypotheses and into professional judgment as a means of minimizing risk to the species. In allowing professional judgment to override site-specific evidence in some cases during 2001, however, the agencies accepted a high risk of error in proposing actions that the available evidence indicated to be of doubtful utility. The committee, as explained in its interim report, found some proposed actions as given in the 2001 biological opinions to lack substantial scientific support. In their biological opinions of 2002, the listing agencies appear to have resolved this issue either by obtaining concessions from USBR through mechanisms that are generally consistent with USBR's goal of delivering irrigation water (for example, through establishment of a water bank) or by redesigning their requirements to bring them into greater conformity with the existing evidence.

RECOMMENDATIONS

Recovery of endangered suckers and threatened coho salmon in the Klamath basin cannot be achieved by actions that are exclusively or primarily focused on operation of USBR's Klamath Project. While continuing consultation between the listing agencies and USBR is important, distribution of the listed species well beyond the boundaries of the Klamath Project and the impairment of these species through land- and water-management practices that are not under control of USBR require that the agencies use their authority under the ESA much more broadly than they have in the past.

Recommendation 1. The scope of ESA actions by NMFS and USFWS should be expanded in several ways, as follows (Chapters 6, 8, 9).

- NMFS and USFWS should inventory all governmental, tribal, and private actions that are causing unauthorized take of endangered suckers and threatened coho salmon in the Klamath basin and seek either to authorize this take with appropriate mitigative measures or to eliminate it.
- NMFS and USFWS should consult not only with USBR, but also with other federal agencies (e.g., U.S. Forest Service) under Section 7(a)(1); the federal agencies collectively should show a will to fulfill the interagency agreements that were made in 1994.
- NMFS and USFWS should use their full authority to control the actions of federal agencies that impair habitat on federally managed lands, not only within but also beyond the Klamath Project.
- Within 2 yr, NMFS should prepare and promulgate a recovery plan for coho salmon, and USFWS should do the same for shortnose and Lost River suckers. The new recovery plans should facilitate consultations under ESA Sections 7(a)(1), 7(a)(2), and 10(a)(1) across the entire geographic ranges of the listed species.
- NMFS and USFWS should more aggressively pursue opportunities for non-regulatory stimulation of recovery actions through the creation of demonstration projects, technical guidance, and extension activities that are intended to encourage and maximize the effectiveness of non-governmental recovery efforts.

Recommendation 2. Planning and organization of research and monitoring for listed species should be implemented as follows (Chapters 6, 8, 10).

- Research and monitoring programs for endangered suckers should be guided by a master plan for collection of information in direct support of the recovery plan; the same should be true of coho salmon.
- A recovery team for suckers and a second recovery team for coho salmon should administer research and monitoring on the listed species. The recovery team should use an adaptive management framework that serves as a direct link between research and remediation by testing the effectiveness and feasibility of specific remediation strategies.
- Research and monitoring should be reviewed comprehensively by an external panel of experts every 3 yr.
- Scientists participating in research should be required to publish key findings in peer-reviewed journals or in synthesis volumes subjected to external review; administrators should allow researchers sufficient time to do this important aspect of their work.
- Separately or jointly for the upper and lower basins, a broadly based, diverse committee of cooperators should be established for the purpose of pursuing ecosystem-based environmental improvements throughout the

basin for the benefit of all fish species as a means of preventing future listings while also preserving economically beneficial uses of water that are compatible with high environmental quality. Where possible, existing federal and state legislation should be used as a framework for organization of this effort.

Recommendation 3. Research and monitoring on the endangered suckers should be continued. Topics for research should be adjusted annually to reflect recent findings and to address questions for which lack of knowledge is a handicap to the development or implementation of the recovery plan. Gaps in knowledge that require research in the near future are as follows (Chapters 5, 6).

- Efforts should be expanded to estimate annually the abundance or relative abundance of all life stages of the two endangered sucker species in Upper Klamath Lake.
- At intervals of 3 yr, biotic as well as physical and chemical surveys should be conducted throughout the geographic range of the endangered suckers. Suckers should be sampled for indications of age distribution, qualitative measures of abundance, and condition factors. Sampling should include fish other than suckers on grounds that the presence of other fish is an indicator of the spread of nonnative species, of changing environmental conditions, or of changes in abundance of other endemic species that may be approaching the status at which listing is needed. Habitat conditions and water-quality information potentially relevant to the welfare of the suckers should be recorded in a manner that allows comparison across years. The resulting survey information, along with the more detailed information available from annual monitoring of populations in Upper Klamath Lake, should be synthesized as an overview of status.
- Detailed comparisons of the Upper Klamath Lake populations (which are suppressed) and the Clear Lake and Gerber Reservoir populations (which are apparently stable), in combination with studies of the environmental factors that may affect welfare of the fish, should be conducted as a means of diagnosing specific life-history bottlenecks that are affecting the Upper Klamath Lake populations.
- Multifactorial studies under conditions as realistic as practicable should be made of tolerance and stress for the listed suckers relevant to poor water-quality conditions in Upper Klamath Lake and elsewhere.
- Factors affecting spawning success and larval survival in the Williamson River system should be studied more intensively in support of recovery efforts that are focused on improvements in physical habitat protection for spawners and larvae in rivers.
- An analysis should be conducted of the hydraulic transport of larvae in Upper Klamath Lake.

• Relevant to the water quality of Upper Klamath Lake, more intensive studies should be made of water-column stability and mixing, especially in relation to physiological status of *Aphanizomenon* and the occurrence of mass mortality; of mechanisms for internal loading of phosphorus; of winter oxygen concentrations; and of the effects of limnohumic acids on *Aphanizomenon*.

• A demographic model of the populations in Upper Klamath Lake should be prepared and used in integrating information on factors that affect individual life-history stages.

• Studies should be done on the degree and importance of predation on young fish by nonnative species.

• Additional studies should be done on the genetic identities of subpopulations.

Recommendation 4. Recovery actions of highest priority based on current knowledge of endangered suckers are as follows (Chapter 6).

• Removal of Chiloquin Dam to increase the extent of spawning habitat in the upper Sprague River and expand the duration over which larvae enter Upper Klamath Lake.

• Removal or facilitation of passage at all small blockages, dams, diversions, and tributaries where suckers are or could be present.

• Screening of water intakes at Link River Dam.

• Modification of screening and intake procedures at the A Canal as recommended by USFWS (2002).

• Protection of known spawning areas within Upper Klamath Lake from disturbance (including hydrologic manipulation, in the case of springs), except for restoration activities.

• For river spawning suckers of Upper Klamath Lake, protection and restoration of riparian conditions, channel geomorphology, and sediment transport; elimination of disturbance at locations where suckers do spawn or could spawn. These actions will require changes in grazing and agricultural practices, land management, riparian corridors, and public education.

• Seeding of abandoned spawning areas in Upper Klamath Lake with new spawners and physical improvement of selected spawning areas.

• Restoration of wetland vegetation in the Williamson River estuary and northern portions of Upper Klamath Lake.

• Use of oxygenation on a trial basis to provide refugia for large suckers in Upper Klamath Lake.

• Rigorous protection of tributary spawning areas on Clear Lake and Gerber Reservoir, where populations are apparently stable.

• Reintroduction of endangered suckers to Lake of the Woods after elimination of its nonnative fish populations.

• Reestablishment of spawning and recruitment capability for endangered suckers in Tule Lake and Lower Klamath Lake, even if the attempts require alterations in water management, provided that preliminary studies indicate feasibility; increased control of sedimentation in Tule Lake.

• Review of all proposed changes in Klamath Project operations for potential adverse effects on suckers; maintenance of water level limits for the near future as proposed by USBR in 2002 but with modifications as required by USFWS in its most recent biological opinion (2002).

Recommendation 5. Needs for new information on coho salmon are as follows (Chapters 7, 8).

• Annual monitoring of adults and juveniles should be conducted at the mouths of major tributaries and the main stem as a means of establishing a record of year-class strength for coho. Every 3 yr, synoptic studies of the presence and status of coho should be made of coho in the Klamath basin. Physical and chemical conditions should be documented in a manner that allows interannual comparisons. Not only coho but other fish species present in coho habitats should be sampled simultaneously on grounds that changes in the relative abundance of species are relevant to the welfare of coho and may serve as an early warning of declines in the abundance of other species. Results of synoptic studies, along with the annual monitoring at tributary mouths, should be synthesized as an overview of population status at 3-yr intervals.

• Detailed comparisons should be made of the success of coho in specific small tributaries that are chosen so as to represent gradients in potential stressors. The objective of the study should be to identify thresholds for specific stressors or combinations of stressors and thus to establish more specifically the tolerance thresholds for coho salmon in the Klamath basin.

• The effect on wild coho of fish released in quantity from hatcheries should be determined by manipulation of hatchery operations according to adaptive-management principles. As an initial step, release of hatchery fish from Iron Gate Hatchery (all species) should be eliminated for 3 yr, and indicators of coho response should be devised. Complementary manipulations at the Trinity River Hatchery would be desirable as well.

• Selected small tributaries that have been impaired should be experimentally restored, and the success of various restoration strategies should be determined.

• Success of specific livestock-management practices in improving channel conditions and promoting development of riparian vegetation should be evaluated systematically.

• Relationships between flow and temperature at the junctions of tributaries with the main stem and the estuary should be quantified; possible benefits of coordinating flow management in the Trinity and Klamath main stem should be studied.

Recommendation 6. Remediation measures that can be justified from current knowledge include the following (Chapter 8).

- Reestablishment of cool summer flows in the Shasta and Scott rivers in particular but also in small tributaries that reach the Klamath main stem or the Trinity main stem where water has been anthropogenically warmed. Reestablishment of cool flows should be pursued through purchase, trading, or leasing of groundwater flows (including springs) for direct delivery to streams; by extensive restoration of woody riparian vegetation capable of providing shade; and by increase of summer low flows.
- Removal or provision for effective passage at all small dams and diversions throughout the distribution of the coho salmon, to be completed within 3 yr. In addition, serious evaluation should be made of the benefits to coho salmon from elimination of Dwinnell Dam and Iron Gate Dam on grounds that these structures block substantial amounts of coho habitat and, in the case of Dwinnell Dam, degrade downstream habitat as well.
- Prescription of land-use practices for timber management, road construction, and grazing that are sufficiently stringent to prevent physical degradation of tributary habitat for coho, especially in the Scott, Salmon, and Trinity river basins as well as small tributaries affected by erosion.
- Facilitation through cooperative efforts or, if necessary, use of ESA authority to reduce impairment of spawning gravels and other critical habitat features by livestock, fine sediments derived from agricultural practice, timber management, or other human activities.
- Changes in hatchery operations to the extent necessary, including possible closure of hatcheries, for the benefit of coho salmon as determined through research by way of adaptive management of the hatcheries.

COSTS

The costs of remediation actions are difficult to estimate without more detail on their mode of implementation by the agencies. Based on general knowledge of costs of research and monitoring at other locations, an approximate figure for the recommendations on endangered suckers over a 5-yr period is $15–20 million, including research, monitoring, and remedial actions of minor scope. Excluded are administrative costs and the costs of remedial actions of major scope (e.g., removal of Chiloquin Dam), which would need to be evaluated individually for cost. For coho salmon, research, monitoring, and remedial projects of small scope over 5 yr is estimated at $10–15 million. Thus, the total for all three species over 5 yr is $25–35 million, excluding major projects such as removal of dams. These costs are high relative to past expenditures on research and remediation in the basin, but the costs of further deterioration of sucker and coho populations, along with crisis management and disruptions of human activities, may be far more costly. A hopeful vision is that increased knowledge, improved management, and cohesive community action will promote recovery of the fishes. This outcome, which would be of great benefit to the Klamath basin, could provide a model for the nation.

1

Introduction

The United States attempts to reduce the rate of extinction within its diverse and valuable biota primarily through the Endangered Species Act (ESA) of 1973. The ESA prohibits or severely limits the intentional or incidental taking of species that are listed as endangered or threatened. The ESA is ecologically practical in requiring that habitat necessary for each life-history stage (critical habitat) of a species be preserved and, if possible, expanded or enhanced. Among the requirements of the ESA, the prohibition of intentional taking is relatively easy to implement, the prohibition of incidental taking raises many practical difficulties because of its conflict with ordinary human activities, and the requirement for protection of critical habitat can be troublesome in the extreme because it often is in direct conflict with customary and valued uses of natural resources.

The ESA has been applied to the upper Klamath River basin of Oregon and California (Figure 1-1) for protection of the Lost River sucker (*Deltistes luxatus)* and shortnose sucker (*Chasmistes brevirostris)* and for the Klamath basin component of a genetically distinct population of coho salmon (*Oncorhynchus kisutch)* that is designated the southern Oregon/northern California coasts (SONCC) "evolutionarily significant unit" (ESU). The listing of these three fish species has, as required by the ESA, led to an intensive effort on the part of federal agencies and others to identify critical habitat and to propose federal actions that would promote recovery of the species. Analysis of the needs of the species has extended necessarily to private lands and to privately held water rights, given that the fishes range well beyond the boundaries of federal land and water management.

FIGURE 1-1 Map of the upper Klamath River basin showing surface waters and landmarks mentioned in this report. Source: Modified from USFWS.

Requirements of the endangered and threatened fishes (see Chapter 9 for the difference between these two designations) came into especially sharp focus during 2001, a year of drought, when federal agencies, in an effort to protect these fishes, all but eliminated the distribution of water from Upper Klamath Lake for irrigation. The severe economic consequences of that decision for some segments of the Klamath basin community brought a sense of crisis to a controversy that had already developed around environmental, cultural, and commercial interests in fish as opposed to agricultural and economic interests in the uses of land and water.

This report presents the results of a study conducted by the National Research Council's (NRC) Committee on Endangered and Threatened Fishes in the Klamath River Basin. The committee was formed at the request of the Department of the Interior and the Department of Commerce, whose agencies are responsible for implementing requirements of the ESA in the Klamath River basin. The committee's tasks were to evaluate the scientific merit of federal agencies' proposals or requirements for protection of the endangered and threatened fishes and to analyze the long-term requirements for recovery of these fishes. The committee's final report, which is given here, presents conclusions and recommendations that bear on the requirements of the endangered and threatened fishes. The committee hopes that its report will assist the federal government both in implementing the requirements of the ESA and in minimizing adverse effects of ESA actions on residents of the Klamath River basin.

OVERVIEW OF THE ENVIRONMENT

For purposes of environmental analysis, it is convenient to divide the Klamath River basin into an upper basin, which extends north and east from the Iron Gate Dam on the main stem of the Klamath River, and a lower basin, which extends south and west to the Pacific Ocean (Figure 1-1). The upper basin is dominated by the activity of large volcanoes and active faulting, which controls the location and shape of broad valleys. These fault-bounded valleys contain all of the large natural lakes and large wetlands of the Klamath basin. Crater Lake, the second deepest lake in North America and one of the most transparent of all lakes, is a notable geographic feature of the upper basin, but is irrelevant to the welfare of the endangered and threatened fishes because of its hydrologic isolation. The upper basin has a relatively dry, high desert climate typical of areas that lie east of the Cascade Range. The widespread volcanic rocks of the upper basin produce numerous springs that are important local sources of water.

Within the lower basin, below Iron Gate Dam, the Klamath River is incised deeply into bedrock, forming a narrow canyon. The mountains that surround the lower Klamath, including the Trinity Alps and Coast Ranges, are rugged, with dense conifer and fir forests and steep tributary streams. The climate is quite variable in the lower basin, but is distinguished by its very high annual rainfall and relatively mild temperatures. Some fertile valleys, including those of the Shasta and Scott rivers, are found in the lower basin.

Because the Klamath River flows directly to the Pacific, it is isolated from other inland waters. This isolation, which was compounded in the past by separation of the upper and lower parts of the Klamath basin, explains the high degree of endemism in the fish fauna of the basin (Chapter 5). Isolation also accounts for the spectacular ecological success, before human intervention, of the endemic fishes of the upper basin, as shown by formerly great abundances of the shortnose and Lost River suckers, which are adapted for living in a naturally variable high desert environment (Chapter 5). Although isolation has been less absolute for anadromous fishes, which occupy the lower basin and mix with other populations in the Pacific Ocean, the homing characteristics of salmonids in combination with regional selective forces have led to the presence of genetically distinct populations of anadromous fishes, including the SONCC population of coho salmon, in the lower Klamath basin and several adjacent drainages (Chapter 7).

With respect to water management, the upper basin has two parts: (1) waters draining to Upper Klamath Lake and (2) Klamath Lake plus all lands lying between it and Iron Gate Dam, including the Lost River basin. There are no lakes of significance to the endangered suckers above Upper Klamath Lake, but the streams and rivers above Upper Klamath Lake, especially the Williamson and Wood rivers and their tributaries, historically were and still are important for spawning of the endangered suckers (Chapter 6). The Lost River historically was isolated from the rest of the upper basin in all but wet years and has lakes that are or were important to endemic fishes. It is now hydrologically connected to the Klamath River through water management.

The issues of importance above Upper Klamath Lake include physical degradation and blockage of tributaries by dams or water-management structures and misdirection of fish through entrainment. Correction of these problems will involve private parties because most water management in this portion of the basin is not under federal control. As explained more fully in Chapter 2, cattle and irrigated crops are important.

Below the Upper Klamath Lake watershed, Upper Klamath Lake, Gerber Reservoir, Clear Lake, and the now small remnants of Lower Klamath Lake and Tule Lake all are affected by water management through the U.S. Bureau of Reclamation's (USBR) Klamath Project, as are the flows of all tributary waters (most notably the Lost River) that lie below all of these water bodies. Water management in this region is largely federal in that USBR delivers water from Upper Klamath Lake to the Klamath Project and also stores and routes water by using the other lakes and waterways. Thus, any loss of fish caused by hydraulic manipulation or water-management structures of the Klamath Project is the responsibility of USBR as it fulfills its contracts for delivery of water. Private water users, however, determine land use and application methods for water delivered by USBR and use privately managed diversion structures and small dams to regulate the routing of water. Thus, both USBR and private water users may affect the suitability of environmental conditions for endangered suckers. Although the details are complex, the general pattern is that water stored in Upper Klamath Lake, Clear Lake, and Gerber Reservoir is diverted for agricultural use, and the unused portion of this diverted water is returned via Tule Lake, Lower Klamath Lake, or the Lost River to the main stem of the Klamath River (Figure 1-2). Approximate quantities of water flow are as shown in Table 1-1.

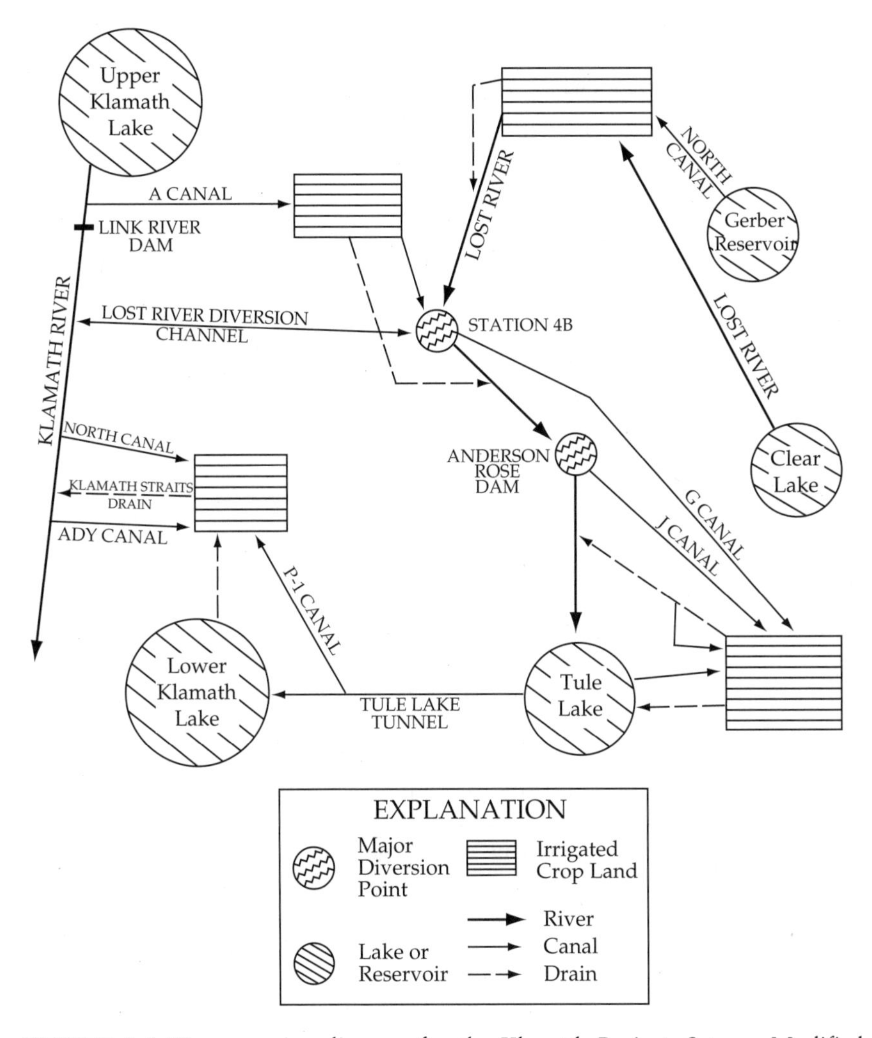

FIGURE 1-2 Water routing diagram for the Klamath Project. Source: Modified from USFWS 2002.

TABLE 1-1 Flows Under Conditions of Average Water Availability in the Upper Klamath Basin[a]

Location	Amount (acre-ft per yr)
Upper Klamath Lake outflow[b]	1,300,000
Outflow April-September	500,000
Directed to Klamath Project	400,000
Directed downstream	900,000
Clear Lake inflow[b]	117,000
Directed to Klamath Project[c]	36,000
Gerber Reservoir inflow[b]	55,000
Directed to Klamath Project	40,000
Total Klamath Project consumptive use, including refuges[b]	350,000
Total Klamath Project returns to Klamath River[b]	100,000
Nonproject irrigation diversions, upper basin[d]	420,000
Total flow at Orleans[e]	6,000,000
Trinity River flow	3,800,000
Total flow at mouth	13,400,000

[a]Approximate only—actual values differ from year to year.
[b]USBR 2000a.
[c]Evaporative losses are especially high in Clear Lake (long retention time and evaporation at about 3.8 ft/yr).
[d]NMFS 2001 (estimated from percentages).
[e]Near the mouth of the Klamath River, but above the Trinity River.

The upper basin contains seven national wildlife refuges and several other public and private preserves, as shown in Figure 1-3. The abundance of refuges and preserves in the upper basin is an indication of its exceptional value for waterfowl and other forms of life that depend on great expanses of shallow water and wetlands. Refuges and preserves around the lakes can be considered a means of conserving or enhancing wetlands that may be relevant to the welfare of endangered suckers.

Near Lower Klamath Lake and Tule Lake, water management is especially complicated in that the refuge lands within the original inundation zones of these two lakes now are used extensively for agricultural purposes according to agreements that were reached during the early history of the refuges (Chapter 2). The two lakes function hydrologically primarily as drainage conduits; they are not allowed to accumulate water because of governmental commitments to continuing agricultural use of the former lake beds. Thus, both lakes now lack the large populations of shortnose and Lost River suckers that once occupied them, although Tule Lake does still support a small population of endangered suckers (Chapter 6).

Also in the upper basin are six main-stem dams (Figure 1-4). The Link River Dam (completed in 1921), which is near the outlet of Upper Klamath

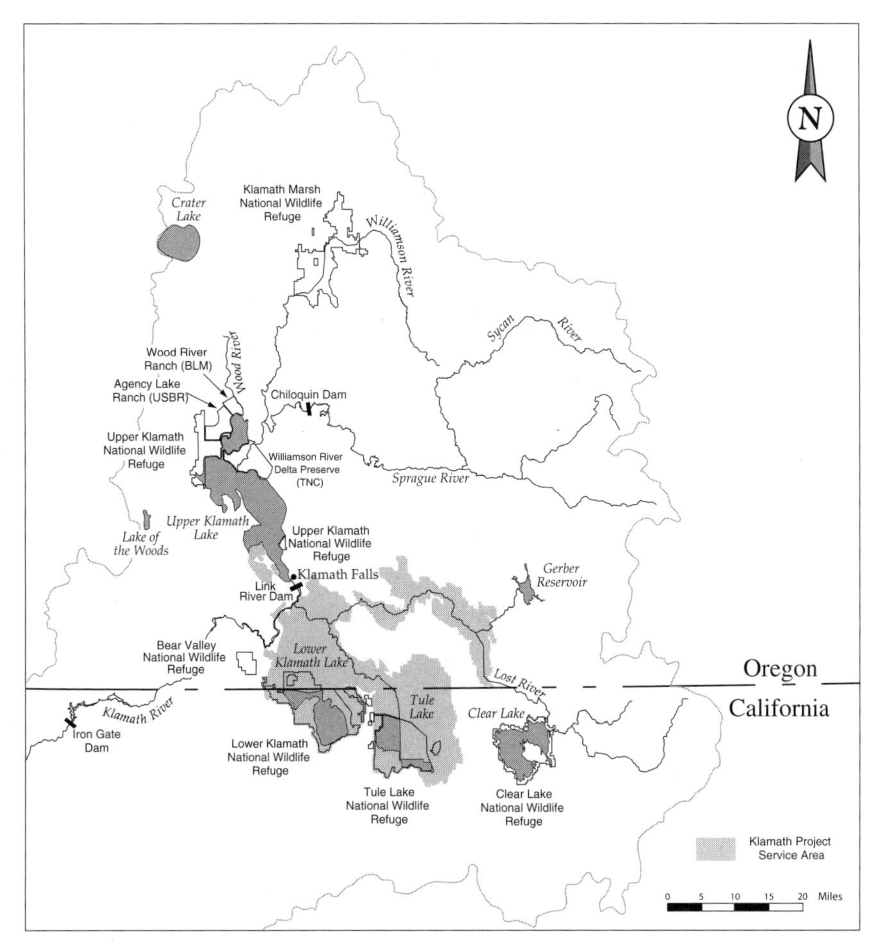

FIGURE 1-3 Map of the upper Klamath basin. Abbreviations: BLM, Bureau of Land Management; TNC, The Nature Conservancy; USBR, U.S. Bureau of Reclamation.

Lake, is used in regulating the level of Upper Klamath Lake for water-management purposes and also produces hydropower. Irrigation water is withdrawn seasonally in large quantities through the A Canal, which is just above the Link River Dam. Principles of operation of the dam are a major point of controversy related to the welfare of the endangered suckers (Chapter 6).

Below the Link River Dam are five additional dams; all the dams except the Keno Dam produce hydropower. All six dams are operated by PacifiCorp, a utility company, through agreements with USBR. Iron Gate Dam, the terminal dam, is used for reregulation of flow to the Klamath River

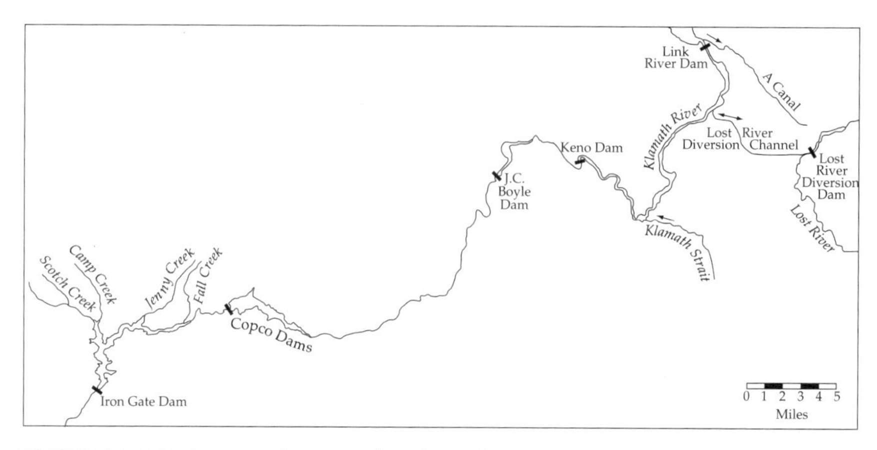

FIGURE 1-4 Main-stem dams on the Klamath River.

main stem. The six dams block access of both endangered suckers and coho salmon to large portions of their historical ranges and can be direct or indirect agents of fish mortality. Through the operation of Link River Dam, endangered suckers have been historically entrained into the A Canal and thus killed (Chapter 6). In addition, the suckers enter the unscreened intakes of the power-production facilities and thus may pass through turbines. Dams also are the means by which ramping of flow (change in discharge over short periods), which is consistent with optimal operation of hydropower production facilities, is achieved; ramping of flow can be detrimental to coho fry, which can become stranded at the river margin when flow decreases rapidly.

In the lower part of the basin (below Iron Gate Dam), the main stem of the Klamath River is the pathway of migration for numerous anadromous fishes and is important for spawning and rearing of some of them (Chapter 7). Flow to the main stem at Iron Gate Dam is reduced and altered seasonally through the operation of the Klamath Project and private water management above Iron Gate Dam and is regulated hourly by PacifiCorp (Chapter 4). Releases can be regulated to some degree by control of storage in Upper Klamath Lake, but irrigation commitments constrain this management flexibility, especially in dry years. Although groundwater flow is substantial in some parts of the Klamath River basin, there appears to be little accrual of groundwater to the Klamath main stem below Iron Gate Dam. Increase in discharge downstream occurs through four large tributaries—the Shasta, Scott, Salmon, and Trinity rivers (Figure 1-1)—and through numerous small tributaries. The large tributaries all are physically altered, and some show severe depletion of flow and are excessively warm because of loss of riparian vegetation and high relative contribution of irrigation return flows to total stream discharge (Chapter 4). As explained in Chapters 7 and 8, the small tributaries now provide some of the best habitat for coho salmon. Land and water relevant to the welfare of the coho and other fishes in the lower basin are primarily under private control. Water-management structures interfere with the movement of fish in this part of the watershed, as they do elsewhere.

The Trinity River, which is the largest tributary of the Klamath River, reaches the Klamath about 43 mi from the estuary. In 1964, the Trinity River Diversion began delivering up to 90% of the upper Trinity's flow out of the basin to the Central Valley Project. This diversion and other changes in the watershed were followed by a severe decline in the anadromous fish populations of the Trinity River. Studies of coho salmon and other fishes of the Trinity River have been conducted separately from those of the Klamath River basin through processes prescribed by the National Environmental Policy Act, which involves an environmental impact statement (EIS) rather than ESA procedures. In December 2000, the EIS resulted in a record of decision (ROD) for the Trinity River (USFWS 2000). The ROD called for increased minimum flows, habitat restoration for the benefit of anadromous salmonid populations, and use of an adaptive management approach involving further study and evaluation of the outcomes of flow and habitat manipulations. As a result of judicial decisions, however, a supplementary EIS is still in progress. Recovery of the Trinity River coho populations is important for recovery of the coho in the Klamath basin as a whole; hydrologic linkages between the two rivers are especially important for the migration of coho (Chapters 4, 7, and 8).

The hydrologic characteristics of the Klamath River main stem and its major tributaries are dominated by seasonal melt of snowpack. Summer storms and release of groundwater from springs also make contributions, but they are smaller in aggregate than the snowmelt effect. The schedule of melting differs from year to year, reflecting climatic variability, but a universal feature of hydrographs is a spring pulse in flow followed by recession to a baseflow condition by late summer. These main features of the hydrograph undoubtedly have influenced the adaptations of native organisms, as reflected in the timing of their key life-history features (see Chapters 5 and 7).

Even though water is now managed (Table 1-1), hydrographs of the Klamath River basin still show the dominant influence of snowmelt and spring precipitation on water flow. For example, Figure 1-5 compares the flow near the mouth of the Williamson River, above which there are no major impoundments, with the flow at Iron Gate Dam, above which a great deal of water management occurs. Flows at the mouth of the Williamson River are affected by privately managed irrigation diversions but, given the large total flow in the Williamson, the hydrograph has predominantly natu-

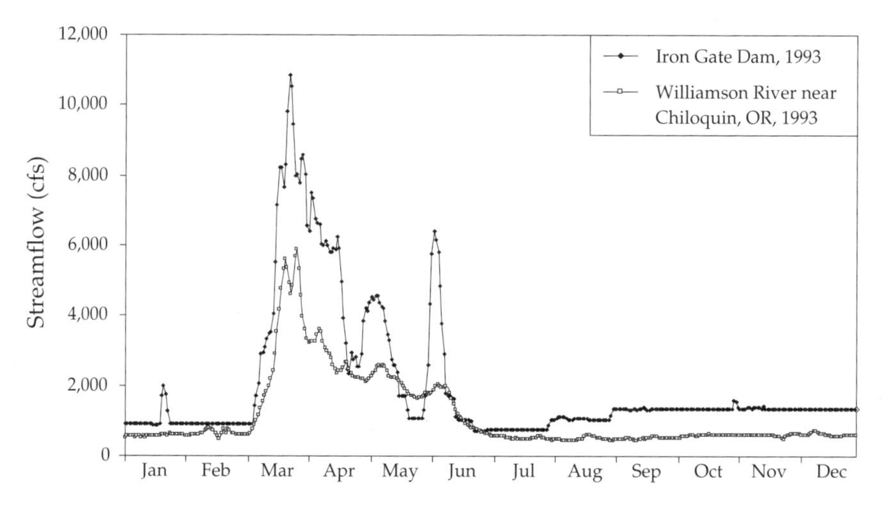

FIGURE 1-5 Flow of the Williamson River, the largest water source for Upper Klamath Lake, and of the Klamath River main stem (at Iron Gate Dam) in a year of near-average water availability. Source: USGS gage records.

ral features. At Iron Gate Dam, the retention of water in reservoirs of the Lost River and in Upper Klamath Lake has the potential to alter the hydrograph more extensively. Alteration is, as expected, more severe during years of drought than years of average flow.

The management of hydrographs, in combination with natural climatic variation, now is a major focus of attention in the analyses of environmental factors that may affect the welfare of the two endangered sucker species and the coho salmon (Chapter 4). Hydrology has environmental effects not only through its direct control of physical attributes of standing and flowing water (mean depth, water velocity), but also because of its indirect control of other aspects of the physical, chemical, and biological environment such as temperature of flowing water, nutrient concentrations in lakes, and extent and type of aquatic vegetation. Even so, numerous influences on the endangered fishes, such as the introduction of nonnative fishes, loss of riparian vegetation, and anthropogenic mobilization of nutrients, involve factors other than hydrology.

THE FISHES

The shortnose and Lost River suckers are large, long-lived fishes of high fecundity. Although they spend most of their lives in lakes, flowing waters are important to them for spawning. Some subpopulations spawn around the perimeter of Upper Klamath Lake, particularly near springs, but fish of both species migrate or attempt to migrate into tributaries for spawning. Shortly after hatching, fry return to the lake, where they occupy very shallow water at first and move to progressively greater depths as they mature. The endangered suckers do not spawn until they are several years old (Chapter 5).

The two endangered sucker species were so abundant before colonization that they served as a major food source for Indian tribes (Chapter 2). After the Klamath basin was colonized, the fish were harvested in large numbers commercially. Because they are large and tend to migrate during spawning, they were highly vulnerable, and their numbers were drastically reduced through harvest. Records of the size of spawning runs and sport fishing indicated during the 1980s that both species had declined to such a point that without special protection they might be extirpated. Fishing for the species was eliminated except for very small numbers of fish allocated for ceremonial purposes to Indian tribes. In 1988, both species were listed as endangered under the ESA (53 Fed. Reg. 27130, 18 July 1988).

It was clear in the 1980s and even earlier that prohibition of fishing, although essential, might not be sufficient to produce recovery of the endangered suckers. Factors that probably have contributed to the suppressed abundances of these species include blockage of migration pathways to spawning areas; entrainment of large numbers of fish by water-management structures; poor water quality, especially in Upper Klamath Lake; physical degradation of habitat; and adverse genetic consequences of scarcity and fragmentation (Chapter 6). Mass mortality of large fish in Upper Klamath Lake, although recorded for over 100 yr, caused particular alarm during the 1990s because of its sequential occurrence in 3 yr (1995–1997). The abundance of large adults appears to have been strongly suppressed by fishing, which was banned after 1987, and by mass mortality caused by poor water quality. Although recruitment of young fish has been documented since the listing of the suckers in 1988, there is no indication of recovery in overall abundances (Chapter 6).

Populations of coho salmon in the Klamath River were substantial when commercial salmon fisheries first developed (Chapter 7). Abundances of most anadromous fishes in the Klamath River basin and other Pacific coast basins have declined drastically since then. Decline of the coho salmon in the Klamath River basin led to federal listing of the SONCC ESU as threatened in 1997 (62 Fed. Reg. 24588, 6 May 1997); California listed the ESU as endangered in 2003.

The coho salmon, except in the case of some early-spawning males, has a 3-yr life history that is divided almost equally between marine and freshwater environments. A fall-winter migration brings the fish up the main stem of the Klamath River. Although some spawning may occur in the main stem, the primary spawning occurs in tributaries (Chapter 7). Young fish remain primarily in the tributaries or tributary mouths during development, after which they move downstream as smolts undergoing physiological transformation that is essential for life in marine waters. The many factors that are known or suspected to have contributed to the decline of the coho salmon include harvest (which is now prohibited), depletion of flows, anthropogenic warming of water, loss of cover, blockage of migration routes, and adverse water quality. In addition, the release of large numbers of hatchery-reared salmonids (coho and other taxa) introduces increased predation and competition.

Physical and chemical conditions in the tributaries are undeniably bad and can be remedied only through extensive remediation on private lands, either with or without facilitation by federal or state agencies. Examples of private efforts to promote recovery of salmon are already available for some parts of the Klamath basin (Chapter 8). USBR has considerable control over flows in the main stem, however, and the degree to which coho can benefit from changes in water management by USBR has been the subject of much controversy. The coho is strongly oriented toward tributaries for most of its life cycle, but the upstream migration of adults and the downstream migration of smolts involve the main stem. Thus, it is reasonable to ask whether regulation of flow in the main stem is important in holding back the recovery of the coho salmon; the evidence is reviewed in Chapter 8.

REQUIREMENTS OF THE ENDANGERED SPECIES ACT

With the listing of the coho and the two endangered suckers, the ESA introduced a new legal framework that has become the dominant factor in resolving water issues in the Klamath River basin, except for the Trinity River, where EIS procedures predominate. The listing of the two sucker species and the coho salmon triggered a suite of ESA regulatory requirements, as follows:

- Section 4 of the ESA requires the listing agency—the U.S. Fish and Wildlife Service (USFWS) for the endangered suckers and the National Marine Fisheries Service (NMFS) for the coho salmon—to designate "critical habitat" for endangered and threatened species unless exceptions, which are narrow, apply.
- Section 4(f) of the ESA requires the listing agency to develop and implement a "recovery plan" for endangered and threatened species.
- Section 7(a)(1) of the ESA requires all federal agencies, through consultation with the listing agency, to use their authority to carry out programs for the "conservation" of endangered and threatened species.

• Section 7(a)(2) of the ESA requires all federal agencies, through consultation with the listing agency, to ensure that actions carried out, funded, or authorized by them do not "jeopardize" the continued existence of endangered and threatened species and do not result in "adverse modification" of their critical habitat.

• Section 9(a)(1) of the ESA prohibits all persons subject to U.S. jurisdiction (including federal, state, tribal, and local governments) from "taking" endangered species unless authorized by the listing agency pursuant to appropriate provisions of the ESA; and section 4(d) allows the listing agency to extend the same level of protection to threatened species.

As explained in Chapter 9, some of these requirements have not been fully implemented for the endangered suckers and threatened coho salmon of the Klamath River basin. Nevertheless, primarily through the prohibition in Section 7(a)(2) against federal agencies causing "jeopardy" to listed species, the ESA, after the listings, has affected USBR's operation of the Klamath Project. Primarily through the jeopardy-consultation procedure of Section 7(a)(2), USFWS has influenced USBR's maintenance of water levels in Upper Klamath Lake for the protection of the endangered suckers and NMFS has influenced USBR's releases from Upper Klamath Lake to the Klamath River main stem for protection of the coho salmon.

The full force of the jeopardy opinions manifested itself in the Klamath River basin on April 6, 2001, when, because USFWS and NMFS had concluded in their consultations with USBR that the proposed operation of the Klamath Project for delivery of irrigation water to USBR's water-contract parties would jeopardize the endangered and threatened species, USBR determined that it could not deliver water through the Klamath Project. The social and economic consequences of that decision focused the attention of many observers on two of numerous conclusions given in the USFWS and NMFS biological opinions: that the continued existence of the species would be jeopardized unless USBR maintained the water levels in lakes that USFWS specified and the main-stem flows that NMFS specified. The ESA required USFWS and NMFS to base the jeopardy findings on the "best scientific and commercial data available" (16 U.S.C. 1536(a)(2)). Some observers questioned whether the two agencies had met that standard; others contended that the decision was fully justified.

Application of the ESA to fishes of the Klamath River basin puts into focus one of the central dilemmas of the ESA: the need to reconcile the ESA's legal framework with its scientific foundations. For example, the ESA demands that USFWS and NMFS make clear distinctions as to whether an action will cause jeopardy, but the scientific process is not fully compatible with such sharp distinctions. Biologists studying the status of a species are likely to speak in conditional terms and only rarely to express the definitive conclusions that would be most useful for application of the ESA. Moreover, the listing and jeopardy-consultation procedures require that USFWS and NMFS use "best available" data, but such data often do not resemble the products of scientific review processes used by leading scientific journals. It may be possible, therefore, for USFWS or NMFS to satisfy the demands of the ESA with an analysis that would not satisfy the demands of scientific review for publication or other peer-review processes common in modern science. This issue is dealt with further in Chapter 9.

INTERESTED PARTIES

The work of the NRC committee does not involve conclusions or recommendations on economic or social issues, but the various interested parties that are present in the basin provide some context for evaluation of the controversies that have developed around the endangered and threatened fishes and thus the focus of scientific research on specific topics. Ultimately, the interested parties must work together on a sustained basis in order to achieve and maintain recovery of the listed species (Poff et al. 2003).

Indians were the first occupants of the basin; they now operate institutionally as the Klamath tribes (a group of related but formerly separate tribes, including the Klamath, Modoc, and Yahooskin Band of the Snakehead Indians) and the Yurok, Karuk, Shasta, and Hoopa Valley tribes. The tribes extensively used the fish of the Klamath River basin for food before the arrival of colonists, and they have cultural traditions involving the fish. The endangered and threatened fishes of the Klamath River basin, and numerous other fishes not now listed as threatened or endangered (see Chapters 5 and 7), are tribal trust species; the U.S. government has an acknowledged obligation to preserve these fishes for use by the tribes. Preservation of the fishes for use obviously implies water rights. The priority date for these water rights is "time immemorial." Thus, in the seniority system for water rights, tribal water rights related to the protection of fishes are senior to all others. Two practical issues, as yet unresolved, are how to translate the protection of fishes into specific amounts of water at specific points in the basin and the degree to which any such commitment would curtail other uses of water. These legal matters are directly relevant to research on the requirements of the endangered and threatened fishes.

The USBR, another interested party, has been working in the Klamath River basin for about a century. In 1905, Congress, Oregon, and California granted USBR authority to create the Klamath Project, which involved the acquisition of extensive water rights in the upper basin, the construction of storage and distribution systems, and extensive drainage of lakes and wetlands around Tule Lake and Lower Klamath Lake so that agriculture could displace the natural aquatic habitats there (USBR 2000b). The project matured over about a half-century and is considered to have taken its full modern operational characteristics in 1960. Thus, the interval between 1960 and today is often taken as the benchmark period for judging proposals for the future.

The USBR, as a federal agency, must follow all requirements of the federal government, even including those not associated with its mission, but it devotes its energy primarily to the orderly distribution of water in support of agricultural water use. The ESA requires, however, that USBR analyze and put into writing its assessments of the effect of Klamath Project operations on endangered and threatened species and that it enter into consultation about the assessments with USFWS (for suckers) and NMFS (for coho). In 2001, USBR issued two assessments (USBR 2001a, b) acknowledging that some aspects of project operations were harmful to the two endangered sucker species and to the threatened coho salmon. USBR proposed changes in operations that it believed would offset some of the adverse effects. The 2001 biological assessments were succeeded by revised assessments issued in 2002 (USBR 2002a), which proposed a plan of operations to extend over the next 10 yr. The revised assessments, which contain some additional proposals for amelioration of potential damage to the endangered and threatened species, are summarized at the end of this chapter.

The USFWS is charged with issuing biological opinions related to the endangered suckers of the Klamath River basin. It has been in consultation with USBR over the two endangered sucker species since the species were listed, and it has reviewed USBR's biological assessment of 2001 and USBR's 10-yr biological assessment of 2002. The role of USFWS is to analyze environmental information and set requirements for protection of the fishes, to issue the analyses as biological opinions, and through the creation of "reasonable and prudent alternatives" (RPAs), to call for changes in Klamath Project operations as it believes necessary to reduce risk to the endangered suckers.

The USFWS endorsed a number of proposals contained in USBR's assessment of 2001, but it judged the USBR proposals for control of water level to be inadequate overall for protection of the endangered suckers. The reasonable and prudent alternative proposed by USFWS in 2001 included prescriptions for higher water levels in lakes. Although USBR's assessment of 2002 (10-yr operating plan) was revised with respect to water levels, USFWS again found it to be inadequate overall and proposed its reasonable and prudent alternative, as described below.

The NMFS responded to USBR's biological assessment of 2001 on coho salmon and to USBR's 10-yr operating plan as given in its 2002 biological assessment. NMFS approved of a number of elements in the biological assessments of 2001 and 2002 but differed with USBR on the matter of minimum flows in the Klamath River main stem below Iron Gate Dam. As required by the ESA, NMFS issued a reasonable and prudent alternative prescribing higher flows in the main stem than had been proposed by USBR in 2001 and 2002.

The USFWS and NMFS ("the listing agencies") have the last word in judging the requirements of the endangered and threatened fishes. Thus, as of 2002, the USBR 10-yr proposal, as given in the 10-yr assessment, was in part rejected, and the listing agencies are requiring several new procedures and practices.

The USFWS also plays a second, very different role as manager of refuges in the Klamath basin. On the downstream end of the Klamath Project, the Lower Klamath Lake and Tule Lake refuges receive drainage water from the Klamath Project. The drainage water is used to manage the two refuges within constraints that are set by water availability and requirements for agricultural use of the land in or surrounding the refuges. In this role, USFWS is not able to demand specific amounts of water or specific timing for delivery of water to benefit the refuges. Instead, it negotiates with USBR and with agricultural interests for water to manage the refuges. Thus, although delivery of water to the refuges is required, it has a lower priority than the agricultural use of water or the agricultural use of land near the two lakes. The two uses of water are connected, however, in that some of the water delivered for agricultural use appears downstream for use by the refuges. Thus, curtailment of water for irrigation on the Klamath Project raises questions about the availability of water for the refuges.

Irrigators were present even before initiation of the Klamath Project and came in increasing numbers to use waters of the Klamath Project and waters in parts of the basin not affected by the Klamath Project. About 43% of consumptive use in the upper basin occurs outside the Klamath Project, and 57% occurs through the project, which irrigates about 220,000 acres. There is also a substantial amount of irrigation along tributaries in the lower basin beyond the boundaries of the Klamath Project. Agricultural uses of irrigation water are numerous (Chapter 2), and include extensive production of alfalfa by use of sophisticated water-distribution systems and reuse of irrigation tail water.

Irrigators have consistently been skeptical of reasoning that suggests a need for changing water management for the benefit of endangered and threatened fishes; their consultants have entered the debate about the merits of various hypotheses underlying proposed changes in water use. The experiences of 2001, when the occurrence of a drought coincided with the USFWS and the NMFS biological opinions to make the delivery of irrigation water from the Klamath Project virtually impossible for the first time since the creation of the project, sharpened the objections but also have increased the interest of the agricultural community in restoration projects that may benefit endangered and threatened species without curtailing the availability of water (Chapters 2 and 4).

General environmental interests in the Klamath River basin are strong, in part because of the extraordinary value of environmental resources in the basin. Environmental interests have worked toward the moderation of consumptive use and the remediation of past damage to environmental resources. The Nature Conservancy, for example, has purchased a large tract of land on the northern shore of Upper Klamath Lake (Figure 1-3), where it is restoring wetlands (Chapter 2).

Oregon and California also are involved in assessing and forming opinions on endangered and threatened fishes in the upper Klamath basin. The states have placed the two endangered sucker species and the coho under special protection and have supported extensive studies, including those related to the Environmental Protection Agency's (EPA) total maximum daily load (TMDL) requirements, which are administered through the states (e.g., Boyd et al. 2001).

Logging, mining, and commercial fishing are important forces within the basin (Chapter 2). Logging and mining, although reduced from their past maximums, have been cited as sources of habitat degradation, but they operate outside the reach of the Klamath Project. Commercial and sport fishing for salmon and subsistence fishing by the tribes have been drastically curtailed in recent decades, first as a result of declining fish populations and then in a regulatory effort to protect the remaining stocks. The change has caused a loss of income and food for inhabitants of the lower basin.

The biological assessments and biological opinions on the endangered and threatened fishes have focused primarily on the operations of the Klamath Project because federal agencies must operate federal facilities in such a way as to avoid jeopardy to endangered or threatened species (Chapter 9). Other potential threats to the endangered and threatened fishes exist outside the range of the Klamath Project, however, and cannot be remedied solely through requirements related to USBR, which lacks direct control over use of land or water outside the area of the Klamath Project.

THE COMMITTEE

The cessation of water deliveries through the Klamath Project during 2001 as required by the jeopardy opinions on coho salmon and the two endangered sucker species of the Klamath River basin motivated the U.S. Department of the Interior and the U.S. Department of Commerce to seek an outside evaluation of the scientific basis of the requirements set by

USFWS and NMFS for higher water levels in Upper Klamath Lake and higher main-stem flows in the Klamath River. These federal agencies therefore asked the NRC to create a committee to be charged with external, independent review of the biological opinions and assessments and of the long-term needs of the endangered and threatened fishes in the Klamath River basin. As a result of the request, the NRC formed the Committee on Endangered and Threatened Fishes in the Klamath River Basin.

The committee was provided with a written statement of task, as given in Appendix A. The task has two components. First, the committee was asked to complete an interim report by early 2002. The interim report was to focus on the scientific strength of the biological assessments and opinions issued in 2001 on threatened coho salmon and endangered suckers in the Klamath River basin. The purpose of the interim report was to allow the federal agencies to consider a preliminary external review as they were writing their biological assessments and opinions for 2002, which they needed to do because the assessments and opinions of 2001 extended for only 1 yr.

Second, the committee was to prepare a final report to be issued in 2003. The scope of the final report includes the biological assessments and opinions of 2002 but also extends to all matters related to the long-term welfare of endangered suckers and threatened coho salmon in the Klamath River basin. Like the interim report, the final report focuses on the scientific basis of actions that are proposed or required by federal agencies for the benefit of the endangered and threatened fishes. Another important aspect of the final report is its analysis of the need for additional studies of specific issues about which there is too little knowledge to support confident proposals for remedial action.

The committee's interim report proved controversial. The committee found strong scientific support for all components of the reasonable and prudent alternatives given by USFWS in 2001 for the endangered suckers except for recommendations on maintenance of higher water levels in Upper Klamath Lake, for which the committee found no empirical support. At the same time, however, the committee found that USBR's recommendations, which could have caused mean water levels in Upper Klamath Lake to be lower than in the recent past, also were without scientific support. Thus, the committee's overall conclusion was that there was no substantial scientific evidence to support deviation from the water levels produced by operational principles that were in effect during the 1990s. Similarly, in reviewing the biological opinion of NMFS on the coho salmon, the committee concluded that all components of the reasonable and prudent alternative were supported scientifically except the one calling for higher flows in the Klamath River main stem. The committee found little scientific support for these recommendations in relation to coho salmon, nor did it find any scientific justification for the proposals of USBR, which would have allowed the river to be operated at lower mean flows than had been the case for specific categories of water availability applicable during the 1990s.

The committee, in drawing conclusions for its interim report, was bound by its charge to evaluate and comment on the scientific strength of evidence underlying various proposals. Its charge kept it from weighing economic concerns or weighing the advisability of minimizing risk by using professional judgment in place of scientific evidence to support particular recommendations. As explained more fully in Chapter 9, agencies charged with ESA responsibilities can be expected to use professional judgment when no scientifically supportable basis is available for a decision, or where they judge the scientific support to be inadequate. Thus, the agencies may recommend practices for which the committee would find virtually no direct scientific support. The committee acknowledges the necessity of this practice in many situations where information is inadequate for development of scientifically rigorous decisions (Chapter 9).

For its final report, the committee adopted some specific conventions for judging the degree of scientific support for a specific proposal or hypothesis; Table 1-2 gives a summary. Any proposal for specific actions of a remedial or protective nature has an implicit or explicit underlying hypothesis that connects the proposed action with a beneficial effect on a threatened or endangered species. The scientific value of such a hypothesis ranges from negligible to very high, depending on the amount of testing to which it has been subjected. At the low end of the scale of scientific strength is an assertion or proposal that is entirely intuitive and thus without scientific support. For example, the catch phrase "fish need water" has been used as an assertion supporting increased water levels in Upper Klamath Lake and

TABLE 1-2 Categories Used by the Committee for Judging the Degree of Scientific Support for Proposed Actions Pursuant to the Goals of the ESA

Basis of Proposed Action	Scientific Support	Possibly Correct?	Potential to be Incorrect
Intuition, unsupported assertion	None	Yes	High
Professional judgment inconsistent with evidence	None	Unlikely	High
Professional judgment with evidence absent	Weak	Yes	Moderately high
Professional judgment with some supporting evidence	Moderate	Yes	Moderate
Hypothesis tested by one line of evidence	Moderately strong	Yes	Moderately low
Hypothesis tested by more than one line of evidence	Strong	Yes	Low

increased flows in the main stem of the upper Klamath River. The statement is true, but it does not constitute a scientifically valid argument for specific flows or specific water levels.

Professional judgment has more value than unsupported intuition. It typically is based on knowledge of the importance of various environmental factors or the requirements of various species in other locations or on general experience with or knowledge of the response of a particular category of organism to specific kinds of environmental challenges.

Professional judgment can be used in three ways, and the distinctions among them are quite important in the case of the Klamath River basin. First, for an issue about which there is no information whatsoever, an agency that is charged with protecting a threatened or endangered species can justify the use of professional judgment. Such agencies are charged with reduction of risk to the species; lacking site-specific information on a particular type of risk, they would logically draw analogies with the same or similar risks in other settings or for other species, or they would use general principles related to the known tolerance of particular species or groups of species. Although such an approach is weak in that the transferability of ecological knowledge from one set of circumstances to another is problematic, there is some scientific basis for it, and barring the feasibility of other approaches, it can be said to have weak but not negligible scientific strength.

Second, a resource agency might use professional judgment to endorse various proposals for action when valid scientific information contradicts it. This use of professional judgment is difficult to justify. The agency may hold to its desire to use professional judgment in preference to empirical information of direct significance to a particular issue on the grounds that something is wrong with the empirical information. Scientifically, however, sound and relevant empirical information always trumps speculation or generalization; an agency could argue the reverse only on the basis of a very conservative approach to risk.

Third, an agency might choose to use professional judgment that is consistent with a small amount of direct evidence. In this case, the use of professional judgment is reinforced rather than contradicted, and scientific support for it can be deemed moderate rather than negligible.

A step beyond professional judgment is the empirical testing of scientific hypotheses involving cause and effect. If a properly designed single line of evidence is developed as a means of testing such a hypothesis, and the hypothesis is not invalidated, scientific support for the hypothesis can be considered moderately strong. Ideally, this approach would be extended by the collection of additional, independent evidence through which the hypothesis could be tested in a different way; barring contradiction between the evidence and the hypothesis, the hypothesis could be considered a theory of considerable strength to be relied on in proposing and pursuing vigorously the action upon which the hypothesis is based.

The committee has used the six-tiered system summarized in Table 1-2 and described above in assessing the scientific basis of actions that have been recommended in the Klamath basin for protection of the endangered suckers and threatened coho salmon. It found its greatest differences with the resource agencies in the second category: instances in which the agencies have used professional judgment that is contradicted by scientifically valid, relevant evidence. In carrying out its task to categorize the scientific support for specific proposals, the committee would characterize any proposal justified by such means as having negligible scientific support. This does not preclude the resource agency from using such an approach, but the justification for it would involve extreme sensitivity to risk, and in this way might be judged not reasonable.

The committee's charge requires that it estimate the costs associated with its recommendations. For the recommendations involving additional research or monitoring, the committee was able to approximate costs based on the experience of the committee members with similar types of research. Even so, the mode of implementation of a particular research program could cause costs to deviate markedly from the committee's estimates. For example, implementation could involve a much broader or narrower geographic scope than suggested by the committee, or it could involve multiple organizations in a way that would increase costs. The committee also was able to estimate, on the basis of general experience, the costs of selected minor restoration activities. The committee did not attempt, however, to estimate costs for major restoration activities. In most instances these activities must be studied for feasibility prior to the time any commitment is made to them, and their final approval and execution may be complicated to an extent that cannot be meaningfully judged by the committee in terms of cost.

SUMMARY OF THE BIOLOGICAL ASSESSMENTS AND BIOLOGICAL OPINIONS OF 2002

The biological assessments issued by USBR in 2001 and the biological opinions issued by USFWS and NMFS in 2001 all expired after 1 yr, so new assessments and opinions were issued in 2002. The assessments and opinions of 2002 differ from those of 2001 in several respects. First, they cover a 10-yr interval rather than a 1-yr interval. In working with 10 yr rather than 1 yr, the agencies are cooperatively attempting to stabilize and add flexibility to management in such a way as to benefit both water use and environmental remediation. At the same time, consultation between the agencies probably will continue, and requirements of USFWS and NMFS probably will be modified within the 10-yr interval as new information becomes available. Reinitiation of consultation is required by ESA Section 7 under some circumstances, and both USBR and NMFS must issue a new biological assessment and opinion in any case because of the ruling of a U.S. District Court (see below). The texts of assessments and opinions of 2002 show that they were influenced to some extent by the committee's interim report. The interim report was not binding on the agencies but provided a basis for additional consultation and appears to have stimulated some new kinds of discussions among the agencies.

Endangered Suckers

The USBR Biological Assessment

The USBR, which in 2001 had prepared two assessments (one for the threatened coho and one for the two endangered sucker species), dealt with all three species in a single document during 2002. This makes sense because water resources at times of scarcity must be shared not only among consumptive uses and listed species but also among the listed species themselves, given that the coho and the suckers occupy different parts of the basin. USBR proposed maintenance of specific water levels in lakes and some other actions previously suggested by USFWS or others, reflecting the consultation process through which gaps between the viewpoints of the agencies are intended to be minimized.

Table 1-3 lists in abbreviated form the commitments that USBR made in its 2002 assessment to accommodate the needs of the endangered suckers. It proposed to manage water levels in Upper Klamath Lake, Clear Lake, and Gerber Reservoir so as to stay within the operating ranges of the 1990s. Specifically, it proposed not to allow water levels to fall below the 1990–1999 minimums for specific water-year categories and not to allow the mean water level for any water-year category to decrease through increased average drawdown. Thus, the water-level proposals in the assessment were responsive to the criticism made by the committee in its interim report (2002) that the USBR proposal of 2001 would have allowed, without any ecological rationale relevant to the suckers, greater mean drawdown within any given water-year category.

A second element of the assessment is a water bank, which USBR proposed to be as large as 100,000 acre-ft. The water bank would provide operational flexibility in meeting multiple needs for water during years of water scarcity and would help USBR to ensure that water-level targets in lakes (or flow requirements at Iron Gate Dam, for coho salmon) would be met.

USBR also proposed a procedure for developing project operations in a particular water year. The procedure would begin in April with classifica-

TABLE 1-3 Summary of Commitments of the USBR Biological Assessments of 2002 that are Relevant to the Two Endangered Sucker Species

Assessment Commitments
Water levels in Upper Klamath Lake, Clear Lake, and Gerber Reservoir:
Maintain water levels at or above 1990–1999 minimums for specific water-year types[a]
Maintain mean water levels at or above 1990–1999 means for specific water-year types
Establish water bank of about 100,000 acre-ft
Use specific procedure for determining annual operations, including 70% exceedance principle for water availability
Coordinate externally and produce annual report on operations
Reduce entrainment and enhance passage in Link River and at other locations
Enhance water supply
Cooperate with USFWS in operation of refuges

[a]Special concerns and procedures are clarified by subsequent memoranda on Clear Lake and Gerber Reservoir (USBR, unpublished memo, February 21, 2003; USFWS, unpublished memo, March 4, 2003).

tion of the year by water-year type—above average, below average, dry, or critical dry (see Chapter 3 for details)—through the use of forecasts from the National Resource Conservation Service (NRCS). A 70% exceedance factor would be used in applying the forecast; that is, forecasts of the availability of water for the Klamath Project would be conservative in that there would be a 70% chance that the forecast would be equaled or exceeded by actual water availability. Having thus classified a developing water year as belonging to one of the four categories, USBR would follow specifications on minimum water levels for the appropriate water-year category. A second, later calculation would facilitate maintenance of water levels in lakes no lower than the average (rather than the minimum) end-of-month elevations for specific water-year types over the interval 1990–1999.

Another component of the assessment was a commitment to an annual report on operations, which would be useful because of the general interest in operations and the difficulty of discovering the details of operations without an interpretive document. Coordination not only with USFWS, as required through ESA, but also with other groups is a component of this portion of the assessment proposal.

The USBR proposed to reduce entrainment of fish by diversions and to increase fish passage in the Link River. Specifically, entrainment of fish at the A Canal is known to be large. Entrainment of fish above a size of about 30 mm would be reduced by installation of a permanent fish screen by a specified date (April 1, 2004). Salvage operations are included, as are measures to promote fish passage at the Link River Dam to be completed in January 2006. Increase in water supply through increased storage capacity and leasing also is a component of the proposals from USBR for 2002, but details are not yet available. Because these measures would require congressional approval and funding, they were not attached to a specific schedule in USBR's assessment.

The USFWS Biological Opinion

In responding to the portion of the USBR assessment dealing with endangered suckers, USFWS, through its biological opinion of 2002, reacted favorably to a number of the USBR proposals, including the water bank and specifically scheduled actions intended to reduce entrainment and improve fish passage. In the text of its opinion, however, USFWS expressed its position that water levels higher than those proposed by USBR would be favorable to the suckers through improvement in water quality and maintenance of habitat (see Chapters 3 and 6). Overall, USFWS found that the operations proposed by USBR would leave the two endangered sucker species in jeopardy and therefore formulated an RPA under which USBR must operate (Table 1-4).

The USFWS concluded that low water levels in the lakes are less favorable than high water levels to the welfare of the suckers. It required that water levels in the lakes not deviate from minimums (for single years) or averages (for groups of years) of the 1990s for specific categories of water years, as proposed by USBR. In addition, USFWS required through its RPA that USBR use a 50% exceedance probability rather than a 70% probability in forecasting water availability. As shown in the USFWS biological opinion, use of a 70% forecast, although favorably conservative for water-management purposes in tending to underestimate water availability, could be unfavorable from the environmental point of view if it were allowed to justify water-level drawdown in lakes more extreme than would be consis-

TABLE 1-4 Summary of Components of USFWS Biological Opinions of 2002 that are Relevant to the Two Endangered Sucker Species of the Klamath River Basin

Component of Biological Opinion[a]
Use 50% rather than 70% exceedance probability for planning water levels in Upper Klamath Lake
Screen power-plant intakes at Link River Dam
Study cause of death and habitat needs of endangered suckers in Upper Klamath Lake
Take actions leading to more favorable water quality and expansion of habitat
Monitor populations of endangered suckers
Produce annual assessment report on suckers
Follow specific implementation schedule

[a]Components shown here are in addition to proposals of the USBR in its biological assessment.

tent with the actual availability of water. Thus, USFWS justified the 50% exceedance requirement for estimates as a means of ensuring that estimates of water availability would not be biased. Currently, it appears that USBR and USFWS are in agreement that April projections can be corrected as appropriate whenever they later appear to have been in error (USFWS 2002; p. 118).

A second element of the RPA was to reduce entrainment of fish at Link River Dam and hydropower intake facilities. USBR had committed to screening the A Canal, but it did not make the same commitment for the power-production facilities at Link River Dam. Thus, the USFWS RPA appears to extend USBR's commitment to screening. This requirement of the RPA raises questions about the feasibility of requiring USBR to manage entrainment for facilities that are operated by PacifiCorp, a power production company. The application of this feature of the RPA to the Link River Dam will depend on the nature of the federal action that USBR takes with respect to PacifiCorp's operation of the facilities. If USBR has sufficient discretionary authority over PacifiCorp's operation within the meaning of ESA Section 7 (carry out, fund, or authorize operations) for the facilities to be properly within the scope of the interagency consultation, the RPA would be an appropriate component of the USFWS biological opinion. If not, USFWS would need to explore application of ESA Section 9 to PacifiCorp and determine whether PacifiCorp would be in violation of the ESA in the absence of screening and other measures that may be developed between USFWS and PacifiCorp (see Chapter 9). Thus, USFWS and USBR still must clarify the status of the Link River Dam operations under Section 7 of the ESA.

Other requirements of the biological opinion are that USBR study the causes of mass mortality of fish and access of endangered suckers to habitat in Upper Klamath Lake, take actions designed to reduce unfavorable aspects of water quality or limitations in sucker habitat, monitor populations of endangered suckers, and produce an annual assessment report. A detailed implementation schedule and requirements for collaborative work of USBR with other parties accompany this element of the RPA.

Threatened Coho Salmon

The USBR Biological Assessment

In its biological assessment of 2002, the USBR made a number of proposals relevant to coho salmon, as shown in Table 1-5. First, USBR committed itself to maintain river discharges no lower than those observed during 1990–1999 for the categories of water years that it uses in water management. It also committed itself to maintain interannual averages no

TABLE 1-5 Summary of Components of USBR Biological Assessments of 2002 that are Relevant to Threatened Coho Salmon of the Klamath River Basin

Assessment Component
Discharge of water from Iron Gate Dam
Above-average and below-average years: monthly flow will be no lower than 1990–1999 year minimums or FERC minimums, whichever is greater
Dry and critical-dry years: monthly flow will be no lower than actual 10-yr averages plus pulse of 10,000 acre-ft in April
Establish water bank of about 100,000 acre-ft
Use specific procedure for determining annual operations, including 70% exceedance principle
Coordinate externally and produce annual report on operations
Enhance water supply

lower and sometimes higher than interannual averages of 1990–1999 for specific categories of years, thus answering the concern expressed in the committee's interim report that a commitment to maintain minimums without a commitment to maintain averages would in fact allow future operations to produce lower averages.

As was the case for water levels of Upper Klamath Lake, Clear Lake, and Gerber Reservoir, USBR proposed to use a 70th percentile exceedance factor applied to the April 1 forecast of NRCS for planning annual operations. For above-average and below-average years, USBR proposed to provide flows no lower than the minimums observed during the 1990s and also no lower than the Federal Energy Regulatory Commission (FERC) minimums if the FERC minimums happen to be higher. For the two drier categories of years (dry and critical dry), USBR proposed to provide flows no lower than the observed averages for the 1990s and also to provide 10,000 acre-ft of additional flow during April to facilitate smolt migration. The use of averages rather than minimums from the 10-yr observation period is a commitment of additional water above what had been committed by USBR in its 2001 assessment, as is the 10,000 additional acre-ft for April.

An additional component of the proposed operating plan for any given year is the establishment and operation of a water bank, which also serves the needs of endangered suckers, ultimately to be as large as 100,000 acre-ft. Mechanisms for water banking could involve offstream storage but also could include reduction in irrigation demand with compensation to irrigators and conjunctive use of groundwater and surface water to provide a buffer that would be especially useful in dry years (Chapter 10).

The USBR proposal also made a commitment to coordination extending beyond the ESA implementation agencies to include the tribes, Pacifi-

Corp, and private water users. Coordination would be supplemented with an annual report documenting the preceding year's activities. Enhancement of water supply, not necessarily limited to the water-banking concept, was also an element of the USBR proposal.

The NMFS Biological Opinion

After consultation with USBR during 2002, NMFS concluded that proposed actions of USBR as presented in its 2002 biological assessment, although containing several constructive components, would leave the threatened coho in jeopardy. Thus, according to the requirements of the ESA, NMFS prepared a biological opinion containing an RPA summarized in Table 1-6. In revising its biological opinion of 2001, NMFS recognized that the Klamath Project accounts for about 57% of the total irrigation-related depletions of flow at Iron Gate Dam. Thus, according to the opinion of 2002, it would not be reasonable to require USBR to provide directly and immediately all increments of flow judged by NMFS to be necessary for improvement of habitat in the main stem of the Klamath River below Iron Gate Dam. Accordingly, NMFS assigned USBR a 57% share in the responsibility for providing flows in the main stem to meet the requirements of the threatened coho as judged by NMFS. In doing so, however, NMFS did not absolve USBR entirely of responsibility for making up the other 43% of flows. The biological opinion requires USBR to facilitate and coordinate a phased effort to provide capacity for the additional flows.

NMFS, as part of the RPA, requires USBR to build a water bank, which USBR has agreed to be its preferred method for meeting its obligation to provide the 57% of flow shortfalls that NMFS will require it to provide for support of the threatened coho salmon (specific flows are shown in Table 9

TABLE 1-6 Summary of Components of NMFS Biological Opinions of 2002 that are Relevant to Threatened Coho Salmon in the Klamath River Basin

Component of Biological Opinion[a]
Apply 57% rule for proportionate USBR direct responsibility for flow at Iron Gate Dam
Use task force to develop the 43% additional flow from nonproject sources
Use phased approach to raising flows and lowering temperatures
Develop water bank (100,000 acre feet) on specific schedule
Adopt water-year types as identified in draft phase II flow study report (Hardy and Addley 2001)
Limit ramping rates below Iron Gate Dam
Conduct designated scientific studies with advice from external experts

[a]Components shown here are in addition to proposals of the USBR in its biological assessment.

of NMFS 2002 and in Chapter 4 of this report). USBR must create a water bank to 100,000 acre-ft capacity by 2006 according to the RPA. A U.S. District Court judge found during July 2003, however, that reliance on the water bank is unjustifiably speculative until more particulars are given. Thus, USBR soon must issue a new biological assessment in consultation with NMFS, which must issue a new biological opinion.

In its recommendation for flows, NMFS gave greatest emphasis to improvement of the conditions for smolt migration, probably because tributary conditions are most important for spawning and rearing, while the main stem performs a critical and irreplaceable function in smolt migration (Chapter 7).

In prescribing flows, NMFS did not follow the method of USBR in assigning specific water years to categories. NMFS used estimates of unimpaired flows from the Hardy Phase II draft report (Hardy and Addley 2001) and the idea that the shape of the natural hydrograph and a natural range of interannual variabilities should be represented as completely as possible in the flows of the main stem. The five categories and their percentiles used by NMFS in its flow prescriptions for the Klamath main stem are as follows: wet years, 10%; above-average years, 30%; average years, 50%; below-average years, 70%; and dry years, 90%. The percentile in each case indicates the proportion of years that would exceed the unimpaired monthly flows. The RPA provides specific dates by which USBR must meet the flow requirements.

NMFS specified upper limits on ramping rates below Iron Gate Dam. The specifications are more stringent and more detailed than those governing previous operations. As in the case of screening plant intakes, however, the direct responsibility for meeting this requirement may lie with PacifiCorp rather than USBR.

According to the RPA of 2002, USBR is required to convene a panel of experts capable of identifying studies that improve the current understanding of relationships between river discharge and welfare of coho salmon. One specific element of the studies is a test of the effect of various flows on thermal refugia in the main stem of the Klamath River.

Overview of the 2002 Biological Assessments and Opinions

The USBR assessment and the accompanying biological opinions of USFWS and NMFS for 2002 reflect considerable constructive interaction among the agencies between 2001 and 2002. There is still a gap between the assessments and the opinions, but the gap has narrowed from 2001 through some carefully considered movement toward consensus among the three agencies. USFWS and NMFS are requiring some substantial actions beyond those proposed by USBR. In general, however, the actions adhere more closely than those given by the listing agencies in 2001 to the relevant available scientific evidence or to professional judgment reinforced by at least some scientific evidence. As explained in this report, USFWS and NMFS in a few instances have made requirements based almost entirely on professional judgment, without direct scientific support, as is their prerogative. In doing so, however, they appear to have made a special effort to frame their requirements in such a way as to cause minimal impairment of Klamath Project operations and, in contrast with 2001, have recognized the inevitable need to include parties other than USBR in modification of environmental conditions for the benefit of the endangered and threatened fishes.

CONTEXT FOR THE COMMITTEE'S REPORT

The NRC committee has evaluated a very extensive accumulation of data collected both in the field and laboratory, historical records of various kinds, opinions and interpretations by individuals intimately familiar with the environmental conditions in the Klamath, and numerical analyses of many kinds. Though the documentation for questions related to endangered and threatened fishes in the Klamath basin is impressive in scope and volume, it must be viewed as a preliminary step toward what eventually can and must be known about the Klamath River basin in support not only of the recovery of endangered fishes but also of the more general restoration of aquatic environments in the Klamath basin. As will be shown by this report, the number of firm conclusions that can be reached about cause-and-effect relationships still is modest, yet these types of conclusions are essential for planning, managing, and predicting the outcomes of actions in the Klamath River basin. The NRC committee sees its own work only as a best effort given the information available; the committee fully expects to see new kinds of data and new tests of ideas yield insights that the committee could not have anticipated based on current information. Effective efforts to cause recovery of the endangered and threatened fishes rest on information, and the committee urges the creation of new information that will place management decisions on increasingly firm ground.

2

Land Use and Water Management

The Klamath River watershed covers 12,000 mi^2 of northern California and southwestern Oregon and extends more than 350 river mi from its headwaters to its estuary at the Pacific Ocean. The watershed derives its unique character largely from its geology and climate (Mount 1995), which are discussed in the first quarter of this chapter. The rest of the chapter describes land uses and resulting changes in the basin since 1848, the beginning of the gold-mining era. The topography, hydrology, ecosystems, and unusual plant and animal communities of the watershed reflect diverse dynamic processes in the landscape of today and in the past. These features of the watershed are tied to the natural resource economies of the watershed, which include logging, grazing, agriculture, mining, and fisheries. The diversity of land uses and landscape features poses a significant challenge to land managers and those seeking to restore the watershed's aquatic communities. As this chapter shows, simple or uniform approaches to restoration of impaired ecosystems are unlikely to succeed in a watershed as diverse as that of the Klamath River.

DESCRIPTION OF THE KLAMATH RIVER WATERSHED

Geologic Setting

The physiography of the Klamath watershed records the oblique convergence between the North American tectonic plate and the plates that underlie the Pacific Ocean. The Juan de Fuca and Gorda Plates, which lie off the shore of Washington, Oregon, and northern California, are being subducted in a northeasterly direction beneath western North America, forming the Cascadia subduction zone (Figure 2-1). A consequence of the subduction is the formation of an extensive north-south oriented chain of volcanoes known as the Cascadia volcanic arc or Cascade Range. The arc includes two of the more prominent volcanoes in the upper Klamath watershed: Mount Shasta and Mount Mazama (the site of Crater Lake). The volcanic arc bisects the Klamath watershed, dividing the upper basin from the lower basin (Figure 2-1). The upper basin, including the large natural lakes and their tributaries, lies in the back-arc of the Cascadia margin. The lower basin—which includes the mountainous, steeper portions of the main-stem Klamath and the Scott, Salmon, and Trinity rivers—lies in the dynamic fore-arc area of the margin. The Shasta River straddles the tectonic boundary between the back-arc and the fore-arc (Figure 2-1); its confluence with the main-stem Klamath occurs in the fore-arc region.

Geophysical and geodetic surveys coupled with geologic mapping efforts have shown that portions of the fore-arc and back-arc regions of the Cascadia margin form discrete crustal blocks, each with its own motion (Wells et al. 1998, McCaffrey et al. 2000). The motion of these blocks and their interactions with each other have dictated the dynamic topography of the region.

Tectonic Setting of Klamath Watershed

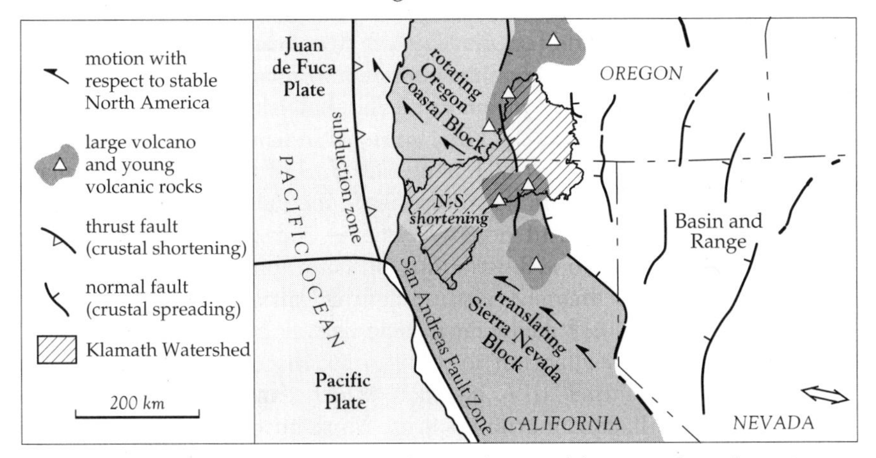

FIGURE 2-1 General tectonic setting for northern California and southern Oregon illustrating the Cascadia subduction zone, the Cascade volcanic arc, the Basin and Range Province, and the Oregon fore-arc and Sierra Nevada blocks. Note that the Klamath watershed occurs at the intersection of these tectonic blocks. Source: Modified from Wells and Simpson 2001.

Within the Klamath watershed region, the back-arc portion of the Cascadia margin is part of the crustal block known as the Basin and Range Province. Although attached to North America, the province is undergoing east-west extension of as much as 1 cm/yr (Bennett et al. 1998, Magill et al. 1982). Right-lateral shear oriented north northwest-south southeast occurs along the western edge of the province and is superimposed on the east-west extension (Bennett et al. 1998, 1999). This shear has formed the distinctive grabens showing north-northwest south-southwest orientation, which appear topographically as fault-bound troughs and valleys of the Klamath Lake area. The crustal extension of the northwestern basin and range in southern Oregon and northern California has been accompanied by widespread Neogene volcanism that has formed the distinctive volcanic tablelands and broad valleys and marshes of the upper tributaries within the Klamath watershed.

Unlike most watersheds, the Klamath watershed has its greatest relief and topographic complexity in its lower half rather than in its headwaters. This unusual physiography stems from the location of the fore-arc region, which encompasses the lower half of the watershed. The Cascadia fore-arc of northern California is arguably the most dynamic landscape in the region (Mount 1995). The regional compression associated with subduction of the Gorda Plate immediately off shore has produced some of the fastest rates of uplift recorded in California. Additionally, the fore-arc occurs at the poorly defined intersection between two large crustal blocks (Figure 2-1): the Sierra Nevada block and the Oregon fore-arc block (Wells et al. 1998, McCaffrey et al. 2000). The Sierra Nevada block includes the Sierra Nevada-Great Valley of central California and the Klamath Mountains and Coast Ranges of Northern California. The block is bounded on the east by the Basin and Range Province and on the west by the San Andreas-Coast Range Fault system (Wells et al. 1998). Geodetic surveys indicate that the block is moving northwest relative to North America and is rotating in a counter-clockwise manner (Argus and Gordon 1990). The Oregon fore-arc block extends from the Cascadia subduction zone on the west to the Basin and Range on the east. Its southern boundary occurs at the transition to the Sierra Nevada block, roughly in the vicinity of the California-Oregon border. The Oregon fore-arc block is rotating clockwise relative to North America (Wells et al. 1998).

The lower Klamath River watershed, which extends from Iron Gate Dam to the Klamath estuary, traverses the northern portions of the Sierra Nevada block along its transition to the Oregon fore-arc block (Figure 2-1). The steep, rugged watersheds of the lower Klamath, coupled with the bed-rock-controlled main stem, reflect the rapid uplift in the region and the constant adjustment of the river to its dynamic landscape (Mount 1995). The patterns of uplift and faulting also control the orientation of most tributaries. Because the main tributaries of the lower Klamath River—the Shasta, Scott, Salmon, and Trinity rivers—are important for salmonids, their individual geologic features are of interest.

The Shasta River watershed is at the junction between the Basin and Range Province, in the Sierra Nevada block within the Cascadia volcanic arc. Its watershed, which originates at Mount Eddy, encompasses about 800 mi^2. Like the Scott River watershed to the west, the Shasta has a large central alluvial valley, steep headwaters on the west, and a steep gorge in the lowermost portion of the watershed. The eastern portions of the watershed are dominated by Tertiary and Quaternary volcanic flows and by debris flows associated with Cascade volcanism. The lower gorge and westernmost edge of the basin are underlain by Paleozoic metamorphic rocks of the Sierra Nevada block. The most conspicuous topographic feature of the Shasta Valley is a large Pleistocene volcanic debris avalanche derived from nearby Mount Shasta that creates the unusual hummocky topography in the upper reaches of the valley (Crandall 1989). The north-south orientation of the valley is associated with large basin and range faults similar to those controlling the formation of the upper basin. The hydrology of the Shasta River watershed, unlike that of the other tributary watersheds of the lower basin, is dominated by discharge from numerous springs.

The Shasta subbasin lies within the extensive rain shadow of the Salmon and Marble mountains. Precipitation averages 12–18 in/yr and is as low as 5 in/yr in the vicinity of Big Springs (Mack 1960). The bulk of this precipitation occurs from October to March as snow. Like the upper Klamath basin, the Shasta subbasin has warm summers (mean daily temperatures commonly exceeding 30°C) and cool winters (mean daily temperatures of 5°C). The average length of the growing season in the basin is about 180 days (Mack 1960). As discussed in Chapter 8, climate may change over the coming decades.

The Scott River watershed lies at the transition between the Cascadia volcanic arc and the fore-arc basin (Figure 2-1). The watershed, which is about 820 mi^2, has headwaters nearly 8,000 ft above sea level in the Salmon Mountains along the west side of the watershed. The Scott joins the Klamath River at river mile 142. The physiography of the watershed shows elements of its neighboring watersheds. Like the Salmon watershed, the headwaters of the Scott are heavily forested and have annual precipitation of 50 in or more, high water yields, and extensive snowpack more than 4,000 ft above sea level. Like the Shasta watershed, the Scott has a large, fault-bound alluvial valley in the middle portions of the watershed that supports extensive agriculture and grazing. This valley, like the eastern portion of the Scott watershed, lies in the rain shadow of the Salmon and Marble mountains; mean annual precipitation is about 20 in. The Scott

River, like the Shasta River, has a steep bedrock gorge downstream of the alluvial valley and above its confluence with the Klamath River. Mean daily temperatures in the valley exceed 32°C during late July or early August (peaks, above 40°C); mean daily temperatures reach 10°C in winter (Rantz 1972, CDWR 2002).

The tributaries of the Scott River strongly affect the hydrology (Mack 1958) and aquatic habitat of the basin. The fourth-order tributaries of the west side of the watershed—including Scott, French, Sugar, Etna, Patterson, Kidder, and Shackleford creeks—are steep-gradient, perennial bedrock tributaries. Several of these tributaries have built coarse-grained alluvial fans where their gradients decrease as they meet the valley floor. In contrast, the East and South Forks of the Scott and the third- and fourth-order creeks of the Scott River Canyon, a tributary of the Scott, enter the river in steep reaches and have no alluvial fans. The relatively dry east side of the watershed has several low-gradient ephemeral tributaries; Moffett Creek is the largest and most important of these.

In its upper reaches and within the canyon, the Scott River is primarily a bedrock river characterized by alternating step-pool and cascade reaches with discontinuous riffle-pool reaches containing narrow alluvial floodplains. Within the Scott Valley, the river has various forms that are controlled principally by grain size, slope, tributary contributions, and channel modifications. In coarse-grained, steep-gradient reaches of the river, the channel appears to be actively braiding. In low-gradient, fine-grained reaches with cohesive banks, the channel alternates between a single-channel meandering river and a multichannel, anastomosing river, albeit with numerous modifications for flood management and irrigation diversions. Some incision within the channelized reach has lowered the channel bed by several feet (G. Black, Siskiyou Resource Conservation District, Etna, California, personal communication, 2002). Sloughs, which indicate historical channel avulsion and cutoff events, apparently were numerous before agricultural development of the valley. Several large sloughs remain in the valley along the west side and receive flow from tributaries and from the main stem during large flow events.

At 750 mi^2 the Salmon River is the smallest of the four major tributaries to the lower Klamath basin (Figure 2-1). The Salmon watershed is steep and heavily forested and, in comparison with its neighboring watersheds, relatively undisturbed. The bulk of the main stem and its tributaries consist of bedrock channels with numerous step-pool and cascade reaches and narrow riparian corridors. The watershed is located entirely within the Cascadia fore-arc region on the Sierra Nevada block. The high uplift rates and the lack of extensional tectonics have prevented the formation of any important alluvial valleys, such as those of the Scott and Shasta drainages. The rugged terrain and the lack of a large alluvial valley have limited some of the land-use activities that have affected anadromous fishes in other tributaries.

The Trinity River is the largest tributary to the Klamath River. At 2,900 mi^2 with an annual average precipitation of 57 in, it is also the largest contributor of runoff and sediment to the Klamath River. It is a rugged, step, and heavily forested watershed. Its eastern portions in the Trinity Alps and Coast Ranges reach elevations in excess of 9,000 ft and support thick winter snowpacks. The bulk of the watershed is below 5,000 ft in elevation and is dominated by conifer and mixed conifer and hardwood forests. The confluence of the Trinity and Klamath rivers is located 43 mi upstream of the mouth and exerts considerable influence over conditions in the lowermost Klamath River and its estuary. The Trinity watershed is located entirely within the Sierra Nevada block, west of the Cascade volcanic arc. The basin lies close to the junction between the Cascade subduction zone and the northernmost San Andreas Fault. The physiography of the watershed is controlled by high rates of uplift and a series of large, seismically active northwest trending faults. The eastern half of the basin is composed of rocks of the Klamath Mountains Geologic Province, while the western half is dominated by rocks of the Coast Range Geologic Province. Both provinces contain rock types that are prone to landsliding and high rates of erosion, particularly when disturbed by poor land-use practices. The high rates of uplift, unstable rock types, and high rates of precipitation produce a naturally dynamic landscape and a river with a variable hydrograph and sediment yields.

Uplift in the Trinity watershed has precluded the formation of extensive alluvial valleys such as those found in the Scott and Shasta watersheds. The upper reaches of the main stem and the tributaries support steep-gradient rivers with numerous cascades. In portions of the main stem and the South Fork, however, low-gradient reaches with narrow alluvial valleys occur. These reaches historically supported dynamic, meandering coarse-grained channels that provided ideal spawning and rearing habitat for salmon and steelhead. The size of the Trinity watershed, coupled with its extensive high-quality spawning and rearing habitat, made the Trinity a productive source of coho salmon and other anadromous fishes (USFWS/HVT 1999).

Climate and Historical Hydropattern

The tectonic setting of the Klamath watershed exerts primary control over its irregular distribution of precipitation. The uplift of the Cascadia fore-arc and the formation of the Cascade volcanic arc have produced an important rain shadow in the upper basin and the Shasta Valley. The upper watershed has a relatively low mean annual precipitation (27 in; Risley and Laenen 1999), about half of which falls as snow. Precipitation in the lower watershed varies greatly and reaches as much as 100 in/yr in the temperate rain forest close to the coast. The rapid uplift of the fore-arc has produced a series of steep mountain ranges with strong orographic effects. Where mountain ranges exceed 5,000 ft above sea level, they maintain large winter and spring snowpacks in wet years and are associated with very high amounts of runoff during warm winter storms.

Annual runoff, as measured near the mouth of the Klamath River, is approximately 13 × 10^6 acre-ft. The upper watershed above Iron Gate Dam, which comprises about 38% of the total watershed area, provides only 12% of the annual runoff of the watershed. The low yields from the upper watershed are a product of its location in the rain shadow of the Cascades, its low relief, and its extensive marshes and lakes that increase hydraulic retention times. In contrast, the tributaries of the lower watershed dominate the total runoff of the Klamath watershed. Their high runoff stems from their high relief and the orographic influence of the Coast Ranges, Trinity Alps, and the Marble, Salmon, and Russian mountains. For example, one relatively small tributary, the Salmon River, supplies runoff about equal to that of the entire upper watershed, but from less than one-fifth of the area (Table 2-1).

TABLE 2-1 Runoff, Yield, and Basin Areas for the Klamath Watershed[a]

Location	Average Annual Runoff, 1,000 acre-ft	Drainage Area, mi^2	Runoff, %	Drainage Area, %	Ratio of Average Runoff to Drainage Area, acre-ft/mi^2
Klamath River below Iron Gate Dam	1,581	4,630	12	38	341
Shasta River near mouth	136	793	1	7	172
Scott River at mouth	615	808	5	7	761
Other tributaries	615	709	5	5	867
Klamath River below Scott River	3,020	6,940	23	57	435
Indian Creek at mouth	360	135	3	1	2,667
Salmon River at mouth	1,330	750	10	6	1,773
Other tributaries	1,350	650	10	5	1,500
Klamath River at Orleans	6,060	8,475	47	70	715
Trinity River at Hoopa	3,787	2,950	29	24	1,283
Other tributaries	3,021	675	23	6	4,476
Klamath River at mouth	12,868	12,100	100	100	1,109

[a]Data compiled from reports of the California Division of Water Resources 2002, representing average current conditions (including depletion caused by consumptive use) and gage records of the U.S. Geological Survey. Periods of record for data vary by site from 22 to 50 yr, principally between 1951 and the present, and include both pre- and post-Trinity River Diversion operations.

The hydropattern, or timing of runoff, varies throughout the watershed. Seasonal runoff from the upper watershed is regulated by the long and complex transport pathways in the basin and, historically, by the natural buffering effect of overflow into the Lost River and Lower Klamath and Tule lakes.

Under unregulated conditions, peak runoff from the upper watershed would typically occur in April and decrease gradually to minimums in late August or early September. Flow regulation and land-use activities in the upper basin have altered the hydropattern. Unlike the upper basin, the lower Klamath basin exhibits two potential flow peaks, depending on the water year. Subtropical storms strike the Klamath watershed with high frequency from late December to early March and are responsible for all peak daily discharges in the Klamath main stem and its tributaries. The short hydraulic retention times of the tributaries to the lower Klamath basin enhance the effect of these storms. The second and more predictable flow peaks are associated with spring snowmelt. The timing of the snowmelt pulse varies, but it usually occurs in April. Historically, the decline in flow from the tributaries to the lower basin was gradual and reached minimums in September. During the low-flow periods in the late summer or early fall when no precipitation occurs, spring-fed tributaries such as the Shasta River and flow from the upper basin constitute the bulk of base flow in the main stem of the lower basin.

Even the Trinity, the largest annual contributor of runoff to the Klamath, historically provided very little flow in the late summer and early fall.

AQUATIC ENVIRONMENTS IN THE UPPER KLAMATH BASIN

The upper Klamath basin encompasses about 5,700 mi^2 (USBR 2000a). Major lakes in the upper Klamath basin include Upper Klamath Lake (now 67,000 acres at maximum lake elevation), Lower Klamath Lake (historical maximum area, 94,000 acres; now about 4,700 acres), Tule Lake (historical maximum area, 110,000 acres; now 9,450–13,000 acres), Clear Lake, and Gerber Reservoir (see Chapter 3).

Upper Klamath Lake, now the largest water body in the Klamath basin, receives most of its water from the Williamson and Wood rivers. The Williamson River watershed consists of two subbasins drained by the Williamson and Sprague rivers. The Williamson River arises in the Winema National Forest, flows to the north through Klamath Marsh, and turns south to Upper Klamath Lake. The Sprague River arises in the Fremont National Forest and flows westward to connect with the Williamson River just below the Chiloquin Dam (Figure 1-1). The Sycan River, a major tributary of the Sprague, drains much of the northeastern portion of the watershed. Both the Williamson and Sprague subbasins are primarily forested (about 70%). Other important land-cover types are shrub and grassland (14%), agriculture (6%), and wetland (6%; Boyd et al. 2002). The Williamson and Sprague together provide over half the water reaching Upper Klamath Lake (Kann and Walker 2001).

The Wood River is the second largest source of water (16%) for Upper Klamath Lake (Kann and Walker 2001). Annie and Sun creeks join to form the Wood River. The watershed drains an area northeast of Upper Klamath Lake and extends from the southern base of the mountains that surround Crater Lake to the confluence of the Wood River with Upper Klamath Lake by way of the northern arm (Figure 1-3), which is often called Agency Lake. Although primarily forested, the Wood River has extensive agricultural lands and wetlands. The balance of the water reaching Upper Klamath Lake is derived from direct precipitation on the lake and flows from springs, small streams, irrigation canals, and agricultural pumps.

Before development of the Klamath Project, Lower Klamath Lake (Figure 1-3) was often larger than Upper Klamath Lake. Flows from the Klamath River, supplemented by springs around the lake, supported a complex of wetlands and open water covering approximately 80,000–94,000 acres in the spring, during high water, and 30,000–40,000 acres in late summer. The open water provided habitat for suckers, and the variable combination of open water and marsh created important habitat for migratory birds along the Pacific Flyway, making it one of the most important aquatic complexes for waterfowl in the West. By 1924, however, development of the Klamath Project eliminated more than 90% of its open water and marsh. Only about 4,700 acres of open water and wetland remain. Draining the lake led to the extirpation of sucker populations that had been in the lake (USBR 2002a), and also eliminated much of the habitat suitable for waterfowl and other birds.

Connections between the Klamath River and Lower Klamath Lake were severed by development, which changed the hydrology of both the lake and the river in ways that are not entirely clear. Before 1917, when railroad construction blocked the Klamath Straits, "water flowed from Upper Klamath Lake, through the Link River into Lake Ewauna, and then into the Klamath River. Between Lake Ewauna and Keno, the river meandered through a flat, marshy country" (Henshaw and Dean 1915, p. 655) for about 20 mi before descending over a natural rock barrier that stretched across the river at Keno. "Water in the river periodically backed up behind the reef at Keno and spread out upstream, flowing into Lower Klamath Lake through Klamath Straits" (Weddell 2000, p. 1). Today, connectivity between Lower Klamath Lake and the rest of the basin is limited to water pumped through Sheepy Ridge from Tule Lake and water from irrigation channels that lead to the Keno impoundment (USFWS 2001, Figure 1-2).

Before the Klamath Project, the lake and wetlands probably retained substantial amounts of early spring precipitation and some of the high flow of the river. "By storing and subsequently releasing this water into the river, Lower Klamath Lake would have augmented the effects of groundwater in shifting the Klamath River hydrograph to the river" (Weddell 2000, p. 7). Lower Klamath Lake was "neither an undrained basin nor a thoroughly drained floodplain. At times, its waters flowed into the Pacific Ocean via the Klamath River, yet this drainage was only partial" (Weddell 2000, p. 8).

Before 1924, suckers appear to have been abundant in Lower Klamath Lake, even after its connection to the river was severed in 1917. Suckers migrated into the lake from Sheepy Creek, a spring-fed tributary on the western edge of the lake, in numbers large enough to support a fishery (Coots 1965, cited in USFWS 2001).

Before the Klamath Project, Tule Lake (Figure 1-3) varied from 55,000 to over 100,000 acres, averaging about 95,000 acres (making it often larger than Upper Klamath Lake). Like Lower Klamath Lake, Tule Lake was connected seasonally to the Klamath River. During periods of high runoff, water from the Klamath River flowed into the Lost River slough and down the Lost River to Tule Lake. The direction of the river's flow is now determined by operators of the Klamath Project, depending on irrigation needs. Most of the former bed of Tule Lake has been drained for agriculture, leaving about 9,450–13,000 acres of shallow lake and marsh.

The fluctuation in surface area of Tule Lake afforded by its connections to the Klamath River may have been critical in maintaining the high aquatic productivity of Tule Lake and its wetlands (ILM 2000). Tule marshes on the north and west sides of the lake supported populations of colonial nesting water birds and summer resident waterfowl. The large fish populations in the lake supported what was probably the largest concentration of nesting osprey in North America (ILM 2000). Much of the historical variability in lake and marsh habitats has been lost as a result of management. Nevertheless, well into the 1960s and early 1970s, Tule Lake National Wildlife Refuge was considered the most important waterfowl refuge in North America; duck populations exceeded 2.5 million at their peaks. Siltation caused by agriculture and loss of wetland productivity has occurred in the last several decades, however, and waterfowl populations have declined (ILM 2000).

Historically, suckers in Tule Lake and the Lost River were abundant enough to support cannery operations along the Lost River (USFWS 2001). After the Klamath Project drained most of Tule Lake for agriculture and diversion dams of the project blocked the access of suckers to spawning areas in the Lost River, sucker populations declined substantially (Scoppettone et al. 1995, USBR 2002a).

The hydrology of Tule Lake and of the Klamath River first changed in 1890, when settlers built a dike across the Lost River slough in an attempt to protect lands near Tule Lake from flooding (USFWS 2001). The dike prevented Klamath River floodwaters from overflowing into the Lost River drainage and ultimately draining into Tule Lake. As is the case with respect to Lower Klamath Lake, the amount of water that flowed from the Klamath River into Tule Lake and the effect of this overflow on the historical hydrograph of the Klamath River are unclear. Estimates of historical Klamath River flows are derived from measurements recorded before Lower Klamath Lake was disconnected from the Klamath River, but the measurements were taken after Tule Lake was disconnected from the river.

The Lost River drains Clear Lake and flows north toward the Klamath River (Figure 1-3). The structure and hydrology of the Lost River have been highly modified by the Klamath Project. Historically, the Lost River was connected to the Klamath River during periods of high flow via the Lost River slough. There is now no direct outlet to the Klamath River, although diversion canals can be used to send water into the Klamath Project (Figure 1-2).

Aquatic habitats have been modified throughout the upper Klamath basin, but the Lost River watershed has been particularly altered by development of the Klamath Project. The Lost River, once a major spawning site for suckers, today supports few suckers (Chapter 6). According to the U.S. Fish and Wildlife Service (USFWS), the Lost River "can perhaps be best characterized as an irrigation water conveyance, rather than a river. Flows are completely regulated, it has been channelized in one 6-mi reach, its riparian habitats and adjacent wetlands are highly modified, and it receives significant discharges from agricultural drains and sewage effluent. The active floodplain is no longer functioning except in very high water conditions" (USFWS 2001, III-2-24). New lakes have been created and old lakes drained, new waterways have been dug and old rivers turned into irrigation ditches, and new sucker habitat has been created while original sucker habitat has been eradicated.

Before 1910, a natural lake, marsh, and meadow complex occupied what is now Clear Lake (Figure 1-3). Water from this lake drained into the Lost River and then to Tule Lake (USBR 2000a). In most years, the Lost River below the present Clear Lake dam ran dry from June through October. To hold back floodwaters from Tule Lake and store seasonal runoff for irrigation later in the season, a dam was constructed at Clear Lake in 1910, impounding the waters of the Lost River and creating a larger lake.

Where Gerber Reservoir now stands (Figure 1-3), 3,500 acres of seasonal wetlands existed before the Klamath Project, but there was no lake. Construction of Gerber Reservoir in 1926 for flood control and irrigation created new sucker habitat and a population of suckers persists there (USBR 2002b, Chapter 5).

AQUATIC ENVIRONMENTS IN THE LOWER KLAMATH BASIN

The lower Klamath River, including the Trinity River, is the largest of the coastal rivers of California (Figure 1-1). The lower Klamath basin historically was dominated by large runs of anadromous fishes with diverse life-history strategies (Chapter 7), some of which penetrated into the headwaters of tributary streams and into the rivers feeding Upper Klamath Lake. Four major tributaries to the Klamath River—the Salmon, Scott, Shasta, and Trinity rivers—were major salmon and steelhead producers. The Shasta River in particular, with its cool summer flows, was once one of the most productive streams of its size for anadromous fish in California (Chapter 7).

Historically, most of the aquatic habitat in the lower Klamath River consisted of streams with moderate to high gradients and cool water in summer, although the main-stem Klamath River may have been fairly warm during late summer. Similar conditions existed in the Trinity River (Moffett and Smith 1950). The flows in tributary streams were high in winter and spring from rain and snowmelt and low in summer. Native fishes of the lower basin are mainly anadromous but also include a few nonanadromous stream fishes (Chapter 7).

Many small tributaries enter the main-stem Klamath between Iron Gate Dam and Orleans. These creeks largely drain mountainous watersheds dominated by forest. Most creeks are affected to some degree by logging, mining, grazing, and agriculture. Water withdrawal leads to reductions in summer base flows in many of these tributaries. Water quality has not been extensively studied, but these tributaries may be particularly important in providing cold-water habitats for salmonids (Chapter 4).

As described below, the watershed has been drastically altered by human activities. The anadromous fishes have been in decline since the 19th century, when dams, mining, and logging severely altered many important streams and shut off access to the upper basin. The declines continued through the 20th century with the development of intensive agriculture and its accompanying dams, diversions, and warm water. Commercial fishing also contributed to the declines.

HISTORY OF LAND USE IN THE KLAMATH BASIN

For at least 11,000 yr, ancestors of the Klamath and Modoc Indians inhabited the upper Klamath basin (OWRD 2000). Most of the year, the Klamath and Modoc tribes lived near creeks, springs, riparian areas, and marshes (Cressman 1956). Their family groups were small, so they were able to extract enough resources for survival on a sustainable basis. Family groups came together during seasons of resource abundance for communal hunts, for celebrations, and to take advantage of seasonal concentrations of suckers and riparian plants (Cressman 1956).

The Klamath Indian name for Lost River suckers is *tchwam*; shortnose suckers are referred to as *kuptu* (L. K. Dunsmoor, Klamath Tribes, Chiloquin, Oregon, personal communication, September 3, 2002). Suckers in general became known to settlers as mullet. Lost River suckers in particular were once a staple food of the Modoc and Klamath tribes; they provided important protein in the spring, when food reserves had been depleted (Cope 1879, USFWS 2002). Gilbert (1898) reported them as the most important food fish in the Klamath Lake area, and Stern (1965) estimated an artisanal harvest of 50 tons/yr, which would correspond to 13,000 fish at an average weight of 3 kg.

The Klamath and Modoc tribes manipulated the wetlands and riparian areas to increase their resources. For example, the Klamath burned riparian areas because women preferred to weave baskets with the supple young stems that sprouted after a fire. They burned wet meadows in fall to increase production of root plants, to lure animals that were attracted to the protein-rich shoots that grew after fire, and to protect their shelters from wild grassland fires. Intensive digging, particularly for roots, also altered riparian areas (C. Burnside, Malheur National Wildlife Refuge, personal communication, 1997).

Four tribes occupied the lower Klamath basin. The Yurok lived along the Pacific coast from about 15 mi south of what is now Crescent City down to Trinidad Bay and up the Klamath River to Bluff Creek, a few miles past the junction with the Trinity River. The Hupa people lived along the Trinity River, where 13 villages were concentrated in a 7-mi reach called Hoopa Valley. The Karuk lived along the Klamath River upstream of the Trinity to a point beyond Happy Camp. Above Happy Camp, the Shasta Nation occupied the upper reaches of the Salmon, Klamath, Scott, Shasta, and McCloud rivers (Beckman 1998).

The Yurok, Hupa, and Karuk were closely allied with the sedentary cultures of the northwest coast; the Shasta showed cultural traits more akin to those of the migratory tribes of the inland West (Beckman 1998). The Yurok, Hupa, and Karuk spoke languages of three very different language groups—Yurok is Algonquian, Hupa is Athapaskan in origin, and Karuk is Hokan and thus associated with old languages of Mexico—but their cultural habits were similar (Beckman 1998). In contrast with the tribes down river, the Shasta did not occupy permanent villages, and their traditions were closer to those of the tribes of the upper Klamath basin.

The Yurok and Hupa, unlike tribes in the drier inland regions, were able to be almost completely sedentary because of salmon runs (Nelson 1988). As Beckman (1998) noted, their resources were so plentiful that they had the free time to nurture the arts and crafts in a way that was uncommon in California and that gave them a hierarchy of status and wealth. Unlike most other California nations, the Yuroks recognized no chiefs and had no organized political society. They were unique in believing in individually owned land; a family's wealth was measured by the amount of land that it owned, and land could be sold. The Hupa were strictly a river people, whereas the Yurok were divided between river and coastal villages. Most Yurok, however, lived along the Klamath River and relied on riverine resources (Waterman 1920), even though they used coastal resources, such as shellfish, surf fish, and seals. Anadromous fish that brought the abundant energy of the Pacific Ocean upstream were the Yurok's, Hupa's, and Karuk's most important resources and were critical resources for the Shasta as well.

Fur Trapping

When fur trappers from the Hudson Bay Company of Canada arrived in the Klamath basin in the 1820s, tribes throughout the basin coexisted in relative peace with them. Trappers were not seeking to establish permanent settlements in the basin that might threaten tribal rights. Rather, in an attempt to discourage Americans from laying claim to the region, Hudson Bay Company's written policy was to trap fur-bearing animals from streams south of the Columbia River to extinction. In July 1827, George Simpson of the Hudson Bay Company stated the policy clearly, writing that the best protection from Americans was to keep the "country closely hunted" (Williams 1971, p. xiv). Peter Skene Ogden, the trapper who opened up much of the basin to white exploration, followed that policy. By the summer of 1828, Ogden wrote of the region that "almost every part of the country is now more or less in a ruined state, free of beaver" (Ogden 1971, p. 98). During the next spring, he wrote that "it is scarcely credible what a destruction of beaver by trapping at this season, within the last five days upwards of fifty females have been taken and on average each with four young ready to litter. Did we not hold this country by so slight a tenure it would be most to our interest to trap only in the fall, and by this mode it would take many years to ruin it" (Ogden 1971, p. 17).

Ironically, it was the removal of beaver by fur trappers that helped create the basis for ranching. When beaver were removed, their dams fell into disrepair and the small wetlands behind the dams were drained and became the fertile meadows that were soon to sustain ranchers' cattle (Elmore and Beschta 1987).

Mining

Although the tribes were able to coexist with trappers, the miners who followed them proved disastrous to the Indian nations. Far more than trappers, miners transformed the basin's rivers and wetlands, partly because of mining activities in the rivers and streams and also because of their indirect encouragement of permanent white settlements. Miners created a new market for food and supplies and thus attracted farmers and ranchers to the region. Many of the settlements in the lower Klamath River basin originated from the mining boom of the middle 1800s (NMFS 2001). Miners also depended upon federal troops and Indian agents to cope with the problems that mining generated; they created a U.S. Army presence in the basin that further destabilized relations with the tribes (Malouf and Findlay 1986).

Mining in the 19th century was particularly destructive of fish habitat along the lower Klamath basin. In 1853, miners discovered a way to excavate gold-bearing placer deposits by using blasts of water to wash away gravel. Mining companies soon diverted creeks into reservoirs that fed water at high pressure to huge nozzles that could deliver water at up to 30,000 gal/min. The jets of water could level entire hillsides and their use rearranged much of the riparian landscapes of California. The waterborne debris was directed into sluices containing mercury, which captured the gold. Before a court ruling halted the practice in 1884, hydraulic miners released 1.6×10^9 yd^3 of sediment into California waterways, while hard-rock miners produced another 3×10^7 yd^3 of tailings, and dredgers left behind about 4×10^9 yd^3 of debris—a total of about 5.6×10^9 yd^3 for the entire state (Krist 2001).

Water was diverted and pumped for use in sluicing and hydraulic operations that resulted in increased turbidity and siltation. Silt from mining harmed benthic invertebrates, covered salmon redds, suffocated salmon eggs, and filled pools that were used by salmon. Wood for equipment and structures, railroad tracks, housing, and fuel was obtained through deforestation, often on steep slopes, and caused erosion, flooding, fires, and loss of animals. Miners also reduced freshwater resources by overfishing, damming, and diverting streams (Malouf and Findlay 1986).

The gold rush brought extensive changes to the Scott River watershed, particularly the main stem and South Fork and Oro Fino, Shackleford, and French creeks. Placer mining began as early as 1851 and expanded to widespread hydraulic mining in 1856 (Wells 1881). Large Yuba dredges that operated in 1934–1950 (Sommerstram et al. 1990) left some of the most visible effects of mining in the basin. They excavated material 50–60 ft below the river bed and created tailings piles more than 25 ft high downstream of the town of Callahan. The processing of the sediment by Yuba dredges left much of the coarsest material (typically boulders) at the top of the piles, effectively armoring the finer sediments. Early surveys in the basin (Taft and Shapovalov 1935) noted the severe damage that the dredging had caused to fish habitat. To support the mining, numerous ditches were constructed along the margins of the valley to intercept tributary flows, and these ditches eventually became sources of irrigation water for early agricultural development.

The Salmon and Trinity rivers were also severely affected by mining. Along the Salmon River, during the late 1800s and into the 1990s, extensive placer gold mining and some hydraulic gold mining were conducted in the main stem and the South and North Forks. The main stem of the Trinity River was severely impaired by placer mining within the channel and by hydraulic mining and extensive dredging.

One of the most problematic effects of the gold rush was the release of mercury into the environment; the consequences continue today. Mercury was critical in the mining and processing of gold; it is estimated that at least 2.6×10^7 lb of elemental mercury were used between 1850 and 1900 in gold mining. Much of the mercury remains in soils and sediments, and some of it has been converted into methyl mercury, which is particularly dangerous for humans because it travels through the food chain into fish and becomes a threat for those who eat fish. In addition to contamination from mercury used in gold mining, mercury contamination comes from mercury mines, some of which were in the Klamath basin. Most of the mines are now abandoned (Krist 2001).

By the late 1850s, gold mining in California was a large-scale industry that required infusions of capital for construction of mills, rail lines, dams, flumes, and smelters. Miners used two major processes to extract gold: stamp mills and hydraulic placer mining. Both methods used a great deal of mercury. Stamp mills pounded gold-bearing ore into dust that then was washed across mercury-coated plates; the gold sank and stuck to the mercury, and the less dense debris was carried away. The mercury-gold amalgam then was heated in furnaces, which vaporized the mercury and left the gold. Some of the evaporating mercury was captured in a condensation chamber for reuse, but much escaped into the air or was crushed by the stamp mill and released into the water. Hydraulic placer mining released even more mercury into the environment—perhaps as much as 1 lb of mercury for every 3 or 4 oz of gold recovered, or about 1.3×10^7 lb of mercury in the 19th and early 20th centuries (estimate by Ronald Churchill of the California Division of Mines and Geology, cited in Krist 2001).

Because salmonids achieve most of their growth in the marine environment, mercury accumulation in adult salmon presents less of a health risk to humans than would mercury accumulation in other kinds of large predatory fish. Nevertheless, mercury contamination may affect the coho salmon themselves. Young salmon are sensitive to mercury released by placer mining (USFWS 1991). Early life stages of coho salmon are harmed by low concentrations of methyl mercury (Buhl and Hamilton 1991, Devlin and

Mottet 1992), and placer mining releases contaminants that can be toxic to early life stages of salmonids (Buhl and Hamilton 1990).

The deleterious effects of mining on salmonid habitat were so rapid and intense that in 1852, only 4 yr after Sutter's discovery of gold in the foothills of the Sierra Nevada, California enacted its first salmon statute, which required "'all good citizens and officers of justice' to destroy man-made obstructions to salmon migration, except those erected by Indians." That statute did little to stem habitat destruction. In the 1880s, all obstructions to salmon migration, including those built by Indians, were banned by state law (Lufkin 2000).

The gold rush struck all California tribes hard (Heizer 1978, White 1991). Within a year after Sutter's 1848 discovery, at least 80,000 miners and others came to California, overwhelming governmental and military authority. In the quarter-century from 1845 to 1870, the Indian population in California declined from about 150,000 to 30,000 largely because of direct and indirect effects of the gold rush (Franzius 1997).

In 1851–1852, 18 treaties were negotiated with California tribes, including the Yurok, Hupa, and Karuk. The treaties set aside 7,466,000 acres of lands for the tribes and promised agricultural and educational assistance. But in 1852, California's new state senators refused to ratify the treaties.

Among the tribes of the lower Klamath basin, violent resistance to miners and to the California legislature's increasingly repressive policies erupted in 1860–1872. The Hupa were more successful than many other California nations in resisting encroachments of settlers on their land. When federal troops entered the Hoopa Valley, the Hupa were able to withstand the troops and force them into a stalemate. On August 12, 1864, the Treaty of Peace and Friendship was signed between the Hupa and the U.S. government; it promised the Hupa a reservation that included about 90% of their original homeland. In 1891, President Harrison signed an executive order joining the Hupa and Yurok reservations. The Karuk and Shasta, however, never gained legal ownership of their homeland. Most land occupied by the Karuk was claimed by the government with little compensation, and much of it became part of the national forest system. Timber development in the 20th century brought some measure of prosperity to the Hupa and Yurok reservations. For example, seven new sawmills were constructed in the Hoopa Valley during the 1950s, and timber income was distributed throughout the tribe. Yet this was also the "Termination Era," when federal Indian policy shifted toward the termination of tribal rights and the breakup of Indian land holdings (Nelson 1988).

As miners, ranchers, and the army came to the Klamath basin in the 1850s, confrontations erupted, culminating in the Modoc Indian War of 1872. In 1864, the Klamath and Modoc tribes and the Yahooskin band of Snake Indians met with federal officials to sign a treaty that relinquished more than 19 million acres of their homeland, reserving about 2.5 million acres for the Klamath Indian Reservation. This land was soon substantially reduced through correction of a federal survey error (Gearheart et al. 1995). The treaty of 1864 specified the Klamath Tribes' exclusive right to hunt, fish, and gather on Klamath Indian reservation lands. Although the Klamath tribes lost their reservation land following termination of the reservation in 1954 (Haynal 2000), they retained their water rights and their right to harvest a number of fish species designated as tribal trust species, reflecting their traditional practices.

Ranching

After the Modoc Indian War, open hostilities between whites and Indians diminished in the upper basin, and white immigration to the basin increased. Early white settlement in the upper Klamath basin centered on ranching rather than farming because without irrigation, precipitation often was insufficient for growing most crops (Blake et al. 2000). The General Allotment Act of 1887 allowed Indian lands to pass into white ownership, and much of the best grazing land on the reservation was bought by whites.

In the upper Klamath basin, as throughout the entire inland portion of the West, cattle increased in abundance during the 1870s and 1880s until by the late 1880s overgrazing became a political and ecological issue. In 1875, the Central Pacific Railroad completed a shipping facility at Winnemucca, Nevada, giving cattle operations relatively rapid access to San Francisco beef markets. With an efficient transportation infrastructure in place, ranchers brought more animals to the open range. When prices were low, few ranchers sold their young cattle, and herd sizes rose while ranchers waited for better prices (Gordon 1883). Overgrazing was the result.

The federal government responded to overgrazing with the Gordon report, the product of a study motivated in part by the disastrous winter of 1879–1880, when extraordinary cold led to high mortality of cattle across the West. Gordon noted that overgrazing meant that wetlands and riparian meadows were becoming critical habitat for cattle, especially in southeastern Oregon. Ranchers fenced riparian areas and planted them with alfalfa for winter feed. That took some of the pressure off the land, but only for a short time (Gordon 1883). The result, as the 1883 edition of *West Shore* magazine reported, was a landscape "almost bare of grass except for a few clumps under the dense, scraggly sage brush" (Lo Piccollo 1962, p. 115).

In the wake of the 1879–1880 disaster, cattle and sheep populations were rebuilt until a combination of dry summers and cold winters occurred in the late 1880s (Simpson 1987). Cattle prices collapsed in 1885 and 1886, and ranchers held their stock from market, hoping for higher prices. In 1889, when the geologist Israel Russell toured southern Oregon, streams throughout the region that Ogden had described as level with the surrounding landscape in the 1820s had begun to incise their channels, and Russell (1903, p. 63) concluded that this was caused by "the introduction of domestic animals in such numbers that the surface covering of bunch grass was largely destroyed, and in consequence the run-off from the hills accelerated."

Government inspectors who were sent to the region warned that overgrazing was ruining the very source of the region's prosperity. The inspectors recommended that the only solution was to provide more grass by draining wetlands and planting them with hay so that there would be less competition for a dwindling resource (Griffiths 1902). Ranchers did exactly that as they began diking and draining wetlands in the 1890s along the borders of Upper Klamath Lake to provide more forage for cattle.

Good government records of numbers of cattle in the upper Klamath basin begin with the 1920s, when 30,000 cattle occupied Klamath County, which makes up only part of the watershed (Walker 2001). In the 1960s, the cattle population in Klamath County peaked at 140,000 head (Figure 2-2); by 1999, there were 120,000.

To accommodate cattle, ranchers turned to flood irrigation of pastures and drainage of wetlands. Early methods of flood irrigation did not always degrade riparian and wetland habitat, but a switch to nonnative species for production of hay in the 1950s required changes in irrigation practices that, while increasing efficiency, severed riparian connections to the landscape (Langston 2003). In 1998, the Environmental Protection Agency's Index of Watershed Indicators estimated that at least 110,000 acres of the watershed

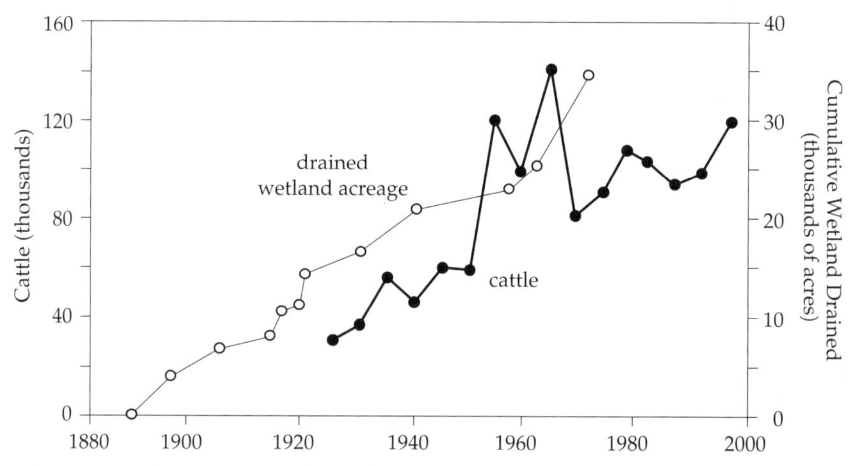

FIGURE 2-2 Changes in numbers of cattle and cumulative acres of drained wetland in Klamath County, Oregon. Source: Modified from Eilers et al. 2001.

had been converted to irrigated pasture or other agricultural activities; Risley and Laenen (1999) estimated an 11-fold increase in acreage of irrigated land between 1900 and the 1990s.

While numbers of cattle were only slightly lower in the 1990s than in the 1960s, the acreage of land being grazed declined much more substantially. The U.S. Bureau of Reclamation (USBR) estimated that by 2000 only 35% of the Upper Klamath Lake watershed was grazed (USBR 2002a). By 2002, nearly 100,000 acres of irrigated agriculture had been retired, and some of this was restored to wetland. Thus, production intensity appears to have increased. Transport of cattle to California during the winter was part of the method for keeping cattle production high while the acreage of irrigated pastureland declined.

The effects of grazing in the watershed were probably profound but are impossible to quantify. Overgrazing in riparian zones can harm fish by degrading riparian vegetation (Chapter 4). Grazing can mobilize nutrients and sediments, both of which are of concern in the upper Klamath basin (Stubbs and White 1993). By 1900, native perennial grasses were being replaced with annual grasses and forbs that, when combined with soil compaction from cattle, may have resulted in higher erosion and greater peak flows (NMFS 2001). For example, on Fishhole Creek, cattle had destroyed streambank vegetation, resulting in erosion and lowered water tables (Thompson et al. 1989). Conditions are similar in the Wood River valley and in some of the Sprague River watershed. Season-long grazing in the past probably contributed to reduction of spawning habitat for trout and suckers in the Sprague River, increased stream temperatures, and increased transport of sediment and nutrients. These changes led the Oregon Department of Environmental Quality to identify the Sprague River as one of the highest-priority streams in Oregon for control of non-point-source pollution (Stubbs and White 1993). Cattle do not always lead to such adverse effects; well-managed riparian pastures can be consistent with good stream conditions.

Irrigated pasture required water diversions from Klamath basin tributaries, and the diversions have played a substantial role in the decline of suckers in the upper basin and of salmonids in the lower basin (Chapters 5 and 7). The Chiloquin Dam on the Sprague River near Chiloquin, Oregon, constructed in 1914–1918 for water diversion and timber milling, is one example.

Timber

Much of the blame for poor watershed conditions is placed on agriculture, but nearly 80% of the Upper Klamath Lake watershed is forested, and much of the forest land has been harvested under federal, tribal, and private management (Gearheart et al. 1995). According to the Oregon State Water Resources Board (cited in Gearheart et al. 1995), over 73% of the forest land in the upper Klamath basin is subject to severe erosion. Therefore, timber management may well have contributed to the decline of suckers and salmonids.

Commercial logging began in the upper basin in 1863 when the U.S. Army constructed a sawmill. The pace of logging accelerated during the late 1910s, when ponderosa pine became an important timber resource for the nation (Langston 1995). By 1918, large amounts of reservation timber were being sold to private parties; by 1920, annual harvest rates had increased to 120 million board ft. Peak lumber production occurred in 1941, when 22 lumber mills processed a total of 808.6 million board ft within the upper basin. Harvest has dropped to about 400 million board ft in recent years (Eilers et al. 2001, Gearheart et al. 1995).

Poorly designed roads and damaging harvest practices on pumice and volcanic soils and on steep slopes probably contributed to loss of fish habitat. When stripped of vegetative cover, steep slopes are subject to erosion. In the lower basin, road construction has increased erosion and also created barriers to fish passage (USBR 2001b). Log storage on the Klamath River below Klamath Falls also has affected fish habitat. After fish kills in the late 1960s, log storage was greatly reduced on the river, but it continues (Stubbs and White 1993).

Forest management and fire suppression over the last century changed forest composition in the Klamath basin. The change may have altered flow regimes in the rivers and nutrient movement in the watershed. Before the 1920s, the upper basin forest was composed largely of old-growth ponderosa pine except at high elevations, and frequent, low-intensity fires minimized understory growth. Logging and fire suppression have led to a much denser understory populated with grand fir (Risley and Laenen 1999). As forest composition has changed, the risk of intense fires has increased substantially. Such fires can contribute damaging amounts of sediments and nutrients to streams and rivers. Moreover, intensive clearcutting may have increased peak flows, and the increased understory and denser forests may have decreased total water yield (Risley and Laenen 1999).

In the lower Klamath basin, timber harvesting began in the 1850s in the Scott River watershed commensurate with the growth in mining. As in most northern California watersheds, logging activity reached a peak in the 1950s (Sommerstram et al. 1990). The construction of roads and trails in the watershed has been a major source of fine sediment in the basin, particularly on decomposed granite soils. About 40% of the Scott River watershed that is underlain by such soils was harvested in 1958–1988; more than 288 mi of logging roads and 191 mi of skid trails were constructed (USFS data, summarized in Sommerstram et al. 1990). Sediments have adversely affected spawning and rearing habitat of coho (West et al. 1990).

Along the Salmon River, logging has been substantial, particularly since the 1950s. Road networks have been identified by the U.S. Forest Service (USFS) and the California Department of Fish and Game (CDFG) as an important source of sediment in the basin, and road crossings have been identified as affecting salmonid habitat (CDFG 1979a). Also, the heavily forested Salmon River watershed is susceptible to large wildfires. Since the early 1900s, more than 50% of the basin has burned, and most of the fires have been intense crown fires (USFS data, summarized in Salmon River Restoration Council 2002). Although poorly funded, federal fuel-management efforts are under way in the basin in cooperation with the Salmon River Watershed Council.

In the Trinity River watershed, logging practices, described as "abusive" by the Secretary of the Interior in a 1981 decision regarding flow releases on the Trinity, has had significant effects on the quality of salmonid habitat on the Trinity (USFWS/HVT 1999). Extensive logging road networks, coupled with highly erosive soils, have produced high yields of fine sediment within the basin. Very large floods on the Trinity River in December of 1964 introduced especially large volumes of fine sediment that caused severe degradation of spawning and rearing habitat in the South Fork and main stem of the Trinity.

Agriculture in the Upper Basin

Serious efforts at irrigation and drainage in the Klamath basin started in about 1882; by 1903 about 13,000 acres in the upper Klamath basin were irrigated by private interests. Land speculators urged USBR to consider the Klamath basin for irrigation, and a USBR engineer estimated in 1903 that irrigation could water 200,000 acres of farmland.

California and Oregon had acquired Lower Klamath Lake through the Swamp Lands Act of 1860, but their efforts to stimulate drainage and reclamation had failed. In 1904 and 1905, California and Oregon ceded the lake back to the federal government for use by USBR. Oregon gave USBR the right to the water of the Klamath River (Jessup 1927). In February 1905, Congress approved the Klamath Project, and work began.

USBR engineers focused their early efforts on Lower Klamath Lake and Tule Lake. The project would dry up these two lakes so that the land under them could be farmed. The government would then construct two new lakes to hold water for irrigation (behind Clear Lake and Gerber dams, Figure 1-3). A dam and canal would divert the Lost River to the Klamath River. Headworks would take water from Upper Klamath Lake into an elaborate irrigation system. USBR would fund construction of irrigation works; people (mostly veterans) would buy land irrigated by those works from the federal government in parcels of up to 80 acres and would pay for the land and improvements over 10 yr. The federal government sold the land, but not the water rights, to Klamath Project irrigators; irrigators were promised use of sufficient water for irrigation each year for a modest fee.

Meanwhile, just three months after Congress authorized the Klamath Project in early 1905, conservationists discovered the basin's extraordinary abundance of avian life. During the summer of 1905, just a few months after Congress approved the Klamath Project, the conservationist William Finley toured the marshlands in the lower Klamath basin. He was awed by what he found, including extraordinary concentrations of pelican rookeries and what he believed to be the greatest feeding and breeding ground for waterfowl on the Pacific Coast. By 1908, Finley had persuaded President Roosevelt to create the Lower Klamath Lake National Wildlife Refuge (Figure 1-3), thus preserving nesting grounds for migratory waterfowl. It was to be one of the largest wildlife refuges ever authorized, one of the first on land of any agricultural value, and the first to be established in a watershed being transformed by USBR. In 1911, President Taft established the Clear Lake National Refuge and in 1928 President Coolidge established Tule Lake National Wildlife Refuge. The Biological Survey would manage the refuges, and land within refuge boundaries would not be made available for settlement.

President Roosevelt's designation created inherent conflicts. The refuges were to be managed by the Biological Survey, which could not function with full independence because the refuges were on land of USBR, which also controlled the water reaching the area. To USBR, wetlands and riparian areas were wastelands waiting for conversion (reclamation) to agriculture (Langston 2003).

President Roosevelt had intended no settlement within the boundaries of the refuge, but USBR interpreted refuge boundaries as encompassing only land covered by water all year. Thus, if USBR drained the lakes and wetlands, it would no longer be refuge land, and it could be sold or leased.

Before draining Lower Klamath Lake, USBR commissioned soil surveys to see whether the area would be good farmland. C. F. Marbut, a government soil scientist with the U.S. Department of Agriculture (USDA), completed a report indicating that the lakebed would be utterly worthless for agriculture. "We can not cite an example of the successful cultivation of a soil of similar character," admitted Copley Amory, an economist with USBR, in response to that discouraging report (Amory 1926, p. 80). Moreover, the report stated, wetlands surrounding the lake would have only a slim chance of supporting agriculture because the underlying peat, once drained, would be subject to smoldering fires and subsidence.

Despite Marbut's report, USBR authorized $300,000 for drainage of Lower Klamath Lake. Conservationists challenged USBR's plans in court, and President Wilson in 1915 reduced the Lower Klamath Lake National Wildlife Refuge from 80,000 acres to 53,600 acres, freeing up the rest for drainage and sale or lease.

The federal government signed an agreement with railroad companies according to which the companies would construct an embankment across the marshes with a gate that would close Klamath Straits. The gates were closed in 1917, cutting off flow of water from the Klamath River into the lake (Jessup 1927). Within a year, the flooded area of the lake decreased by about 53%, from 76,600 acres to 36,000 acres; within 5 yr, most of the waters of the lake had evaporated (Weddell 2000). USBR entered into contracts in 1917, first with California-Oregon Power Company, selling it water rights to the river for power generation, and then to a drainage and land-speculation company, the Klamath Drainage District. The shrinkage of the lake greatly reduced waterfowl populations. The peat beds of the wetlands began to burn and collapse, farm efforts failed, and, by 1925, homesteaders were going bankrupt. By 1925, nearly everyone involved agreed that the project was a failure.

After USBR had drained Lower Klamath Lake, it leased what remained of the refuges for grazing. The ornithologist Ira Gabrielson (1943, p. 13) described the situation in 1920:

> The water table on the lake has been lowered several feet by closing the gates which control the inflow from the Klamath River. This action, made under agreement with the water users' association, has uncovered large areas of alkali flats without thus far benefiting the settlers adjoining the lake or opening up additional land suitable for agriculture. Its future as a refuge is seriously jeopardized. This is an understatement of the wildlife tragedy involved in the loss of one of the two greatest waterfowl refuges then in existence.

Near Tule Lake National Wildlife Refuge, water from drained wetlands was being pumped into headwater ditches, used for irrigation, and then collected in the Tule Lake Sump on the refuge, where it was allowed to evaporate. Farmers wanted the land under the sump for farming, but the Tule Lake Sump was overflowing with irrigation return flows as more and more farmers irrigated reclaimed lands.

A reclamation engineer, J. R. Iakish, proposed to pump the irrigation return flows from the Tule Lake Sump through a 6,600-ft tunnel beneath the ridge to Lower Klamath Lake to put out the fires and restore the wetland. Such a plan, Iakish argued, would create more farmland by draining the sump and more wetland for birds by putting out the fires on Lower Klamath. In 1941, the tunnel was finished, and in the next year, water flowed once again into Lower Klamath Lake. Some of the Lower Klamath Lake wetlands began to refill, and some of the abandoned farmlands were reclaimed when developers figured out how to use the irrigation wastewater, in conjunction with deep drains, to leach alkali out of soils. Lower Klamath, people argued, could indeed be farmed profitably, so waters intended for restoration were instead used for farming (Blake et al. 2000).

In 1946, USBR authorized new allotments on lands north of Tule Lake (shrunk by use of the tunnel) and held a lottery drawing for World War II veterans. The federal government urged thousands of veterans to apply for these new homesteads, promising them as much water as they would ever need for irrigation. Some of the land on the refuges was given to veterans. A total of 22,000 acres was leased to farmers for agriculture in what became known as the lease-land program. For example, nearly half the 39,000 acres of the Tule Lake National Wildlife Refuge became cropland (Kemper 2001). Japanese and Japanese-American citizens who had been interned at the Tule Lake Camp during World War II were the first to farm much of this land, and their labor helped make it ideal farmland for returning veterans.

Agriculture in the Lower Basin

During the early 1900s, farmers and ranchers removed riparian vegetation and valley forests along the lower Klamath River and its tributaries (CDFG 1934). For example, the U.S. Army Corps of Engineers, in conjunction with the National Resource Conservation Service (then known as the Soil Conservation Service), conducted a series of projects on the main stem and tributaries of the Scott River, including removal of riparian vegetation on the middle reaches of the valley, drainage of remaining wetlands, and construction of a series of flood-control and bank-stabilization projects (Scott River Watershed CRMP Council 1997). Today, the Scott Valley supports more than 30,000 acres of farms and irrigated pasture (CDWR, Red Bluff, CA, unpublished material, 1993; Scott River Watershed CRMP Council 1997). The principal crops are alfalfa (33,000 acres) and grain (2000 acres). There are 153 registered diversions in the Scott Valley; 127 are listed by the Siskiyou County Resource Conservation District (SRCD) as active. Fish screens have been installed on 65 of the diversions; another 38 have been funded but not yet built.

In the Shasta River watershed, after the gold rush in the late 1800s, most of the land cover of the Shasta Valley was converted for agriculture and range. About 28% of the watershed is irrigable land that supports a mix of alfalfa, irrigated pasture, and some grain (CDWR 1964). Non-irrigable land supports range and limited dryland farming. The mix of agricultural uses has remained relatively constant in the basin. Mining and timber harvesting are limited and do not substantially affect the river.

Significant urbanization, however, is taking place in the watershed. Most development is occurring in the vicinity of Yreka, the county seat of Siskiyou, and Montague, in the northern portions of the Shasta Valley. There is also increasing pressure to develop in the upper watershed, particularly around the town of Weed and near Lake Shastina (Dwinnell Dam).

FISHING AND ATTEMPTS TO REGULATE LOSS OF FISH

Mining, timber management, dams, and agriculture have degraded fish habitat, but overharvesting also has affected fish populations (Chapters 5 and 7). In the upper Klamath basin, tribal harvests of suckers for family consumption were augmented by commercial harvests beginning in the 19th century, including a cannery that processed Lost River suckers captured from the Lost River near Olene, Oregon, in the late 1890s (53 Fed. Reg. 27130 [1988]). Before the drainage of Tule Lake and Sheepy Creek in the 1920s, suckers were taken in large numbers from Sheepy Creek for consumption by both humans and livestock (Coots 1965). A recreational snag fishery for suckers developed as early as 1909; it focused on fish that were moving into tributary rivers to spawn and secondarily on fish attempting to spawn around the edges of Upper Klamath Lake. The snag fishery remained unregulated until Klamath suckers were declared game species in 1959.

Commercial harvests of salmon intensified with the development of canning technology. Commercial harvesting of salmon began later in the Klamath River basin than in other basins in California and the Pacific Northwest partly because of the inaccessibility of much of the terrain. Nevertheless, by the early 20th century, habitat destruction combined with commercial harvests had resulted in serious salmon depletion on the Klamath River (Pacific Watershed Associates 1994). Cobb (1930) estimated that the peak of the Klamath River salmon runs occurred in 1912; Snyder (1931, p. 7) observed substantial declines in the 1920s. As Snyder observed, "in 1912 three [canneries] operated on or near the estuary and the river was heavily fished, no limit being placed on the activities of anyone."

Millions of juvenile coho salmon, Chinook salmon, and steelhead are released into the Klamath and Trinity rivers each year by the Iron Gate and Trinity River hatcheries, which were built to mitigate the salmonid losses created by large dams. These hatcheries were intended to maintain fisheries for coho and Chinook salmon, but they may have had adverse effects on wild populations of salmonids in the basin (Chapters 7 and 8).

WETLAND TRANSFORMATIONS

Even before the Klamath Project, the actions of humans in the upper basin were concentrated on wetlands. Cattle ranching had been concentrated on the margins of wetlands, extensive efforts to drain wetlands began in 1889, and drainage accelerated with the Klamath Project; restoration began in the 1990s. Figure 2-3 shows the cumulative drained acreage by year for Upper Klamath Lake. The drop in drained wetland acreage after 1990 reflects wetland restoration efforts in the upper basin. In Tule and Lower Klamath lakes, original wetlands were estimated at 187,000 acres; about 25,000 acres remain (Gearheart et al. 1995).

Reclamationists and farmers drained wetlands by building dikes to isolate them hydrologically, constructing a network of drainage ditches within them, and pumping surface water and shallow groundwater (Snyder and Morace 1997, Walker 2001). One effect of lowering the water tables in this way was an increase in aerobic decomposition of peat soils, which liberated nutrients and removed organic deposits. Disking and furrowing can introduce oxygen into the soils, and increase the rate of peat decomposition and nutrient release. Cattle grazing, in contrast, can compact drained soils and slow their decomposition (Walker 2001).

Some scientific work in the upper basin suggests that drained wetlands can become a substantial source of phosphorus (Snyder and Morace 1997), which can lead to increased nutrient loading in the Upper Klamath Lake (Bortleson and Fretwell 1993, Walker 2001). Extensive efforts to restore wetlands, partly to improve nutrient retention, have taken place in the upper basin in the last two decades. Above Upper Klamath Lake, an area of about 101,136 acres has been removed from irrigated agriculture and con-

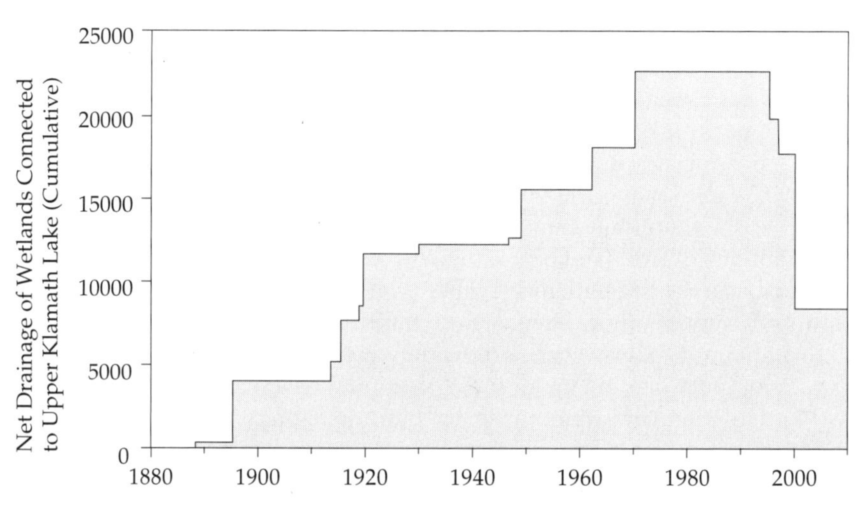

FIGURE 2-3 Net loss, through drainage, of wetland connected to Upper Klamath Lake. A decrease beginning in the 1990s indicates the effects of restoration. Source: Modified from Boyd et al. 2001, p. 48.

verted to artificial wetlands since the 1980s (E. Bartell, The Resource Conservancy, Inc., Fort Klamath, Oregon, unpublished report, 2002). The effects of these conversions on water quality are unclear.

Although wetlands of different types often are lumped in analyses of wetland change in the basin, different kinds of wetlands may have different effects on water quality. Geiger (2001) argues that wetlands in the littoral zone of Upper Klamath Lake may have had particularly important effects on water quality because they were connected to the lake and contributed humic substances that may have played a role in suppressing algae (see Chapter 3). Drainage efforts and subsidence have had pronounced effects on those wetlands. For example, the littoral wetland of Upper Klamath Lake once comprised 51,510 acres of the total lake area (46.2% at maximum elevation). By 1968, after diking and draining, littoral wetland had decreased to 17,370 acres (22.4% of total lake area). The littoral wetland area was reduced by 66.3%, and the wetland area at minimum storage volume (4,136 ft vs the earlier minimum of 4,140 ft) had shrunk from 20,320 acres to 0 acres (Geiger 2001). Some 34,140 acres of former wetland now is isolated behind dikes on Upper Klamath Lake. A total of 17,553 acres of former wetlands behind dikes is now being reclaimed, but subsidence has meant that, even after being restored, these areas remain disconnected from the lake and do not function as the littoral wetlands once did. Once dikes are removed, subsided areas become open-water habitat rather than littoral wetlands (Geiger 2001). Even so, reconnection of the littoral perimeter with open water would lead to the return of processes and functions that have been lost through severance of much of the littoral zone from the offshore areas of the lake.

The conversion of wetlands and associated channelization of riparian habitat have had deleterious effects on sucker habitat (Chapters 5 and 6). For example, sucker larvae historically moved through a meandering Williamson River into the delta area and the adjacent shoreline areas of Upper Klamath Lake. Since 1940, the Williamson River has been channelized, and the delta and adjacent shoreline have been diked and drained, leaving little of the wetlands and riparian vegetation (Klamath Tribe, Chiloquin, Oregon, unpublished material, 1993, cited in Gearheart et al. 1995). As a result, nursery areas have been greatly reduced. Larvae reach the lake sooner, exposing them to poor water quality at an earlier age and for longer.

Substantial wetlands remain in the basin. Klamath Marsh, a 60,000-acre basin underlain by pumice, is one example; 37,000 acres is protected as a federal wildlife refuge. A total of 23,000 acres of the Sycan Marsh was purchased by The Nature Conservancy in 1980 and is undergoing restoration. The largest wetland still connected to Upper Klamath Lake is Upper Klamath Marsh, a federal wildlife refuge on the northwest edge of Upper

Klamath Lake; this refuge is the remnant of an emergent and open-water marsh system that once covered 60,000 acres of the Wood River valley (Gearheart et al. 1995).

THE ECONOMY OF THE KLAMATH BASIN

This section provides an overview, without conclusions or recommendations, of the structure of the economy of the Klamath basin on the basis of data from the Bureau of Economic Analysis (BEA) and the IMPLAN (impact planning) modeling process (Minnesota IMPLAN Group, Inc.). It is divided into discussions of the upper and lower basin economies, which differ substantially. Special attention is given to sectors of the economy oriented toward natural resources, including agriculture in both the upper and lower basin and commercial fisheries in the lower basin. It should be noted that this analysis only includes economic and employment values associated with commodities and services that are traded in markets. Nonmarket values, such as those associated with existence of species, preservation of environmental quality or maintenance of a particular lifestyle, are not reflected directly in the economic values reported here.

Upper Basin

The upper Klamath basin includes parts of five counties in Oregon and California. Almost all the Oregon portion of the basin is in Klamath County, and the basin covers most of the county, including the county seat, Klamath Falls (population about 21,000), which is the major regional population center. In California, the basin covers the northwest corner of Modoc County, not including the county seat, Alturas (population, about 3,000), and the northeast corner of Siskiyou County, including the county seat, Yreka (population, about 7,500). The economy of the upper Klamath basin, which is home to about 120,000 people, in 1998 produced $4 billion worth of output, added $2.3 billion in value to purchased inputs, and had almost 60,000 jobs (Weber and Sorte 2002). This section, which is adapted by permission from Weber and Sorte (2002), describes the upper basin economy.

Over the last 30 yr, full- and part-time employment in the upper Klamath basin has increased from 40,000 to 60,000 jobs, while employment in Oregon as a whole has more than doubled. The composition of the regional economy has changed dramatically over that time. The sectors that grew most rapidly were wholesale trade and services (Table 2-2). Employment in several other sectors declined: military, transportation and public utilities, and manufacturing. Employment in farming, mining, and federal civilian employment grew, but increased more slowly than the regional average

TABLE 2-2 Structural Change in the Upper Klamath Basin Economy, 1969–1999

	Employment		Employment		Employment	
Sector	1969	% of Total	1999	% of Total	Change 1969–1999	% Change
TOTAL FULL- AND PART-TIME EMPLOYMENT	40,392	100.0	60,101	100.0	19,709	48.8
Wage and salary employment	31,751	78.6	44,257	73.6	12,506	39.4
Proprietors' employment	8,641	21.4	15,844	26.4	7,203	83.4
Farm proprietors' employment	2,466	6.1	2,723	4.5	257	10.4
Nonfarm proprietors' employment	6,175	15.3	13,121	21.8	6,946	112.5
Farm employment	4,144	10.3	4,592	7.6	448	10.8
Nonfarm employment	36,248	89.7	55,509	92.4	19,261	53.1
PRIVATE EMPLOYMENT	27,563	68.2	44,926	74.8	17,363	63.0
Agricultural services, forestry, fishing and other	1,090	2.7	1,678	2.8	588	53.9
Mining	70	0.2	71	0.1	1	1.4
Construction	1,442	3.6	2,528	4.2	1,086	75.3
Manufacturing	7,171	17.8	5,883	9.8	–1,288	–18.0
Transportation and public utilities	3,084	7.6	2,474	4.1	–610	–19.8
Wholesale trade	876	2.2	2,388	4.0	1,512	172.6
Retail trade	6,291	15.6	10,213	17.0	3,922	62.3
Finance, insurance and real estate	1,965	4.9	3,573	5.9	1,608	81.8
Services	5,574	13.8	16,118	26.8	10,544	189.2
GOVERNMENT AND GOVERNMENT ENTERPRISE	8,685	21.5	10,583	17.6	1,898	21.9
Federal, civilian	1,665	4.1	1,856	3.1	191	11.5
Military	2,369	5.9	30	0.5	–2,049	–86.5
State and local	4,651	11.5	8,407	14.0	3,756	80.8

Source: Modified from Weber and Sorte 2002.

over the last three decades. Because of the more rapid growth in other sectors, the share of jobs in farming declined from 10.3% to 7.6%. Thus, over the last three decades, the basin's economy has grown slowly, has become more specialized in sectors that are growing rapidly in Oregon as a whole (services and wholesale trade), has shown little proportionate change in some slowly growing sectors (farming and federal civilian employment), and has become less specialized in other slow-growth sectors (manufacturing and transportation, public utilities).

Table 2-3 presents estimates of some basic economic indicators of the regional economy and their distribution among sectors for 1998. The four sectors with the largest shares of output in 1998 were wood products, consisting of forestry, logging, and manufacturing of wood products (15.5%); agriculture, consisting of food, beverage, and textile manufacturing (11.1%), construction (8.1%), and health care and social assistance (7.8%). The four sectors with the largest shares of value added were wood products (11%), retail trade (8.8%), real estate (8.7%), and public administration (8.6%). The four sectors with the largest employment shares were retail trade (11.1%), agriculture (10.7%), educational services (10.1%), and health care and social assistance (9.9%). These measures provide a perspective on the distribution of the regional economic activity among sectors. None of them identifies, however, how much the regional economy depends on each sector.

Table 2-4 summarizes the contribution of each sector to total regional employment and is based on an analysis using the upper Klamath basin input-output model. Such models use estimates of exports from each industry and multipliers for each sector to generate estimates of the dependence of the regional economy on each sector's exports. The procedure used to derive the estimates in Table 2-4 is described in Waters et al. (1999). The table compares the employment in a sector with employment that depends on a sector's exports. The jobs under "Sectoral Employment" are within the sector. The jobs under "Export-Dependent Employment" are from all sectors that depend on the exports from a sector. As an example, there were 4,328 jobs in the wood-products manufacturing sector, but 7,018 jobs in the region were dependent on wood products exports.

Of these, 3,089 jobs were directly dependent on the export of wood products from the county where they were produced; these jobs were related to direct purchases from wood-products firms by households, firms, and governments outside the region. In addition, 2,126 jobs were indirectly dependent on wood-products exports; these jobs were created when wood-products firms purchased inputs (such as logs) from firms in the county and when the suppliers purchased from other businesses in the county. Yet another 1,803 jobs were induced by exports of wood products; these jobs were in retail trade, real estate, and health care and were created when

TABLE 2-3 Output, Value Added, and Employment in the Upper Klamath Basin, 1998

Industry	Output		Value Added		Employment	
	$ Million	(%)	$ Million	(%)	Jobs	(%)
Agriculture and related[a]	320	7.9	169	7.3	5,964	10.0
Forestry and logging	30	0.7	16	0.7	248	0.4
Mining	4	0.1	2	0.1	33	0.1
Construction	327	8.1	119	5.1	3,357	5.7
Manufacturing—food, beverages, textiles and related	128	3.2	20	0.9	407	0.7
Manufacturing—wood products, paper, furniture and related	598	14.8	241	10.3	4,328	7.3
Manufacturing—high technology and related	17	0.4	3	0.1	94	0.2
Manufacturing—other (for example, sheet metal products)	113	2.8	35	1.5	844	1.4
Transportation and warehousing	263	6.5	139	6.0	2,257	3.8
Utilities	128	3.2	80	3.4	429	0.7
Wholesale trade	142	3.5	97	4.2	2,036	3.4
Retail trade	235	5.8	205	8.8	6,568	11.1
Accommodation and food services	163	4.0	92	4.0	4,785	8.1
Finance and insurance	197	4.9	138	5.9	2,179	3.7
Real estate, rental, and leasing	279	6.9	202	8.7	1,535	2.6
Other services	186	4.6	84	3.6	3,733	6.3
Information	100	2.5	55	2.3	1,241	2.1
Administrative and support services, and so on	28	0.7	16	0.7	936	1.6
Arts, entertainment, and recreation	31	0.8	19	0.8	1,133	1.9
Health care and social assistance	316	7.8	194	8.3	5,859	9.9
Professional, scientific, and technical services	38	0.9	26	1.1	865	1.5
Educational services	182	4.5	170	7.3	6,010	10.1
Public administration	200	5.0	200	8.6	4,551	7.7
Inventory valuation adjustment	7	0.2	7	0.3	0	0
Total	4,032	100.0	2,327	100.0	59,390	100.0

[a]Technically, this is agriculture, fishing, and related. However, the IMPLAN database for the upper Klamath basin identifies only 48 of 5,964 jobs (0.8%) in fishing. Thus, the sector is renamed "agriculture and related."
Source: Weber and Sorte 2002.

TABLE 2-4 Export Based Employment, Upper Klamath Basin, 1998[a]

	Sectoral Employment		Export-Dependent Employment				
Sector	No. Jobs	%	Direct	Indirect	Induced	Total	Dependency Index (%)
Agriculture and related	5,964	10.0	4,531	1,052	1,004	6,587	11.1
Forestry and logging	248	0.4	243	144	52	439	0.7
Mining	33	0.1	27	5	9	41	0.1
Construction	3,357	5.7	2,809	1,128	1,139	5,076	8.6
Manufacturing—food, beverages, and related	407	0.7	374	865	288	1,527	2.6
Manufacturing—wood products, paper, furniture, and related	4,328	7.3	3,089	2,126	1,803	7,018	11.8
Manufacturing—high technology, and related	93	0.2	30	24	11	65	0.1
Manufacturing—other (for example, sheet-metal products)	844	1.4	728	320	272	1,320	2.2
Transportation and warehousing	2,257	3.8	1,103	518	619	2,240	3.8
Utilities	429	0.7	36	26	27	89	0.2
Wholesale trade	2,035	3.4	352	76	104	532	0.9
Retail trade	6,568	11.1	423	22	82	527	0.9
Accommodation and food services	4,785	8.1	1,541	189	227	1,957	3.3
Finance and insurance	2,179	3.7	139	35	43	217	0.7
Real estate, rental, and leasing	1,535	2.6	95	50	26	171	0.3
Other services	3,733	6.3	1,110	238	235	1,583	2.7
Information	1,241	2.1	143	49	48	240	0.4
Administrative and support services	936	1.6	48	6	7	61	0.1
Arts, entertainment, and recreation	1,133	1.9	27	5	3	35	0.1
Health care and social assistance	5,859	9.9	371	65	122	558	0.9
Professional, scientific, and technical services	865	1.5	77	10	23	110	0.2
Educational services	6,010	10.1	4,546	86	1,208	5,840	9.8
Public administration	4,551	7.7	4,551	34	1,492	6,077	10.2
Households (social security)	—	—	11,952	1,947	3,185	17,084	28.8
Total	59,390	100.0	38,345	9,020	12,029	59,394	100.0

[a]Export includes any activity that brings dollars to the Klamath economy. The dependency index is the percentage of jobs that are dependent on payments to households from outside the lower Klamath basin.
Source: Weber and Sorte 2002.

households respent income earned in all of jobs generated directly and indirectly by export of wood products. The spending and respending of money brought into the region by export of wood products generated a total of 6,922 jobs.

Table 2-4 indicates the dependence of the basin's regional employment on two natural-resources sectors. Agriculture (agriculture and related plus food-products manufacturing) supports 13.7% of the region's jobs, and wood products (forestry and logging plus wood products manufacturing) supports 12.5%.

Table 2-4 also identifies the dependence of the regional economy on two sectors that often are the focus of local economic development efforts. Although the tourism sector (accommodation and food services; arts, entertainment, and recreation) is responsible for 10% of the total jobs in the region, it contributes only 3.4% of the export employment base. Retail trade, the sector with the largest employment share (11.1%), provides only 1% of the export employment base.

Table 2-4 also shows that regional employment is more dependent on income of households outside the region than on any single sector. Household income from government transfer payments (for example, social security), dividends, commuters' income, rental payments, and other sources of income originating outside the region supported 17,084 jobs (28.8%) in 1998.

The dependence of the basin's economy on federal and state government and educational institutions also is evident in Table 2-4. Almost one-fifth of the jobs in the region depend on federal and state funding for such services as education and other public services. Public administration, which supports 10.1% of jobs, includes federal and state payments to local governments (for example, federal payments in lieu of taxes, federal forest payments, and state-shared cigarette and highway revenues) and to government personnel (in USFS, USDA, and USFWS, for example). State and federal funding of educational services (such as K-12 schools, the community college in California, and the Oregon Institute of Technology) and tuition payments by nonresidents support 9.8% of the region's jobs.

There were 2,239 farms in the upper Klamath basin in 1997 (Table 2-5). A farm is defined as "any place from which $1,000 or more of agricultural products were produced or sold, or normally would have been sold, during the census year" (USDA 1999, p. VII). Farms thus include many places that do not depend significantly on farm income. Indeed, as shown in Table 2-5, 29% of farm operators work more than 200 days per year off the farm, and only 60% consider farming their primary occupation. Just over half the farms (57%) have more than $10,000 in annual sales.

Farms averaged 896 acres; 78% had some irrigation, and 27% of the region's farmland is irrigated. Most farms (82%) are sole proprietorships,

TABLE 2-5 Characteristics of Upper Klamath Basin Farms and Farm Operators, 1997

Farm Characteristic	Klamath, OR	Siskiyou, CA	Modoc, CA	Upper Basin Total
Number of farms	1,066	733	440	2,239
Land in farms (acres)	713,534	628,745	662,927	2,005,206
Average size of farm (acres)	669	858	1,507	896
Farms with sales >$10,000 (%)	54	55	69	57
Farms with irrigation (farms)	851	556	337	1,744
Irrigated land (acres)	243,205	139,534	159,219	541,958
Market value of agricultural products sold ($000)	100,622	74,244	63,797	238,663
Net cash return from agricultural sales for the farm unit ($000)	20,104	16,389	11,249	47,742
Average net cash return per farm ($)	18,859	22,359	25,556	21,323
Government payments received ($000)	817	1,420	666	2,903
Farms receiving payments (%)	16	21	25	19
Average government payments per farm receiving payments ($)	4,750	9,467	6,055	6,720
Farms with hired labor (farms)	380	259	206	845
Farms with hired labor (%)	37	35	47	38
Number of hired farm workers (workers)	1,779	2,795	1,664	6,238
Workers working 150+ days (%)	37	17	21	24
Hired farm labor payroll ($000)	9,745	11,309	6,169	27,223
Average annual pay per hired farm worker ($)	5,478	4,046	3,707	4,364
Sole-proprietor farms (%)	83	82	82	82
Farm operators living on farm operated (%)	82	78	72	78
Operators with farming as primary occupation (%)	58	61	65	60
Farm operators working more than 200 days off-farm (%)	33	27	23	29

Sources: USDA 1999, Weber and Sorte 2002.

and 78% are operated by the person living on the farm. About one-third of the farms (38%) hire farm workers. The average annual pay per hired farm worker was $4,364. About one-fourth (24%) of the 6,238 farm workers worked 150 days or more in 1997.

Net cash return per farm from agricultural sales in the upper Klamath basin averaged $21,323 in 1997. Net cash return equals the value of agricultural products sold minus operating expenses (not including depreciation). Almost one-fifth of the farms (19%) received government payments in 1997, which averaged $6,720.

Table 2-6 reports the value of agricultural production by commodity for each upper Klamath basin county and for the region. The regional value of total agricultural production in 1998 was estimated to be $283 million. Cattle, hay, and pasture accounted for 58% of the value of production, but potato production was also important (15%).

Farm income in the upper Klamath basin, as elsewhere, varies considerably from year to year and from county to county. BEA provides county-level estimates of realized net income from farming, farm proprietors' income, and farm-labor income. Realized net income is equal to total cash receipts from marketing plus other income (including government payments, such farm-related income as custom work and rent, and imputed rent for farm dwellings) minus total production expenses (including depre-

TABLE 2-6 Value of Agricultural Production (Thousands of Dollars) in Upper Klamath Basin, 1998, by County

Commodity	Klamath, OR	Siskiyou, CA	Modoc, CA	Upper Basin Total	Share of Total Value of Production, %
Alfalfa hay	30,726	25,203	12,825	68,754	24.3
Cattle	32,850	23,635	9,000	65,485	23.2
Potatoes	14,217	19,323	7,866	41,406	14.6
Pasture and range	n/a	13,005	7,560	20,565	7.3
Other hay	4,856	3,713	3,588	12,157	4.3
Barley	5,225	3,280	2,187	10,692	3.8
Onions	n/a	2,862	2,464	5,326	1.9
Wheat	1,660	2,805	859	5,324	1.9
Dairy	13,112	2,442	n/a	15,554	5.5
Horseradish	n/a	n/a	896	896	0.3
Sugarbeets	3,832	n/a	3,284	7,116	2.5
Nursery products	n/a	17,271	n/a	17,271	6.1
Other	1,000	5,319	5,973	12,292	4.3
Total	107,478	118,858	56,502	282,838	100

Abbreviations: n/a, not applicable.
Source: Oregon State University Extension Service, California Agricultural Statistics Service.

ciation). In 1997, realized net income in the upper Klamath basin was $30 million, and incomes were positive in all counties. In 1998 (not shown in Table 2-5), realized net farm income in the upper Klamath basin was less than in 1997 (about $1.2 million), and in Klamath County it was negative (–$7 million). BEA estimates farm labor income at $24 million for 1997 and $26 million for 1998 (the 1997 Census of Agriculture estimates hired farm-worker payroll at $27 million).

Farm employment is not as variable as farm income. BEA estimates that there were 2,601 farm proprietors in 1997 and 2,702 in 1998. The Census of Agriculture reports only 2,239 farm operators in 1997 (Table 2-5, USDA 1999). BEA estimates full- and part-time farm wage and salary employment at 1,812 in 1997 and 1,491 in 1998. The Census of Agriculture reports more than 4 times as many hired farm workers (6,238) in the upper Klamath basin in 1997 (Table 2-5, USDA 1999). The Oregon Employment Department estimate of total agricultural (worker) employment in Klamath County in 1997 was 1,490, twice the BEA estimate of 784, suggesting that BEA substantially undercounts farm workers.

The Klamath Reclamation Project provides water to 63% of the 2,239 farms and to 80% of the irrigated farms in the upper Klamath basin (Table 2-7). The Klamath Project contains 36% of the region's irrigated acreage. Farms served by the Klamath Project produce almost half (45%) the value of agricultural sales in the region.

Lower Basin

Except for regulation of releases at Iron Gate Dam, USBR's Klamath Project is disconnected from the lower basin, but the economic implications of measures that may be necessary to facilitate the recovery of coho and benefit other fishes along the Klamath main stem may be considerable for the lower basin.

As explained in this chapter, irrigation-based economies are important in the Shasta and Scott rivers and in the Trinity River, which has been studied specifically with reference to water transfers that generate economic benefits outside the watershed. Changes in irrigation practices and facilities may be necessary for the benefit of the coho and other species, and any such changes in the lower basin would need to be carried out with the cooperation of private water providers and private landholders. As will be shown in Chapters 7 and 8, present timber management and mining practices may also be inconsistent with the welfare of the coho salmon and may require modification, which could affect both public entities and private parties. Commercial fishing is involved economically in the restrictions on take, which are a disadvantage in the short term, and in efforts at restoration, which potentially provide long-term benefits.

TABLE 2-7 Farms in the Klamath Reclamation Project and in the Upper Klamath Basin

Irrigated Farms, 1997		Irrigated Acres, 1997 (1,000s)		Value of Sales, 1997 ($000)	
Basin	Project	Basin	Project	Basin	Project
1,744	1,400	542	195	$238,663	$108,539

Sources: USDA 1999; and Tables 1 and 2 from Burke 2002.

The lower Klamath basin includes parts of three counties in northwestern California: Del Norte, Humboldt, and Trinity. The Klamath River flows from the upper basin in Klamath County, Oregon, into Modoc and Siskiyou counties, California, and then to the lower basin in northern Humboldt County. It continues through southern Del Norte County before reaching the Pacific Ocean near Requa, California. Although the Klamath River itself does not flow through Trinity County, the county is drained mostly by the Trinity River, which is the largest tributary of the Klamath River. The basin does not include Crescent City, the county seat in Del Norte County, or the region's most populous area, Humboldt Bay (including Eureka and Arcata) in Humboldt County. Because demographic and economic statistics are gathered for government jurisdictions, the analysis that follows includes all three relevant counties. Humboldt County dominates the region demographically and economically; it has three-fourths of the region's population and over three-fourths of its full- and part-time jobs. The economy of the lower Klamath basin, which is home to about 167,000 people, in 1998 produced $5.9 billion worth of output, added $3.3 billion in value to purchased inputs, and had more than 84,000 jobs.

Much of the information given below is derived from a report by Sorte and Wyse (in press) and like information on the upper Klamath basin, is based on longitudinal data from BEA; profiles of farm numbers, type, and production from the 1997 Census of Agriculture (USDA 1999) and California County agricultural commissioners' reports; and information from a proprietary input-output economic IMPLAN model constructed by the Minnesota IMPLAN Group, Inc. The IMPLAN model was edited by using agricultural-production data from the California Agricultural Statistics Service, employment data from BEA's Regional Economic Information Service, and fisheries data from Hans Radtke and Shannon Davis of The Research Group, Corvallis, Oregon. Because a number of data sources were used, there is some variation in the categories used to aggregate the industrial sectors and to estimate the number of jobs in each sector.

From 1969 to 1999, full- and part-time employment in the lower Klamath basin increased by 171% from 49,000 to 84,000 jobs. Over the same period, employment in California increased by 211%, and U.S. employment by 180%. As in the upper basin, the composition of the regional economy changed substantially over this time. A summary of the changes is provided in Table 2-8. In the lower basin, the sectors that grew most were construction and services. The share of jobs in construction grew from 2.9% to 5.4% of the total; jobs in services grew from 16.6% to 29.9%. Modest growth occurred in agricultural services, forestry, fishing, and other; retail trade; and finance, insurance, and real estate. Employment declined in the mining, manufacturing, and military sectors. Lower than average growth occurred in the farming, transportation and public utilities, wholesale trade, and federal civilian sectors.

Table 2-9 gives estimates of some basic economic indicators and their distribution among sectors for 1998. This table, which is based on data from Minnesota Implan Group's Input-Output IMPLAN Model, varies slightly from Table 2-8, which is based solely on Bureau of Economic Analysis data. The sectors with the largest shares of output in 1998 were combined wood products including forestry and logging and manufacturing—wood products, etc. (19.8%), construction (8.4%), retail trade (6.8%), and combined agriculture including agriculture, fishing and related and manufacturing—food, etc. (6.5%). The four sectors with the largest shares of value added were wood products (12.4%), retail trade (10.4%), educational services (9.8%), and health care and social assistance (9.4%).

Retail trade (12.8%), educational services (12.2%), and health care and social assistance (11.8%) had the greatest shares of jobs in the economy.

As noted for the upper-basin economy, output, value added, and employment measures indicate the magnitude and distribution of economic activity among sectors in a region. The magnitude of economic activity in a sector, however, does not necessarily reflect the extent to which the sector sustains economic activity in the region.

Table 2-10 summarizes the contribution of each sector to total regional employment, and is based on an analysis that used the Lower Klamath Basin Input-Output Model, which was developed for this report. The jobs under the sectoral employment columns are within the sector, whereas the jobs in the export-dependent columns are from all sectors that depend on the exports from a sector. For example, there were 5,017 jobs in the construction sector but 6,941 jobs in the region depended on construction exports (for example, building homes for retirees from outside the region or construction roads for federal or state governments). Of those, 3,886 jobs depended directly on the exports of construction services from the region; these jobs were related to direct purchases from construction firms from household, firms, and governments outside the region. In addition, 1,687 jobs depended indirectly on construction exports; these jobs were created when construction firms purchased inputs (for example, building materials)

TABLE 2-8 Structural Change in the Lower Klamath Basin Economy, 1969–1999

Sector	Employment		Employment		Employment Change	
	1969	%	1999	%	1969–1999	%
TOTAL FULL- AND PART-TIME EMPLOYMENT	49,107	100.0	84,192	100.0	35,085	71.4
Wage and salary employment	40,867	83.2	64,298	76.4	23,431	57.3
Proprietors' employment	8,240	16.8	19,894	23.6	11,654	141.4
Farm proprietors' employment	917	1.9	1,166	1.4	249	27.2
Nonfarm proprietors' employment	7,323	14.9	18,728	22.2	11,405	155.7
Farm employment	1,517	3.1	2,320	2.8	803	52.9
Nonfarm employment	47,590	96.9	81,872	97.2	34,282	72.0
PRIVATE EMPLOYMENT	37,780	76.9	66,238	78.7	28,458	75.3
Agricultural services, forestry, fishing, & other	1,176	2.4	2,859	3.4	1,683	143.1
Mining	105	0.2	50	0.1	–55	–52.4
Construction	1,429	2.9	4,531	5.4	3,102	217.1
Manufacturing	12,747	26.0	7,986	9.5	–4,761	–37.3
Transportation and public utilities	2,936	6.0	3,077	3.7	141	4.8
Wholesale trade	1,219	2.5	1,968	2.3	749	61.4
Retail trade	7,667	15.6	15,498	18.4	7,831	102.1
Finance, insurance, and real estate	2,332	4.7	5,098	6.1	2,766	118.6
Services	8,169	16.6	25,171	29.9	17,002	208.1
GOVERNMENT AND GOVERNMENT ENTERPRISE	9,810	20.0	15,634	18.6	5,824	59.4
Federal civilian	1,007	2.1	1,235	1.5	228	22.6
Military	1,495	3.0	473	0.6	–1,022	–68.4
State and local	7,308	14.9	13,926	16.5	6,618	90.6

TABLE 2-9 Output, Value Added, and Employment in Lower Klamath Basin, 1998

Sector	Output		Value Added		Employment	
	$ Million	Share (%)	$ Million	Share (%)	Jobs	Share (%)
Agriculture, fishing & related	186.034	3.2	120.316	3.6	4,055	4.9
Mining	9.013	0.2	4.583	0.1	56	0.1
Construction	496.378	8.4	179.237	5.4	5,017	6.0
Manufacturing—food, beverages, textiles & related	192.714	3.3	31.680	1.0	934	1.1
Forestry and logging	180.966	3.1	75.554	2.3	1,286	1.5
Manufacturing—wood products, paper, furniture & related	983.314	16.7	335.536	10.1	5,175	6.2
Manufacturing—high technology and related	12.676	0.2	4.280	0.1	67	0.1
Manufacturing—other	129.025	2.2	43.022	1.3	936	1.1
Transportation and warehousing	258.081	4.4	111.594	3.4	2,730	3.3
Utilities	300.116	5.1	132.448	4.0	853	1.0
Wholesale trade	181.920	3.1	124.520	3.8	2,196	2.6
Retail trade	399.214	6.8	346.514	10.4	10,623	12.8
Accommodation and food services	208.678	3.5	115.597	3.5	6,486	7.8
Finance and insurance	264.670	4.5	194.334	5.9	2,887	3.5
Real estate, rental, and leasing	389.385	6.6	283.234	8.5	1,444	1.7
Other services	252.268	4.3	139.735	4.2	6,997	8.4
Information	133.216	2.3	64.861	2.0	1,179	1.4
Administrative and support services, and so on	44.278	0.8	26.773	0.8	1,225	1.5
Arts, entertainment, and recreation	31.921	0.5	19.466	0.6	1,189	1.4
Health care and social assistance	522.509	8.9	312.115	9.4	9,799	11.8
Professional, scientific, and technical services	126.789	2.2	87.384	2.6	2,812	3.4
Educational services	354.574	6.0	326.439	9.8	10,162	12.2
Public administration	238.337	4.0	238.337	7.2	5,113	6.1
Inventory valuation adjustment	0.301	0.0	0.301	0.0	0	0.0
Total	5,896.375	100.0	3,317.859	100.0	83,220	100.0

TABLE 2-10 Export Based Employment, Lower Klamath Basin, 1998

	Sectoral Employment		Export Dependent Employment				
Sector	No. Jobs	%	Direct	Indirect	Induced	Total	Dependency Index
Agriculture, fishing, and related	4,055	4.9	2,815	461	662	3,937	4.7
Mining	56	0.1	55	14	25	95	0.01
Construction	5,017	6.0	3,886	1,687	1,368	6,941	8.3
Manufacturing—food, beverages, textiles & related	934	1.1	617	491	210	1,319	1.6
Forestry and logging	1,286	1.5	641	549	291	1,482	1.8
Manufacturing—wood products, paper, furniture, and related	5,175	6.2	4,525	3,126	2,393	10,044	12.1
Manufacturing—high technology and related	67	0.1	31	18	15	64	0.1
Manufacturing—other	936	1.1	913	403	393	1,708	2.1
Transportation and warehousing	2,730	3.3	849	425	418	1,691	2.0
Utilities	853	1.0	262	349	274	885	1.1
Wholesale trade	2,196	2.6	367	72	110	549	0.7
Retail trade	10,623	12.8	2,570	95	457	3,123	3.8
Accommodation and food services	6,486	7.8	1,676	170	202	2,048	2.5
Finance and insurance	2,887	3.5	705	96	188	988	1.2
Real estate, rental, and leasing	1,444	1.7	227	90	61	377	0.5
Other services	6,997	8.4	3,819	402	574	4,794	5.8
Information	1,179	1.4	184	78	56	317	0.4
Administrative and support services, and so on	1,225	1.5	229	90	61	380	0.5
Arts, entertainment, and recreation	1,189	1.4	255	20	26	301	0.4
Health care and social assistance	9,799	11.8	3,500	565	841	4,906	5.9
Professional, scientific, and technical services	2,812	3.4	809	76	192	1,077	1.3
Educational services	10,162	12.2	9,645	112	2,605	12,362	14.9
Public administration	5,113	6.1	5,113	43	1,487	6,643	8.0
Inventory valuation adjustment	0	0.0	0	0	0	0	0.0
Households (for example, social security)[a]	0	0.0	12,268	2,032	2,891	17,191	20.7
Total	83,220	100.0	55,960	11,460	15,800	83,220	100.0

[a]Households do not represent an industry sector with employees so have zero sectoral employment. They do spend a portion of their transfer payments on goods and services (e.g., food and health care) within the region, so they have an induced effect on the economy.

from firms within the region and when the suppliers purchased from other businesses in the region.

Another 1,368 jobs were induced by exports of wood products; these jobs were in sectors like retail trade, real estate, and health care that were created when households respent income earned in all the jobs generated directly and indirectly by exports of wood products. The spending and respending of money brought into the region by exports of construction generated a total of 6,941 jobs.

Table 2-10 shows that the lower-basin economy depends on the natural-resources sectors, although not to the same extent as that of the upper basin. The combined agricultural sectors support 6.3% of the region's jobs, and the combined wood products sectors support 13.9%. Together, these two natural-resources sectors make up about 20.2% of the lower-basin economy. In the upper basin, the agricultural sector supports 14% of the region's jobs, and wood products supports 12.5%, for a total of about 27% of the economy. Table 2-10 also identifies the dependence of the lower-basin regional economy on four other sectors that often are the focus of local economic development efforts, particularly in rural economies oriented to natural resources. Specifically, these are the sectors that include substantial activity related to tourism associated with visitors from outside the region, such as retail trade, accommodation and food services, other services, and arts, entertainment, and recreation, which together contribute 12.5% of the export employment base (slightly more than in the upper basin). Still, these tourism sectors remain primarily service sectors. For example, the retail-trade sector's share of sectoral employment is 12.8%, and it provides just 3.8% of the export employment base.

The lower basin's employment, like the upper basin's, depends heavily on income to households. Household income from government transfer payments (such as social security), dividends, commuters' income, rental payments, and other sources of income originating outside the basin is the most important part of the export base. In 1998, 17,191 jobs, or 20.7%, depended on those payments.

The dependence of the basin's economy on federal and state government and educational institutions is also evident in Table 2-10. Almost one-fourth of the jobs in the region depend on federal and state funding for services, such as education and other public services. Public administration supports 8.0% of all jobs in the basin; this sector includes federal and state payments to local governments (such as federal payments in lieu of taxes, federal forest payments, and state-shared cigarette and highway revenues) and to government personnel (USFS, USDA, and USFWS, for example). State and federal funding for educational services plus tuition payments by nonresidents support 14.9% of the region's jobs.

Two important industries based on natural resources, agricultural crop and livestock production and fisheries, are aggregated and summarized in the tables as the agriculture, fishing, and related sector. Because they are both so strongly affected by water resources in the Klamath basin, some additional review of these industries follows.

Using the same definition of a farm as in the upper basin, there were 974 farms in the lower Klamath basin in 1997, that is about 40% of the number of farms in the upper basin (Table 2-11). As noted in the discussion regarding the upper basin, farms include many places that do not depend on their farm operations as their major source of income. Indeed, as shown in Table 2-11, 35% of farm operators work more than 200 days/yr off the farm, and only 51% consider farming their primary occupation. Fewer than half the farms (45%) have more than $10,000 in annual sales. Farms averaged 653 acres; 39.5% had some irrigation and 3.7% of the region's farmland is irrigated. Over half the farms (61%) are sole proprietorships, and 72% are operated by the person living on the farm. About one-third of the farms (35%) hire farm workers. The average annual pay per hired farm worker was $6,754. Thus, the number of farm workers in the lower basin is about one-third the number in the upper basin, but the average pay per worker is greater in the lower basin. About half (44%) the 2,183 farm workers worked 150 or more days in 1997.

Net cash returns per farm from agricultural sales in the lower Klamath basin averaged $23,016 and were similar to those of the upper basin ($21,323) in 1997. Net cash returns equals the value of agricultural products sold minus operating expenses (not including depreciation). Very few farms (3.1%) received government payments in 1997, which averaged $2,000.

Table 2-12 reports the value of agricultural production by commodity for each of the counties in the lower Klamath basin and for the region. The regional value of total agricultural production in 1998 was estimated to be $114 million, compared with $283 million in the upper basin. Dairy and nursery products are the principal agricultural products of the region, together accounting for 75.6% of the value of agricultural-commodity production. Cattle and livestock products are also important; they account for 13.7% of the value of agricultural commodity production.

Fishing is an important part of the culture of the lower-basin culture and the economy. Table 2-13 provides information on catch and value for the fishing industry in 1997–2001. The catch information reflects only ocean-related commercial fishing, not fishing in rivers. The lower Klamath basin input-output model explicitly considers ocean fishing in the agriculture, fishing, and related sectors because the catch is sold directly for processing or consumption. River fishing is included only indirectly in the model; that economic activity and other activities related to fish in the

TABLE 2-11 Characteristics of Lower Klamath Basin Farms and Farm Operators, 1997

Farm Characteristic	Del Norte County	Humboldt County	Trinity County	Lower Basin Total
Number of farms	66	792	116	974
Land in farms (acres)	13,303	584,538	118,252	716,093
Average size of farm (acres)	202	738	1019	653
Farms with sales >$10,000 (%)	41	49	22	45
Farms with irrigation (farms)	24	301	70	395
Irrigated land (acres)	6,323	17,630	2,212	26,165
Market value of agricultural products sold ($000)	20,797	75,475	1,797	98,069
Net cash return from agricultural sales for the farm unit ($000)	5,229	17,700	–489	22,440
Average net cash return per farm ($)	79,234	22,320	–4,216	23,016
Government payments received ($000)	0	54	6[a]	60
Farms receiving payments (%)	0	3.4	2.6	3.1
Average government payments per farm receiving payments ($)	0	2,000	2,000[a]	2,000
Farms with hired labor (farms)	32	279	32	343
Farms with hired labor (%)	48	35	28	35
Number of hired farm workers (workers)	652	1,345	186	2,183
Workers working 150+ days (%)	47	47	18	44
Hired farm labor payroll ($000)	5,579	8,690	476	14,745
Average annual pay per hired farm worker ($)	8,557	6,461	2,559	6,754
Sole proprietor farms (%)	64	59	78	61
Farm operators living on farm operated (%)	70	73	80	72
Operators with farming as primary occupation (%)	50	53	43	51
Farm operators working more than 200 days off farm (%)	33	35	38	35

[a]Information not disclosed by the Census of Agriculture because few farms (three) received assistance. Average payment of $2,000 was estimated because in 1992 the average was $2,236 for Trinity County and in 1997 the average was $2,000 for Humboldt County.

Source: USDA 1999.

TABLE 2-12 Value of Agricultural Production in the Lower Klamath Basin, 1998

	Value of Agricultural Production, $000				
Commodity	Del Norte $	Humboldt $	Trinity $	Lower Basin Total $	Share of Total Value of Production %
Dairy	10,578	39,028	0	49,606	43.5
Nursery products	13,322	23,277	37	36,636	32.1
Cattle and livestock products	3,495	11,074	1,088	15,657	13.7
Hay and pasture	1,351	8,179	463	9,993	8.8
Vegetables	75	676	32	783	0.7
Sheep, lambs, and wool	38	116	8	162	0.1
Fruit and nuts	435	91	105	631	0.6
Other	472	20	49	541	0.5
Total	29,766	82,461	1,782	114,009	100.0

Source: California Agricultural Statistics Service.

Klamath River main stem are reflected primarily in the tourism sectors. Thus, the actual effects of fish migration through the Klamath basin are difficult to estimate accurately. As Table 2-13 indicates, commercial fishing had a value of $12.4 million in 2001, which was less than in prior years and continues to steadily decline.

In relative terms, commercial fishing accounts for about 10% of the value of agriculture in the lower basin. The most valuable components of the catch are groundfish ($5.5 million in 2001) and crab and lobster ($4.1 million in 2001). Salmon (Chinook) landings were valued at about $0.2 million in 2001.

The economic effects of eliminating or reducing any of the ocean fisheries in the lower-basin economy can be calculated with the same procedure used earlier to determine the export dependency indexes. Using the detailed multi-sector version of the Lower Klamath Basin Input-Output Model, which is based on the 1998 IMPLAN model, to be consistent with the upper basin analysis, the effect of removing all the salmon catch in 2001 ($107,887), assuming that the catch is exported from the region, is a total loss to the regional economy of $164,507. This effect, though relatively small in comparison to the commercial fishing industry or the total regional economy, did extend across 193 of the 204 sectors in the regional economy. Commercial fishing has a multiplier of approximately 1.5 on both employment and output in the region. Thus, for every dollar or job directly involved in commercial fishing there is approximately another fifty cents or

TABLE 2-13 Fisheries Characteristics of Ports of Eureka (Humboldt County) and Crescent City (Del Norte County)

	Round Pounds			
Species Group	1997	1998	1999	2000
Groundfish	16,246,794	13,888,084	12,036,198	10,116,024
Pacific whiting	13,958,624	12,614,230	2,881,997	10,988,772
Salmon (troll chinook)	16,675	26,450	34,500	26,450
Crab and lobster	6,454,585	7,425,668	7,122,922	4,764,952
Shrimp	12,441,711	1,460,207	3,658,543	2,170,063
Coastal pelagic	176,167	161,285	46,246	14,168
Highly migratory	2,222,487	727,022	647,952	823,779
Halibut	9,007	477	891	289
Sea urchins	63,624	2,357	36,532	3,735
Other	1,822,974	564,703	597,413	841,699
	53,412,648	36,370,483	27,063,194	29,793,910

Source: Hans Radtke and Shannon Davis, unpublished.

half a job lost as suppliers or businesses that sell to those working in fishing, or for the suppliers or businesses experiencing reduced sales. The current economic effects of the commercial salmon catch may significantly understate the potential contribution of the salmon fishing to the economy of the lower Klamath basin. Salmon landings at the ports of Eureka and Crescent City have declined by more than 95% since the 1970s. If the average 1976–1980 landings from the two ports of 2,547,000 round lb could be reached, and they were sold at 2001 prices of $1.47 per lb, the combined output from the salmon fishery would be $3,744,090. The estimated value-added component of that level of output in 2001 dollars would be $2,476,908. Returning to that level of output would require an estimated 67 direct jobs in the commercial fishing sector. The multiplied effect of these jobs on commercial fishing to businesses that supply the fisheries sector and from household expenditures in service sector businesses could be an additional 30 jobs, for a total of 97 jobs. These estimates of the economic effects of increased salmon harvest assume the catch is exported outside the region and that the effects are not reduced by changes that might be necessary to achieve the increases (e.g., shifting water from irrigated agriculture to increase stream flows).

In summary, the economics of the upper and lower basins display characteristics common to many rural economies, including heavy reliance on natural resources sectors, such as agriculture and wood products. Together, the entire basin showed economic activity valued in 2002 at $10.5 billion. Of that, about 26% (or $2.7 billion) was derived from sectors based

	Value (Nominal), $				
2001	1997	1998	1999	2000	2001
8,708,018	9,309,576	6,615,305	6,308,414	6,631,668	5,461,928
5,081,398	581,399	391,780	115,275	764,851	170,967
73,600	21,298	41,427	61,577	42,795	107,887
1,719,814	11,132,662	12,193,371	13,210,063	9,403,268	4,073,747
3,447,869	5,020,462	951,542	1,982,483	1,172,213	1,236,641
148,548	93,398	39,260	11,365	7,879	52,975
1,414,603	1,870,065	764,542	630,488	841,564	1,155,138
8	17,866	790	1,669	723	16
22,595	35,352	825	26,438	3,224	12,279
388,929	509,044	227,912	217,430	262,536	138,378
21,005,382	28,591,122	21,226,754	22,565,202	19,130,721	12,409,956

on natural resources. Reliance on such sectors is slowly declining across both the upper and lower basins.

OVERVIEW

The Klamath basin is exceptionally diverse geomorphically because it has been strongly influenced by both crustal movement and volcanism. Geomorphic diversity in the basin has produced a wide variety of aquatic habitats, including extensive wetlands, large shallow lakes, swiftly flowing main-stem waters, and various tributary conditions. The watershed is not densely populated but shows strong anthropogenic influences of several kinds. Management of water for irrigation, which has been in progress for more than a century, has altered the basic environmental conditions for aquatic life, including the hydrographic features of flowing waters, the distribution and extent of wetlands, and the extent and physical characteristics of the lakes that were found originally in the basin. Of the total economic activity in the Klamath basin ($10.5 billion), about 26% is derived from natural resources, including mostly agriculture, wood products, and ocean fishing. Irrigation and agricultural practices have blocked or diverted fish from migration pathways, caused adverse warming of waters, and augmented nutrient transport from land to water. Commercial fishing also has left a mark through depleted stocks of some species and, although now controlled, may have had legacy effects that are difficult to reverse. Timber harvest and mining along tributaries have caused, and in some cases continue to cause, severe physical impairment of aquatic habitats. Although aquatic habitats now are regarded as valuable for the maintenance of native species, remediation of damage to habitat presents great difficulties because of the extent and diversity of changes that have occurred in the basin over the last century.

3

Current Status of Aquatic Ecosystems: Lakes

INTRODUCTION

Natural lakes that were suitable for occupation by suckers before land-use development and water management included Upper Klamath Lake, Lower Klamath Lake, Tule Lake, and Clear Lake (Figure 1-3). All of these lakes have been changed morphometrically and hydrologically and are now used in the Klamath Project water-management system for storing and routing water. Gerber Reservoir is also part of the water-management system, but its location was previously occupied by a marsh rather than by a lake. Other lakes relevant to the welfare of suckers include those lying behind five main-stem dams that, except for Keno Dam, incorporate hydroelectric production facilities (Figure 1-3, Table 3-1). The last in the sequence of main-stem dams, Iron Gate Dam, provides reregulation capability for the main stem of the Klamath River as explained below.

Of the lakes used for storage and routing, Upper Klamath Lake, Clear Lake, and Gerber Reservoir support the largest populations of listed suckers (see Chapter 6 for a detailed treatment of the suckers), and these three lakes have been the main focus of ecological and limnological analysis related to the welfare of suckers. Upper Klamath Lake has been studied especially intensively because it potentially would support the largest population of suckers and shows the greatest number of environmental problems, as indicated by episodic mass mortality of adults and probable hardships in all life-history stages. Clear Lake and Gerber Reservoir afford a useful comparison with Upper Klamath Lake because the sucker populations there have not suffered mass mortality and are generally more stable

TABLE 3-1 Basic Information on Lakes of Upper Klamath Basin[a]

Lake Name	Size Before 1900 (acres)		Size Since 1960 (acres)		Volume[b] (acre-ft)	Mean Depth[b] (ft)	Hydraulic Residence Time[b] (days)
	Minimum	Maximum	Minimum	Maximum			
Lakes and reservoirs used for water storage and routing							
Upper Klamath[c]	78,000	111,000	56,000	67,000	603,000	9	180
Lower Klamath[d]	85,000	94,000	4,700	4,700	<20,000	<4	<70
Clear Lake[d]	15,000	15,000	8,410	25,700	527,000	20	1,600
Tule Lake[d]	55,000	110,000	9,500	13,000	50,000	4	180
Gerber Reservoir[d]	n/a	n/a	1,100	3,900	94,000	24	600
Reservoirs used for power production							
Keno[e]	n/a	n/a	2,470	2,470	18,500	7	6
J.C. Boyle[e]	n/a	n/a	420	420	1,700	4	1.2
Copco No. 1[e]	n/a	n/a	1,000	1,000	46,900	47	12
Copco No. 2[e]	n/a	n/a	40	40	70	2	0.02
Iron Gate[e]	n/a	n/a	950	950	58,800	62	16
Total	233,000	330,000	82,120	116,710	1,420,000	—	—

[a]Lake Ewauna, which is named on some maps, is part of the Keno impoundment; Agency Lake (Figure 1-3) is treated here as part of Upper Klamath Lake.
[b]At maximum depth. Mean depths and hydraulic residence times typically are lower than shown in table, which is based on maximum volume.
[c]From Welch and Burke 2001, Table 2-1. Current maximum corresponds to water level of 4143.3 ft above sea level. Area and volume data from USFWS (2002).
[d]From USBR 2002a, Table 4.1.
[e]From PacifiCorp 2000, pp. 2-16 to 2-17; Keno has no turbines.
Abbreviations: n/a, not applicable.

than the populations of Upper Klamath Lake. The hydroelectric reservoirs on the main stem have been studied sparingly and are of less interest than other lakes from the viewpoint of listed suckers.

The lakes shown in Table 3-1 do not serve as habitat for coho salmon, which are blocked by Iron Gate Dam from entry into the upper Klamath basin. Limnological characteristics of the waters behind Iron Gate Dam are potentially important to the coho salmon, however, in that waters released from the dam have a large influence on the water-quality characteristics of the Klamath River main stem, especially near the dam. Reflecting the relative amounts of research or monitoring and the apparent ranking of lakes with respect to their importance for the endangered and threatened fishes, this chapter devotes most of its attention to Upper Klamath Lake, some to the other lakes that are used for storage and routing of water, and some to waters above Iron Gate Dam that hold non-reproducing populations of listed suckers and have the potential to affect coho downstream; the remnants of Tule Lake and Lower Klamath Lake provide little lacustrine habitat at present, but offer potential for restoration.

UPPER KLAMATH LAKE

Description

Upper Klamath Lake is the largest body of water in the Klamath basin and is one of the largest lakes in the western United States (about 140 mi^2). The lake and its drainage lie on volcanic deposits derived in part from the nearby Crater Lake caldera, which took its present form as a result of the eruption of Mount Mazama (about 6,800 BP). The lake also shows a strong tectonic influence, however, as is evident from a pronounced scarp along its southwestern edge (Figure 3-1). Although Upper Klamath Lake has a very low relative depth (ratio of depth to mean diameter), it has substantial pockets of water over 20 ft deep (maximum, 31 ft at a water level of 4,141.3 ft above sea level; USBR 1999 as cited in Welch and Burke 2001). The northern, southern, and eastern portions of the lake and Agency Lake, which is connected to Upper Klamath Lake and is here treated as part of it, are uniformly shallow; they offer water little deeper than 6 or 7 ft at mean summer lake elevation (4,141.3 ft above sea level). Even though specific runoff for the watershed of Upper Klamath Lake is relatively low (about 300 mm/yr), the hydraulic residence time of Upper Klamath Lake is only about 6 mo because the lake is shallow (there is considerable interannual variability). The flat bathymetry of the lake also causes its surface area to be quite sensitive to changes in water level.

Before the construction of Link River Dam, which was completed in 1921, the water level of Upper Klamath Lake fluctuated within a relatively

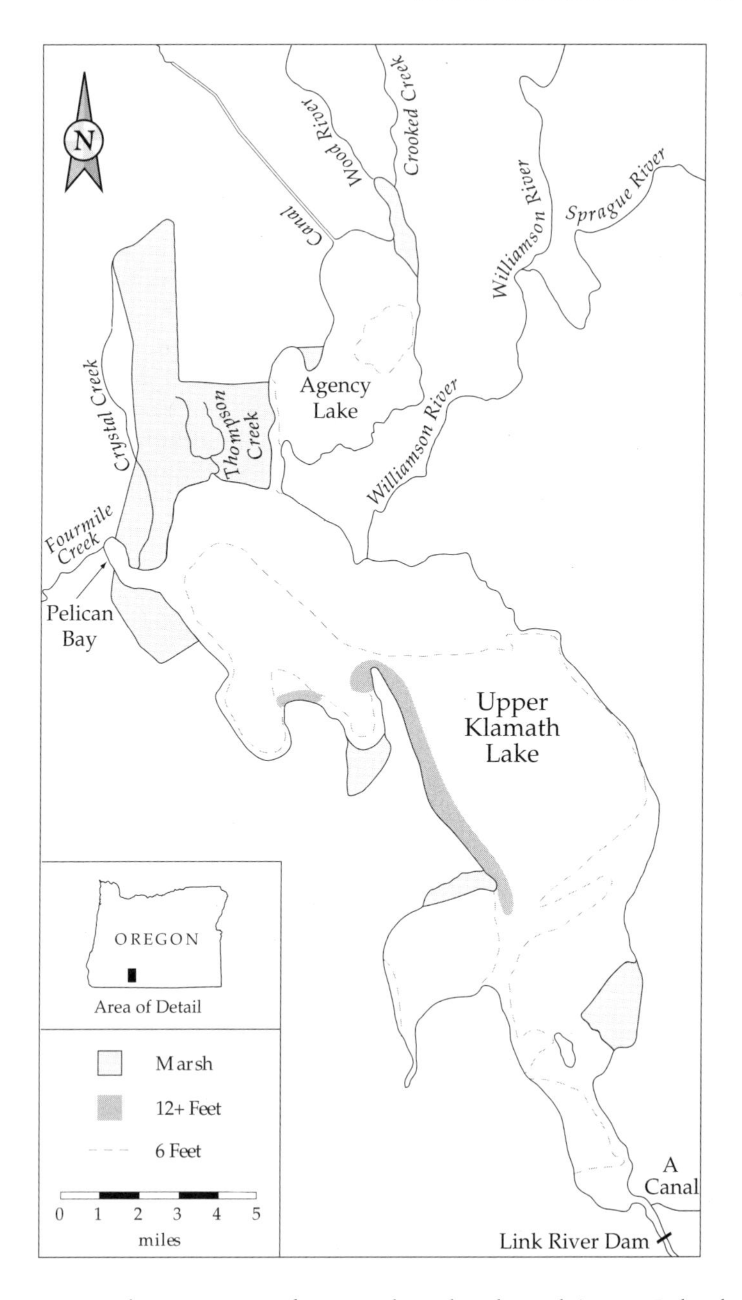

FIGURE 3-1 Bathymetric map of Upper Klamath Lake and Agency Lake showing depths at the mean summer lake elevation of 4,141 ft above sea level. Contours are from data of U.S. Bureau of Reclamation (1999) as reported by Welch and Burke (2001). Source: Welch and Burke 2001.

narrow range (about 3 ft), as would be expected for a natural hydrologic regime (Figure 3-2). Although irrigation was under way in the basin at that time, there was no means of using the lake for storage. Water level in the lake was determined by a lava dam at the outlet (4,138 ft above sea level; USFWS 2002). Even under drought conditions, the lake level remained above the level of the natural outlet, except briefly during oscillations caused by wind (USFWS 2002).

When Link River Dam was constructed, the natural rock dam at the outlet of Upper Klamath Lake was removed so that the storage potential of the lake could be used in support of irrigation. Thus, since 1921, lake levels have varied over a range of about 6 ft rather than the natural range of about 3 ft (Figure 3-2). Drawdown of about 3 ft from the original minimum water level of the lake has occurred in years of severe water shortage (1926, 1929, 1992, and 1994). The operating range of the lake in the context of mean depth and contact between the lake and its wetlands has raised numerous questions about the environmental effects of water-level manipulations, especially under the most extreme operating conditions (USFWS 2002).

The U.S. Bureau of Reclamation (USBR 2002a) has proposed operating Upper Klamath Lake over the next 10 yr according to guidelines that reflect recent historical operating practice (Figure 3-2; Chapter 1). The open question for researchers and for the tribes and government agencies charged with evaluating the two endangered suckers is whether the USBR proposal for future operations is consistent with the welfare of listed suckers in Upper Klamath Lake. In a biological opinion issued in response to USBR's proposals, the U.S. Fish and Wildlife Service (USFWS 2002) has concluded that operations should involve limits on water levels that are more restrictive than those proposed by USBR. USFWS has temporarily accepted the water-level criteria proposed by USBR (2002a), but has required a revised approach to predicting water availabilities in any given year (Chapter 1).

The USBR 10-yr plan is based on a commitment of USBR not to allow Upper Klamath Lake to fall in any given year below the minimum water levels that were observed in 1990–1999 for four hydrologic categories of years and not to allow the interannual mean water levels for these categories to fall below recent interannual means (1990–1999). Figure 3-2 shows the March 16–October 30 operating range based on interannual means for each of the four hydrologic categories. The database for the definition of the categories included water years 1961–1997 (USBR 2002a, p. 39). The calculations were based on the outflow from Upper Klamath Lake for April–September. Years above the mean outflow, which is 500,400 acre-ft, are designated "above average." Those within one standard deviation below the mean are designated "below average"; the expected long-term frequency for these years is 34% (on the basis of a normal distribution). Curve-fitting was not suitable for evaluating years of lower flow, however.

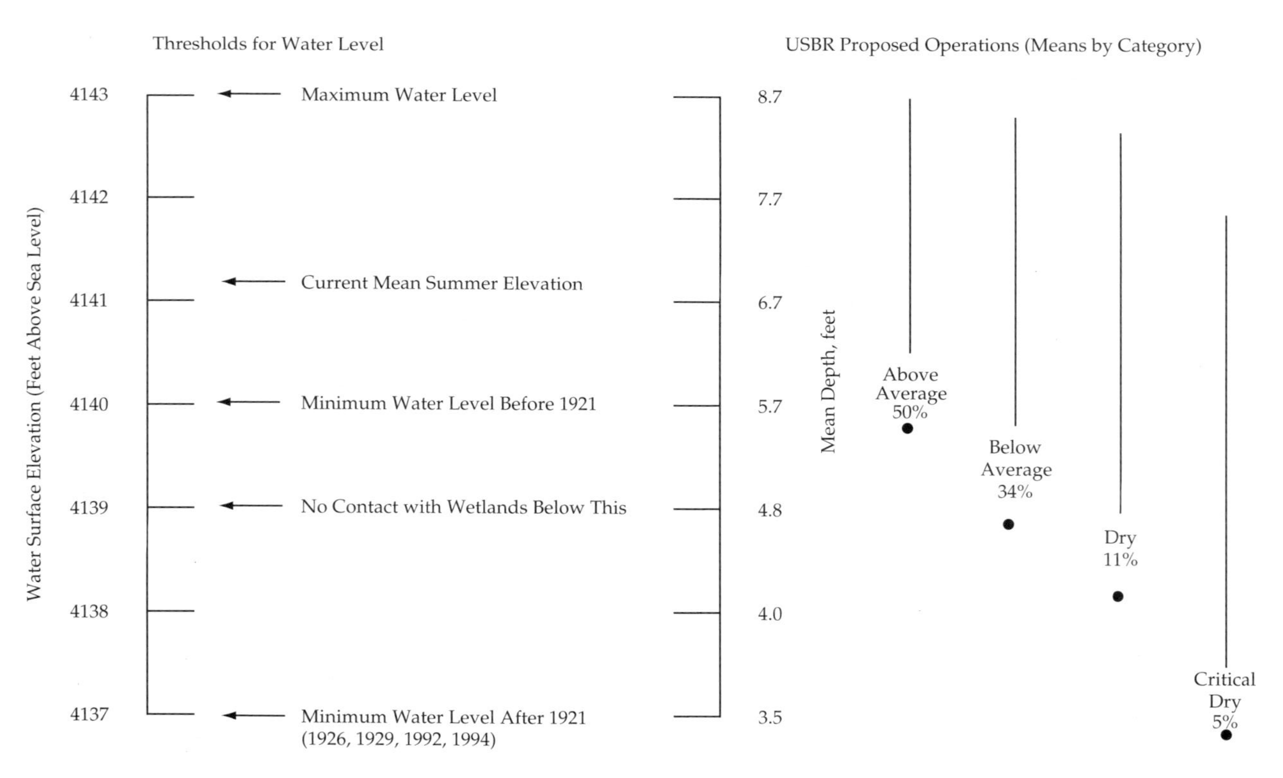

FIGURE 3-2 Water level of Upper Klamath Lake and mean water levels proposed by USBR for years of varying water availability. The vertical lines on the right show interannual mean operating range for an entire season (March 16–October 30). Minimums for specific years can reach below the interannual means; absolute minimums are shown as dots. Source: USBR 2002a.

Two extreme years, 1992 and 1994, were designated "critical dry" and account for about 5% of the total. By difference, a fourth category, designated "dry," is defined; it accounts for about 11% of years.

For each category of years, the maximum water levels occur in the spring. Water levels typically begin relatively high as of mid-March and then rise slightly, after which they fall because of the cumulative effects of drawdown and, after June, the reduced volume of runoff (Figure 3-3). Operations for the four hydrologic categories differ most notably in their lower extremes, which occur after July. In comparison with a baseline condition, which USBR defines as lacking Klamath Project operations but with all project facilities in place, proposed operations typically produce water levels that are above the baseline between March and the end of June and below the baseline during the last half of the summer or fall (USBR 2002a).

Upper Klamath Lake receives most of its water from the Williamson River (including its largest tributary, the Sprague River) and the Wood River. Additional water sources include precipitation on the lake surface, direct drainage from smaller tributaries and marshes, and springs that bring water into the lake near or beneath the water surface. The waters of the lake have only moderate amounts of dissolved solids (interseasonal median, about 100 µS/cm) and the same is true of alkalinity (interseasonal median, about 60 mg/L as calcium carbonate). As described below, the lake is naturally eutrophic, but concentrations of nutrients in the water column may have increased over the last several decades. The fish community of the lake could be described as a diverse array of nonnative species superimposed on a previously abundant but now reduced group of native fishes,

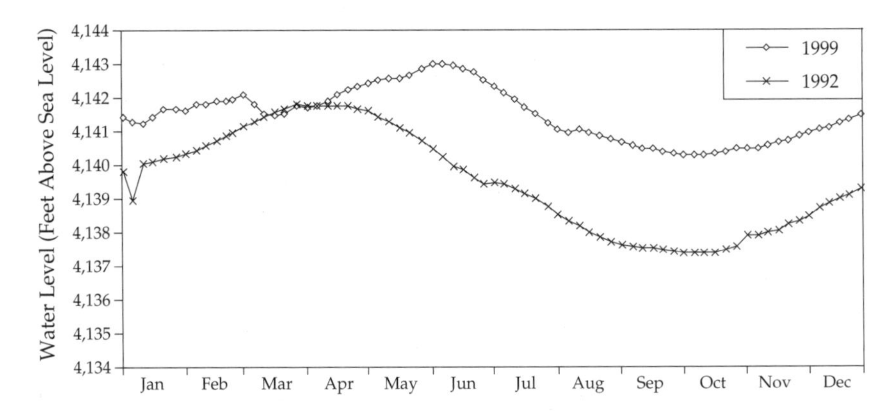

FIGURE 3-3 Water level in Upper Klamath Lake in year of near-average mean water level (1999) and year of extremely low water level (lowest 5%; 1992).

most of which are endemics (Chapter 6). The biota in general has undergone considerable change in the last few decades.

Upper Klamath Lake has several large marshes at its margins. The area of the marshes has been greatly reduced (loss of about 40,000 acres from the lake margin; USFWS 2002). The remaining marshes are most strongly connected to the lake at high water and are progressively less connected at lower water levels down to about 4,139 ft above sea level, at which point they become disconnected.

Poor water quality in Upper Klamath Lake causes mass mortality of listed suckers and may suppress the suckers' growth, reproductive success, and resistance to disease or parasitism. Potential agents of stress and death include high pH, high concentrations of ammonia, and low dissolved oxygen (USFWS 2002). Extremes in these variables are explained by the presence of dense populations of phytoplankton (primarily the cyanobacterial taxon *Aphanizomenon flos-aquae*), especially in the last half of the growing season (Kann 1998, Welch and Burke 2001). Because phytoplankton populations annually reach abundances exceeding 100 µg/L of chlorophyll *a*, the lake can be classified as hypertrophic (or, equivalently, hypereutrophic) according to standard criteria for trophic classification of lakes (OECD 1982: peak chlorophyll over 75 is hypertrophic). Hypertrophic lakes often show extremes in chemical conditions resembling those observed in Upper Klamath Lake.

The main subjects of interest with respect to Upper Klamath Lake proper (discounting the tributaries, which are dealt with in the next chapter) include factors that have been suspected by researchers or by government agencies of being potentially harmful to the endangered suckers. Where water quality is concerned, the causes of the current trophic status of the lake are of great interest, as is the current predominance of a single algal species, *Aphanizomenon flos-aquae,* in the phytoplankton. Within the suite of variables affected by trophic status, special attention must fall on pH, ammonia, and dissolved oxygen, all of which have the potential to be directly or indirectly harmful to the welfare of the endangered suckers. For all water-quality variables, associations between water level and water quality are of special interest because USBR has the potential to modify operations so as to control water level. Finally, physical habitat, especially as affected by water level, is of concern and will be dealt with here.

Nutrients and Trophic Status of Upper Klamath Lake

Nutrient limitation of phytoplankton in lakes usually is seasonal and almost always is associated with nitrogen, phosphorus, or both of these elements. Typically, phosphorus and nitrogen are readily available during winter because demand is low. In spring, the most available forms are taken up, and nutrient limitation often ensues. If the most readily available forms are available in quantities above about 10 µg/L, there is a strong implication that no limitation is occurring (e.g., Morris and Lewis 1988); at lower concentrations, nutrient limitation is possible but may be delayed by internal storage. Nutrient limitation often is relieved in the fall by deep, continuous mixing of the water column, declining irradiance, and lower metabolic rates caused by lower temperatures.

Nitrogen limitation can be defeated by some taxa of bluegreen algae (cyanobacteria) capable of fixing nitrogen (converting N_2 to NH_3). Nitrogen gas (N_2) is present in considerable quantity in water, and the overlying atmosphere acts as a large reservoir that can replenish removal of nitrogen gas by nitrogen fixation. The heterocystous bluegreen algae—which have a special cell, the heterocyst—fix nitrogen readily, although the fixation process requires high intensities of light (Lewis and Levine 1984). Heterocystous bluegreen algae do not grow well in some situations, however, for reasons that are only partly understood (Reynolds 1993). Thus, nitrogen depletion sometimes can occur without inducing growth of nitrogen fixers. Nitrogen fixers grow well in most warm, fertile waters of high pH. When phosphorus is abundant in such waters but nitrogen is scarce, nitrogen fixers have a competitive advantage and often become dominant elements of the phytoplankton. This is the situation in Upper Klamath Lake. For the phytoplankton as a whole in Upper Klamath Lake, nitrogen is limiting (see below), but *Aphanizomenon* has circumvented nitrogen limitation through nitrogen fixation and thus dominates the community.

Typically, the most effective way to control phytoplankton abundance in lakes is to restrict phosphorus supply. Restriction of nitrogen supply is not as effective, because it may lead to the development of nitrogen fixers that are able to offset restrictions in nitrogen supply. Thus, the most obvious way of attempting to control phytoplankton populations in Upper Klamath Lake is to restrict phosphorus supply. As explained below, Upper Klamath Lake presents special difficulties for strategies involving control of phosphorus.

Phosphorus in Upper Klamath Lake

The watershed of Upper Klamath Lake is geologically rich in phosphorus (Walker 2001). Springs have a median phosphorus content of about 60 µg/L as soluble reactive phosphorus, which Boyd et al. (2001, citing Walker 2001) take as an estimate of the background discharge-weighted mean phosphorus concentration. This may be an underestimate, given that springs typically have little or no particulate phosphorus or soluble organic phosphorus, both of which would be present in natural runoff from the watershed. In contrast, watersheds of granitic geology often have discharge-weighted mean total P concentrations of 20 μg/L or less (inorganic P about 5 μg/L), provided that they are not disturbed by human activity (e.g., Schindler et al. 1976, Lewis 1986).

Because background concentrations of phosphorus reaching Upper Klamath Lake are quite high, the lake probably supported dense populations of phytoplankton before land-use development. Early observations indicate that the waters were green, and thus eutrophic, at a time when water quality would have been changed little from the natural state. If, as suggested by Boyd et al. (2001), phosphorus reaching the lake would have had originally a discharge-weighted mean phosphorus concentration of about 60 μg/L, phosphorus in lake water would have been somewhat below 60 μg/L (because of sedimentation of some phosphorus) in the absence of internal loading (net increase originating from sediments). On the basis of empirical relationships between chlorophyll *a* and phosphorus (OECD 1982), the mean chlorophyll *a* in the growing season with total phosphorus at 60 μg/L would have been in the vicinity of 20 μg/L, which would have corresponded to short-term maximums of 40–60 μg/L, or about 20% of the current maximums. The concentrations of phosphorus in the lake could have been higher, however, if substantial internal loading from sediments occurred under natural conditions, in which case chlorophyll could also have been higher.

Monitoring of phosphorus entering the lake has shown that the current discharge-weighted mean phosphorus concentration in waters entering Upper Klamath Lake is near 100 μg/L, about 40% of which is considered to be anthropogenic (Boyd et al. 2001). Concentrations in the lake during spring are only about 50 μg/L (Boyd et al. 2001, Figure 2-6; there is considerable variation from year to year); the difference between the supply water and the concentrations in spring is accounted for by sedimentation of the particulate fraction of incoming phosphorus and by mechanisms that convert incoming soluble phosphorus to particulate phosphorus that can undergo sedimentation. The currently observed total phosphorus concentrations in spring, if not supplemented by any other sources, would support mean algal abundances during the growing season corresponding to chlorophyll *a* at 20 μg/L or less, according to equations developed by the Organization for Economic Co-Operation and Development (OECD 1982).

When the growing season begins (in about May), Upper Klamath Lake shows a steady rise in concentrations of total phosphorus culminating in summer concentrations of 200–300 μg/L (Boyd et al. 2001, Figure 2-6; there is considerable variation from year to year). These concentrations greatly exceed the discharge-weighted mean concentrations in inflowing water (about 100 μg/L) and also greatly exceed the concentrations in the lake during spring (about 50 μg/L, Figure 3-4). Thus, the great increase in

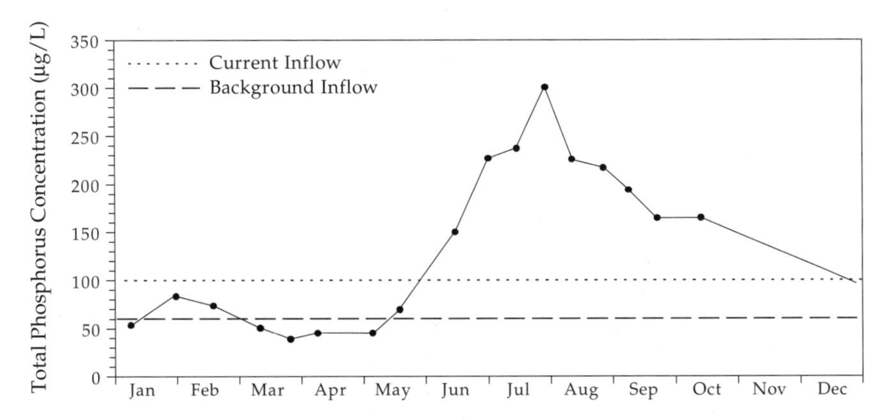

FIGURE 3-4 Total phosphorus concentrations in Upper Klamath Lake during 1997 (an arbitrarily chosen year) and approximate discharge-weighted mean total phosphorus for inflow for background and for current conditions. Source: Data from Walker 2001.

concentrations of phosphorus during the growing season must be attributed to an internal source (sediments).

Concentrations of soluble phosphorus in sediments of Upper Klamath Lake were studied by Gahler and Sanville (1971), as reported by Bortleson and Fretwell (1993). Sediment samples taken at one location in 1968–1970 showed a median soluble phosphorus concentration in the interstitial waters of about 7,000 µg/L, or about 25 times the maximum concentrations observed in the overlying lake water (another location showed less extreme deviation from lake water). Thus, for at least some portions of the lake, sediment pore waters contain substantially more soluble phosphorus than the overlying lake water and can serve as an internal source of phosphorus if the phosphorus leaves the sediments. This is a common situation in fertile lakes.

The efficiency with which phosphorus is released from sediments varies greatly according to the conditions in a particular lake. There are four potential mechanisms of release: (1) If the sediments are disturbed by wind-driven currents or by other means (organisms or degassing), interstitial phosphorus can be transferred to the water column simply by agitation. (2) Decrease in the redox potential (increase in availability of electrons) in the surficial sediments caused by intensive microbial respiration, as would be the case for highly organic sediment, can cause biogeochemical changes that result in accelerated release of mineralized or soluble organic phosphorus from the sediments to the overlying water, even if the sediments are immobile. (3) High pH at the sediment surface may cause release of adsorbed phosphorus from sediments, with or without agitation of sediments.

(4) In shallow lakes, phytoplankton cells may, under calm conditions, sink to the sediment surface, where phosphorus is more concentrated than in the water column, and then be resuspended either by wind or by buoyancy control mechanisms after assimilating phosphorus, thus bringing phosphorus from the sediments to the water column. Internal loading in Upper Klamath Lake is caused by one or more of these four mechanisms, which are not mutually exclusive.

Chlorophyll concentrations in Upper Klamath Lake increase in parallel with concentrations of total phosphorus in the water column from May to July (Boyd et al. 2001). Thus, the data indicate that phytoplankton are assimilating an internal phosphorus load leading to an increase in their biomass. The growth process culminates in concentrations of phytoplankton chlorophyll *a* typically near or above 200 µg/L (Boyd et al. 2001). At such high abundances, phytoplankton approach the maximum sustainable biomass based on light availability (self shading) rather than nutrients (Welch and Burke 2001). The specific limit for phytoplankton biomass based on light rather than nutrients depends on physical conditions in a lake and physiological characteristics of the dominant algae (Wetzel 2001).

Because internal loading increases the phosphorus inventory of the water column in Upper Klamath Lake, thus sustaining high populations of bluegreen algae, its mechanisms are of special importance to the nutrient economy and trophic status of the lake and therefore to water-quality conditions that affect fish.

The simplest mechanism of release of phosphorus from the sediments is disturbance of the sediments. As proposed initially by Bortleson and Fretwell (1993), that mechanism is highly feasible in Upper Klamath Lake because of the lake's low relative depth (a low ratio of depth to area), which is an indication that sediments will easily be mobilized by strong winds, at least over the large expanses of shallow water. Thus, decomposition processes in the sediments may liberate phosphorus from particulate form, and this phosphorus can be transferred to the water column simply by wind-generated sediment movement. Release of gas bubbles from the sediment or invertebrate activity (bioturbation) can produce similar effects. The role of sediment movement in mobilizing phosphorus in Upper Klamath Lake is unknown, but the ability of the wind to move sediments readily over much of the lake bottom is generally acknowledged (Bortleson and Fretwell 1993).

Release of phosphorus from sediments also can occur without any movement of the sediments. If there is a substantial concentration gradient of soluble phosphorus between the sediment pore waters and the overlying water, the potential exists for diffusion of phosphorus from the pore waters to the overlying lake water and distribution of the released phosphorus by eddy diffusion or bulk mixing of the water column. The key requirements for the process include presence of a substantial concentration gradient (which exists in at least some places in Upper Klamath Lake, as indicated by the study cited above) and absence of any physical or chemical barrier to diffusion of soluble phosphorus.

It is well known that iron in the ferric state can bind phosphorus, thus restricting its movement from sediments to water (Mortimer 1941, 1942). Loss of the precipitated (ferric) iron from the surface of lake sediments occurs when sediments are anoxic for long intervals, by conversion of iron to a soluble (ferrous) state. Loss of ferric iron facilitates exchange between the sediment pore waters and the overlying water and releases phosphorus bound by ferric iron. The result can be release of large amounts of phosphorus from the sediments (internal loading). The release of phosphate from sediments caused by changes in the oxidation state of iron is most likely in lakes that show prolonged anoxia at the sediment-water interface. Unlike deeper lakes, Upper Klamath Lake does not remain stratified for the entire growing season, but rather for periods of only days or at most weeks at a time, so a key role for the redox mechanism seems less likely than it would in some other lakes, but it cannot be ruled out.

The adsorption of phosphate by ferric complexes is influenced by pH. Phosphate may pass from a sediment surface to the overlying water if the pH is high (> 8; literature reviewed by Marsden 1989), even without conversion of ferric to ferrous iron. Thus, internal loading in Upper Klamath Lake may involve iron and phosphate under oxic conditions at the sediment surface if pH is high. This mechanism is considered by some researchers to be of special importance in Upper Klamath Lake (summary in Boyd et al. 2001).

Biogeochemical mechanisms (loss of oxygen and high pH) involving release of phosphorus from sediments typically are described in terms of abiotic reactions involving iron, but there is some evidence that bacterial metabolism also accounts for binding or release of phosphorus at the sediment-water interface (Davison 1993). Bacteria also control the oxidation conditions on the sediment surface.

Phosphorus mobilization from sediments of Upper Klamath Lake also may involve direct contact between the algae and the sediments. *Aphanizomenon* contains pseudovacuoles that function as buoyancy-control mechanisms. Under some circumstances, which may coincide with nutrient deficiency, the algae may show higher specific gravity than at other times and thus show an increased tendency to sink. Because nutrients typically are more available in deep water than in shallow water, sinking, which would be notable primarily under calm conditions, can allow algae to reach nutrient reserves that otherwise are not available. In Upper Klamath Lake, a small amount of sinking could allow a substantial fraction of the algal population to have direct contact with the sediments, where phosphorus supplies are rich. Thus, algae may be mobilizing phosphorus through direct contact with the sediments (cf. Ganf and Oliver 1982).

Nitrogen in Upper Klamath Lake

The total nitrogen load to Upper Klamath Lake has been calculated for total-maximum-daily-load (TMDL) purposes as 663,000 kg/yr (Boyd et al. 2001, Walker 2001). Thus, the mass ratio of nitrogen to phosphorus for loading under present circumstances is about 3.6:1. This ratio is extreme in the sense that mass transport of nitrogen and phosphorus from watersheds to lakes typically involves mass ratios well in excess of 5:1 (OECD 1982). Although human activities tend to cause higher relative enrichment with phosphorus than with nitrogen, even disturbed watersheds typically have much higher nitrogen transport than phosphorus transport.

The ratio of nitrogen to phosphorus typically is evaluated with respect to phytoplankton growth by reference to the Redfield ratio, which is an empirically determined value for the relative amounts of nitrogen and phosphorus that are needed by phytoplankton for growth (Harris 1986). The Redfield ratio is 16:1 on a molar basis and 7.5:1 on a mass basis. In environments that show ratios far above the Redfield ratio, strong and persistent phosphorus limitation is expected. Where the reverse is true, all taxa of algae are likely to be nitrogen-limited except those capable of nitrogen fixation. Thus, where the nitrogen:phosphorus ratio is low, as it is in Upper Klamath Lake, the nutritional conditions are ideal for dominance by nitrogen-fixing bluegreen algae, such as *Aphanizomenon flos-aquae*. The fixation of nitrogen by *Aphanizomenon flos-aquae* has the effect of raising the nitrogen:phosphorus ratio by adding atmospheric nitrogen to the lake through the fixation process. While the nitrogen:phosphorus ratio still is low, a rise in this ratio due specifically to *Aphanizomenon* has increased the ability of the lake to produce algal biomass.

Explanations of Dominance by Aphanizomenon

A recent analysis showed that akinetes, which are resting cells of *Aphanizomenon flos-aquae*, are concentrated in recently accumulated sediments but not in sediments of an earlier era corresponding to predisturbance conditions (Eilers et al. 2001). Eilers et al. concluded that the strong dominance of the algal flora in Upper Klamath Lake by heterocystous bluegreen algae is a byproduct of human presence. Historical observations of phytoplankton, as summarized by Bortleson and Fretwell (1993), are consistent with the paleolimnological conclusions. A brief overview of the chronology of observations on phytoplankton is as follows (condensed from Bortleson and Fretwell 1993): In 1906, ice from Upper Klamath Lake was deemed unsuitable for consumption because of high organic matter and green color; in 1913, summer phytoplankton samples showed diatoms dominant and bluegreen algae accounting for only 12% of cells; in 1928, water samples showed abundant algae but no dominance by bluegreens; in 1933, *Aphanizomenon* was reported for the first time but not as a dominant; in about 1939, *Aphanizomenon* was abundant but not dominant; in 1957, *Aphanizomenon* was 10 times more abundant than in 1939 but not yet overwhelmingly dominant; and in the 1960s and later, *Aphanizomenon* constituted almost a monoculture during most of the growing season.

It would be tempting to attribute the low ratio of nitrogen to phosphorus reaching Upper Klamath Lake to anthropogenic augmentation of phosphorus supply. From the TMDL mass-balance analysis, however, it is clear that Upper Klamath Lake probably had an even lower ratio of nitrogen to phosphorus in its predisturbance state (Boyd et al. 2001) because it has an unusually rich geologic source of phosphorus. Thus, nutritional conditions in Upper Klamath Lake favorable to nitrogen-fixing bluegreen algae such as *Aphanizomenon* are not new. The combination of high phosphorus concentrations under background conditions and the low ratio of nitrogen to phosphorus would have created ideal nutritional conditions for the growth of bluegreen algae before human alteration of nutrient loading, yet *Aphanizomenon* blooms appear to be a byproduct of human activity.

The conditions in Upper Klamath Lake prior to anthropogenic change could have involved some factor that prevented the population growth of bluegreen algae, even though nutrient conditions favored nitrogen-fixing algae such as *Aphanizomenon*. It has been suggested, for example, that organic acids (designated here as limnohumic acids and consisting mainly of humic and fulvic acids) present in wetland sediments are capable of chemically suppressing the growth of bluegreen algae (Eilers et al. 2001, Geiger 2001), although the phycological literature on limnohumic acids contains little indication of such effects (Jones 1998, but see also Kim and Wetzel 1993). Drainage of wetlands and hydrologic alteration in the watershed of Upper Klamath Lake probably has reduced the transfer of limnohumic acids to the lake. It is unknown, however, whether limnohumic acids or other substances derived from wetlands would have been present in sufficiently high quantities to inhibit the growth of bluegreen algae under the original conditions of the lake or why this inhibition would have been operating selectively on *Aphanizomenon*, given that other algae were abundant.

Another possibility, apparently not proposed for Upper Klamath Lake (although listed by Geiger 2001), has to do with light climate as influenced by limnohumic acids. A record from 1854 (unpublished document of the state of Oregon, as given by Martin 1997) states suggestively that the water of Upper Klamath Lake "had a dark color, and a disagreeable taste occasioned apparently by decayed tule." Limnohumic acids, which can originate in large quantities from some types of wetlands (especially those of low alkalinity), absorb light strongly at short wavelengths (Thurman 1985) and may substantially affect the light climate of phytoplankton (Jones 1998). For example, Morris et al. (1995) and Williamson et al. (1996) showed that the depth of 1% light declined from 12 m to 2 m as dissolved organic carbon (mostly limnohumic acids) increased from 2 to 10 mg/L in a series of 65 lakes of varied latitude. An increase in absorbance of such magnitude could substantially cut the amount of light reaching phytoplankton. Some diatoms are better adapted to deal photosynthetically with low light availability than most bluegreen algae (Reynolds 1984), but the high light requirement of nitrogen fixation may be even more important. Among the bluegreens, the Nostocales (including *Aphanizomenon*) have especially high light requirements (Weidner et al. 2002, Havens et al. 1998). Thus, a change in light climate rather than a change in nutrient loading or other chemical effects could have been responsible for the shift from diatoms to bluegreen algae. This is only one of several possibilities, however.

Yet another possibility has to do with biotic changes in Upper Klamath Lake. *Aphanizomenon* grows relatively slowly and so is especially vulnerable to grazing, as shown by Howarth and colleagues in marine environments (Howarth et al. 1999, Marino et al. 2002, Chan 2001; see also Ganf 1983). It is conceivable that the intensity of grazing by zooplankton on algae has been altered by the introduction of fishes that are efficient zooplanktivores. In the absence of so many efficient planktivores, zooplankton populations could have been much higher and thus capable of working selectively against *Aphanizomenon* and other nitrogen fixers. Contradicting this hypothesis is the abundance of a large and efficient zooplankton grazer, *Daphnia* (Kann 1998). In fact Kann (1998) proposes that *Daphnia* may promote *Aphanizomenon* by grazing preferentially on its competitors.

Although it seems fairly certain that *Aphanizomenon* has come into dominance in Upper Klamath Lake through human influences, the causal mechanisms of this undesirable change in phytoplankton dominance remain unclear.

Seasonal Development of Algal Biomass

Regular sampling of phytoplankton biomass at multiple stations in 1990–1998 has provided a substantial amount of information on the time course and interannual variability of biomass development of *Aphanizomenon* in Upper Klamath Lake (Kann 1998, Welch and Burke 2001). As is typical of phytoplankton populations, the phytoplankton of Upper Klamath Lake, of which over 90% is *Aphanizomenon* at peak algal abundance, shows a burst of growth in spring followed by decline. The progression of abundance is irregular, however, in that an initial period of rapid growth may be interrupted or delayed, and a period of general decline may lead to renewed growth (Figure 3-5).

The growing season for phytoplankton in Upper Klamath Lake begins generally in April. Wood et al. (1996) proposed that water temperature would show the most direct control on the rate of increase in early spring, when other conditions for growth are favorable, and thus might be a good predictor of the elapsed time between the beginning of the growing season and any particular biomass threshold that might be considered an algal bloom. This concept was investigated by Kann (1998), who showed a statistically significant association between degree days and elapsed time between the beginning of the growing season and the time coinciding with development of a specific biomass. According to Kann's analysis, days elapsed between April 1 and a biomass threshold of 10 mg/L of wet mass could be predicted with fairly high confidence ($r^2 = 0.69$) from degree days between April 1 and May 15. At the lower end of the interannual growth rate spectrum, the threshold was reached after 150 days; at the upper end, after 170 days. A relationship with lake volume in May was also tested and was suggestive but not statistically significant and it depends heavily on an outlying data point for 1992, without which there is no hint of a trend related to lake volume in May. A larger dataset might show a weak but significant relationship on the basis that a lower mean depth might lead to faster warming, but interannual variation in weather introduces considerable variation not related to lake depth.

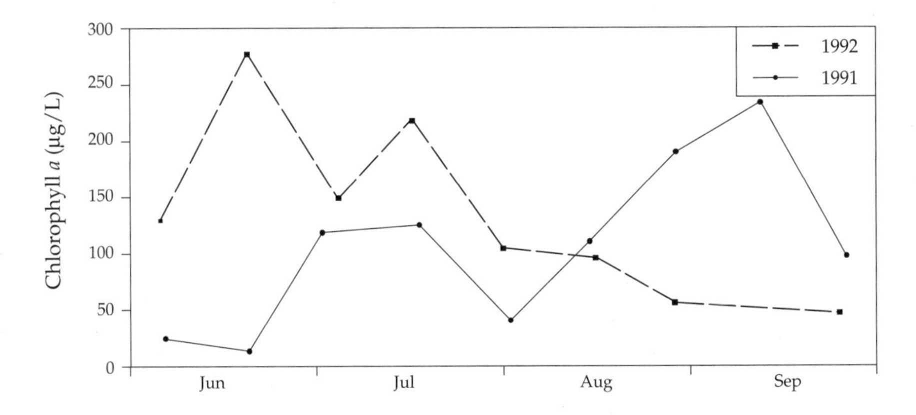

FIGURE 3-5 Change in chlorophyll *a* (lakewide averages, volume-weighted) over growing season for 2 consecutive years showing the potential interannual variability in development of chlorophyll maximums. Source: Redrawn from Welch and Burke 2001.

Kann (1998) and Welch and Burke (2001) have placed considerable emphasis on the relationship between water temperature and the first occurrence of a threshold biomass of *Aphanizomenon* equal to 10 mg/L of wet mass in spring. The relationship is well supported by data, but it has virtually no application to the occurrence or timing of extreme water-quality conditions. The threshold of 10 mg/L of wet mass corresponds to chlorophyll *a* at about 20–30 μg/L, which is only about 10–20% of the maximum abundance of *Aphanizomenon* as it reaches its annual peak. Although temperature influences growth in early spring, it later loses its influence because temperature stabilizes and the full development of the bloom to harmful proportions depends on other factors, as acknowledged by Welch and Burke (2001). Thus, the relationship between temperature and growth rate of *Aphanizomenon* in early spring seems to be a dead end with respect to anticipating the timing of the ultimate biomass maximums or their magnitude.

Of direct interest in connection with extremes of water-quality degradation during summer are the mean and maximum biomasses for suspended algae (primarily *Aphanizomenon*) that the lake shows in a given year. As shown in Figure 3-6, neither peak biomass nor mean biomass during the growing season has any empirical relationship with water level in Upper Klamath Lake.

Welch and Burke have modeled the abundance of *Aphanizomenon* on the basis of light availability with the assumption that nutrients are available in sufficient quantities to produce very high biomass (which is demonstrably correct). Light availability is affected by mean depth. As a water column gets deeper, the mean light availability for individual cells circulating in the water column declines because cells spend a higher proportion of time at greater depth, where light is less available. The modeling led Welch and Burke to conclude that maximum algal biomass of *Aphanizomenon* in Upper Klamath Lake would be quite sensitive to mean depth of the lake (Welch and Burke 2001, p. 3-15). This conclusion is inconsistent, however, with measurements of algal biomass, which show no such relationship. Thus, the model predictions are contradicted by field observations, and the latter must be given greater weight.

Modeling of the type used by Welch and Burke is useful in directing research but often produces misleading predictions because modeling usually requires various assumptions. In the case of modeling related to light, for example, the estimation of light exposure for cells must assume uniform distribution of biomass throughout the water column at all times. Because *Aphanizomenon* is capable of buoyancy regulation, it may have a nonuniform vertical distribution during calm weather. Furthermore, although Upper Klamath Lake is not stratified throughout the growing season, as deeper lakes are, it is stratified for substantial intervals during which the

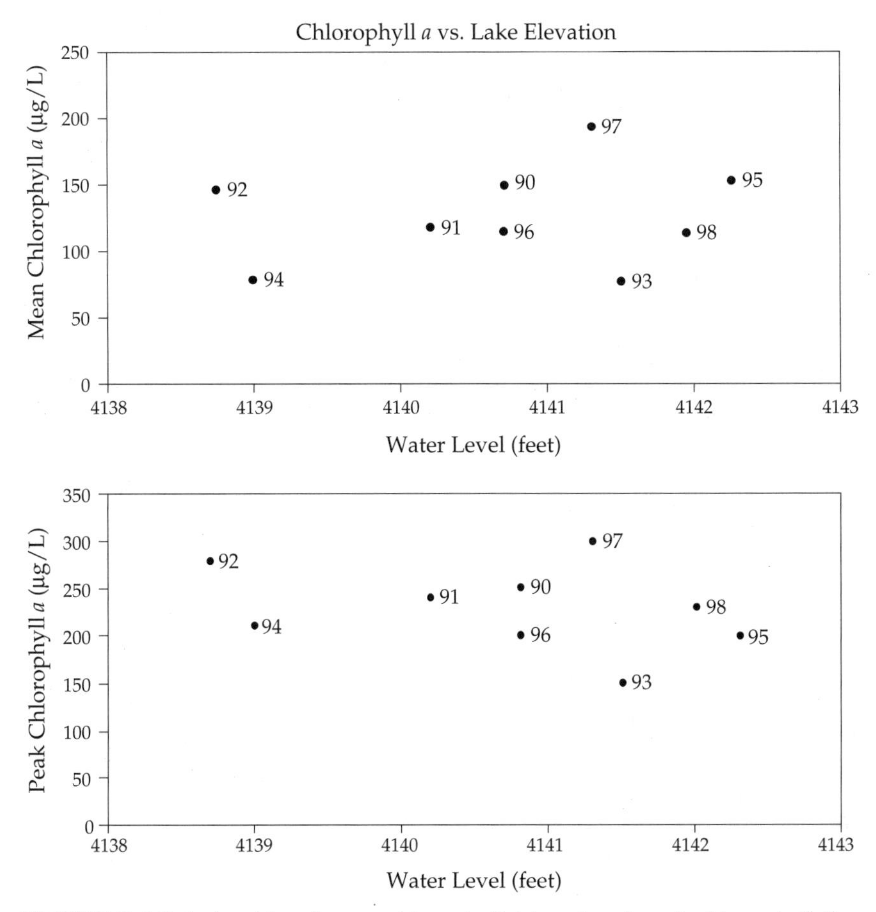

FIGURE 3-6 Relationship of mean chlorophyll (above) and peak chlorophyll (below) to water level in Upper Klamath Lake (median level for July and August). Source: Data from Welch and Burke 2001.

effective depth from the viewpoint of phytoplankton in the surface layer is less than the actual depth of the lake. Many other assumptions were necessary in modeling and could be a cause of divergence between model predictions and observations. At any rate, modeling cannot yet be used as a basis for predicting peak biomass of *Aphanizomenon* from water level in Upper Klamath Lake.

pH

Algal biomass, which typically is measured as chlorophyll concentration, is closely related to pH in Upper Klamath Lake (Kann 1998, Walker

2001). This relationship is consistent with the expected rise in pH caused by high rates of photosynthesis in aquatic environments generally (Wetzel 2001). Thus, high algal abundance sustained by light and abundant nutrients is the proximate cause of high pH during the growing season in Upper Klamath Lake.

The photosynthetically induced high pH of Upper Klamath Lake has been used in formulating a hypothesis related to the control of internal phosphorus loading in Upper Klamath Lake (Boyd et al. 2001, Walker 2001). According to this hypothesis, designated here as the pH-internal loading hypothesis, internal loading occurs primarily under oxic conditions at the sediment-water interface and involves desorption of phosphorus from ferric hydroxide complexes at the sediment-water interface through the replacement of phosphate with hydroxyl ions at high pH. Thus, high pH is proposed as a direct cause of the phosphorus enrichment of Upper Klamath Lake through internal loading during the growing season. As explained above, however, the importance of other mechanisms of internal loading cannot be ruled out, especially because internal loading substantially increases phosphorus concentrations before the lake reaches its peaks of algal abundance that are the cause of peaks in pH.

If high pH is the main cause of internal phosphorus loading, which in turn supports extremes of algal biomass in Upper Klamath Lake, internal loading might be lower if the pH of the lake were lower. Thus, external loading might be connected causally to internal loading by way of pH; this hypothesis is the basis of some recommendations in the TMDL analysis of Upper Klamath Lake (Boyd et al. 2001). The hypothesis is, however, still highly speculative.

The pH of Upper Klamath Lake also may be directly significant to fish, which can be damaged or killed by high pH. For example, Saiki et al. (1999) showed that a mean 24- to 96-h LC_{50} for the two listed sucker species in both larval and juvenile stages was 10.3–10.7. Sublethal effects would be expected below this threshold for exposures of 1 day or longer and have been demonstrated in juvenile shortnose suckers at a pH of near 9.5 (Falter and Cech 1991). Any means of suppressing extreme pH could benefit the suckers, although the degree of potential benefit is not clear. Because pH does not peak during episodes of mass mortality of suckers, however, it seems unlikely that pH contributes to mass mortality (Saiki et al. 1999). Also, because peaks of pH are transitory because of 24-h cycling of pH, impairment of fish by high pH in Upper Klamath Lake is difficult to evaluate.

As mentioned above, the immediate cause of the highest pH values in Upper Klamath Lake is photosynthesis. Furthermore, the abundance of algae, as estimated from chlorophyll *a*, is strongly correlated with pH. Thus, suppression of algal abundance would lead to a suppression of photosynthesis, which in turn would lead to a suppression of pH and, most important, elimination of the highest pH values. Kann and Smith (1999) suggested on the basis of a probabilistic analysis that a target chlorophyll *a* concentration of 100 µg/L would probably lead to a effective suppression of high pH.

The connection between pH and water level in Upper Klamath Lake has been of great interest because water level can be regulated to some degree. Welch and Burke (2001) argued on the basis of modeling that higher water levels would produce lower extremes of pH, which would potentially benefit the suckers. Their projection of pH with modeling was based on the presumption that chlorophyll *a* can be modeled in relation to water level. As mentioned above, however, observations of chlorophyll *a* in relation to water level are not as predicted by the model; there is no relationship between means or extremes of chlorophyll *a* and lake level based on monitoring during the 1990s. Thus, there is no reason to expect a relationship between pH and water level, given that pH is controlled by algal abundance. In fact, the monitoring data show no relationship between pH and water level (Figure 3-7; percentiles other than the one shown also fail to demonstrate a relationship between water level and pH). Even though they predict more favorable pH at higher lake levels, Welch and Burke (2001) acknowledge that there is no empirical relationship between pH and lake level as judged by information collected during the 1990s. The authors open the possibility of more complex relationships between lake level and pH. Any such relationship remains hypothetical, and the weight of current

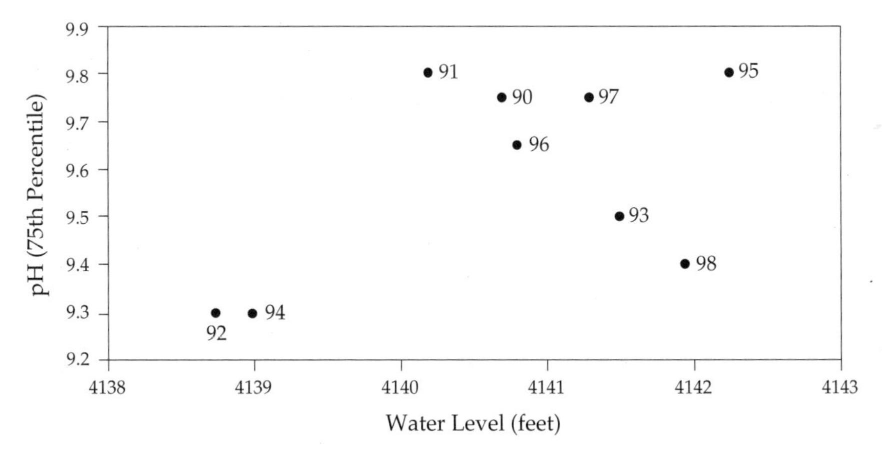

FIGURE 3-7 Relationship between water level (median, July and August) and pH in Upper Klamath Lake. The pH data are water-column maximum pH for 7 monitoring sites distributed across Upper Klamath Lake, shown as 75th percentile for all dates. Source: Data from Welch and Burke 2001.

evidence does not support the argument that higher lake levels will mitigate problems associated with high pH.

One deficiency in the information on pH is lack of consideration of diel cycling in pH (a small amount of information is given by Martin 1997). In highly productive waters such as those of Upper Klamath Lake, pH changes extensively in a 24-h cycle; maximums occur in the afternoon hours, and minimums just before sunrise. The amplitude of pH cycles commonly exceeds 1 pH unit in fertile waters. Thus, evaluation of pH would be more complete if the pH cycle were taken into account.

Overall, pH is regulated by algae, and if the abundance of algae could be reduced, the extremes of pH could be moderated. It is likely that the abundance of algae has been increased by human actions either directly or indirectly, in which case pH under current conditions would be expected to peak substantially above the pH that was present before changes in land use in the basin. Potentially undesirable effects of high pH include direct damage to fish and amplification of internal loading, which is probably the largest source of phosphorus for Upper Klamath Lake. It is not yet clear how much harm high pH is causing suckers (especially in contrast with dissolved oxygen, for example), nor is it clear that internal loading of phosphorus, which can occur by a number of mechanisms, would be strongly suppressed by reduction in pH.

Ammonia

Ammonia has been proposed as a toxicant that potentially affects the endangered suckers of Upper Klamath Lake. Although ammonia is a plant nutrient with no adverse effects on organisms at very low concentrations, it is toxic at high concentrations. Toxicity typically has been associated with the unionized component of ammonia in solution. Thresholds of protection incorporated into various state regulations for warm-water aquatic life usually are in the vicinity of unionized ammonia (expressed as N) at 0.06 mg/L. Toxicity studies on the endangered suckers showed, however, that they are more tolerant of ammonia than many other species of fish (unionized ammonia LC_{50} for 24–96 h, 0.5–1.3 mg/L; Saiki et al. 1999).

Under oxic conditions, ammonia either is removed from the water column by autotrophs (which use it nutritionally) or is oxidized by nitrifying bacteria that convert it to nitrate. Thus, in the absence of a strong point source of ammonia, it is typical to have low concentrations of ammonia in inland waters that are oxic. In the absence of oxygen, however, ammonia produced by decomposition can accumulate, given that its conversion to nitrate or uptake by autotrophs does not occur under these conditions.

Upper Klamath Lake stabilizes in summer when wind speeds are low, as explained below in connection with the discussion of oxygen. At such times, ammonia accumulates in the lower water column as oxygen is depleted. Mixture of the ammonia into the entire water column could produce toxicity. Unionized ammonia seems a less likely cause of mass mortality of fish in Upper Klamath Lake than dissolved oxygen, however, because mass mortality continues after ammonia concentrations have declined (Perkins et al. 2000b), and because the suckers show relatively high tolerance to ammonia.

Dissolved Oxygen

Low concentrations of dissolved oxygen coincide with mass mortality of large suckers in Upper Klamath Lake. The suckers are relatively resistant to oxygen depletion (LC_{50} 1.1 to 2.2 mg/L; Saiki et al. 1999), but their tolerance limits are exceeded under some conditions in Upper Klamath Lake (Perkins et al. 2000b). Unlike extreme pH or high ammonia concentrations, low dissolved oxygen persists for days while mortality occurs. Thus, low dissolved oxygen appears to be the direct cause of mortality.

Most lakes of middle latitude are dimictic; that is, they mix completely in spring and fall but stratify stably during summer and are covered with ice continuously or intermittently in winter. Lakes that are exceptionally shallow in relation to their area, however, are polymictic; that is, they mix many times during the growing season. The shallowest lakes, which can mix convectively at night even in the absence of wind, are designated continuous polymictic lakes (Lewis 1983). Lakes that are too deep to be mixed entirely by free convection every night (about 2–3 m; MacIntyre and Melack 1984) but too shallow to sustain stratification throughout the growing season are intermediate in the sense that they develop and sustain stratification for intervals of calm weather, especially if there is no net heat loss, and mix completely when wind strength increases or substantial heat is lost; they are called discontinuous polymictic lakes. Upper Klamath Lake is a discontinuous polymictic lake, as shown by its episodes of stratification interrupted by extended intervals of full mixing. The dynamics of water-column mixing and stratification in Upper Klamath Lake are not well documented, however, because water-quality surveys have been separated by too much time to allow resolution of the alternation between mixing and stratification in the lake.

A discontinuous polymictic lake shows alternation of the two very different conditions associated with mixed and stratified water columns. While the water column is unstratified, the lake shows minimal vertical differentiation in oxygen or other water-quality variables. When the lake stratifies, however, depletion of oxygen begins in the lower part of the water column, where contact with atmospheric oxygen is lacking and there is not enough light for photosynthesis. Because Upper Klamath Lake is highly productive, its waters have high respiratory oxygen demand that quickly leads to the depletion of oxygen in the lower water column whenever the lake is stratified (e.g., Welch and Burke 2001, Horne 2002).

An empirical relationship has been shown between relative thermal resistance to mixing (RTR, an indicator of stability) and wind velocity during July and August for Upper Klamath Lake (Welch and Burke 2001). Thus, the expectation that intermittent stability is under the control of weather has been verified for Upper Klamath Lake. Further work on the dynamics of mixing would probably be useful for understanding changes in water quality in the lake. Future work should be based on stability calculations rather than RTR, however. Stability can be calculated from morphometric data on the lake, water level, and the vertical profile of density (Wetzel and Likens 2000). Stability depends on water depth and distribution of density with depth, both of which are more irregular in Upper Klamath Lake than would be ideal for use of RTR, which is a shortcut method of estimating stability that overlooks any changes in depth. The advantage of using true stability rather than RTR is that it may show more clearly relationships between stability and factors of interest to the analysis of mixing. The relationships already demonstrated are important, however.

Loss of stability after a period of high stability in Upper Klamath Lake is associated with low concentrations of dissolved oxygen and high concentrations of ammonia throughout the water column and with depression of algal abundances. To some extent, those changes can be understood simply as a byproduct of mass redistribution in the water column. For example, ammonia is expected to accumulate in deep water during stratification because it is a byproduct of decomposition and accumulates where oxygen is scarce or absent; it is distributed throughout the water column by destratification. Likewise, water that is depleted of oxygen near the bottom of the lake, when mixed with the upper portions of the water column, causes a decrease in oxygen concentrations in the entire water column until photosynthesis and reaeration processes at the surface combine to raise oxygen concentrations throughout the water column.

Some of the events that follow destratification in Upper Klamath Lake cannot be explained simply in terms of the redistribution of mass from a stratified water column. Concentrations of ammonia decline rather rapidly after destratification, as expected from the processes of nitrification (oxidation of ammonia to nitrate by bacteria) and autotrophic assimilation (uptake by algae). Low concentrations of dissolved oxygen, however, persist for many days rather than being offset by reaeration and photosynthesis, as might be expected. Furthermore, algal populations show substantial and prolonged decline. The prolonged decrease in oxygen appears to be the main cause of mass mortality of the endangered suckers during transition from a stratified to a fully mixed water column accompanied by the most severe decrease in dissolved oxygen (Perkins et al. 2000b). Therefore, it is important to understand why oxygen concentrations fail to recover.

The likely proximate cause of the extended decrease in oxygen concentrations after destratification is algal death. Stratification of the water column appears to produce conditions that are harmful to the algae. The mechanism of harm is still indeterminate. It could involve, for example, death of the algae that are trapped in the lower portion of the water column when stratification occurs; these algae would lack light and might be exposed to harmful chemical conditions as the lower water column becomes anoxic. Oxygen can be depleted quickly from the lower water column of Upper Klamath Lake, partly because the oxygen demand of sediments is very high (Wood 2001). One would expect that the buoyancy control of *Aphanizomenon* would allow the algae to escape these problems, but perhaps not. Alternatively, the occurrence of calm weather, which probably accompanies the development of stratification, could lead to extensive stranding of buoyant filaments of *Aphanizomenon* at the surface. This type of stranding is known to occur in dense populations of bluegreen algae. When population densities are high, the light climate is poor, and the vacuolate bluegreens often show buoyancy regulation as a means of maintaining the higher mean position in the water column, thus avoiding shading. When the water column is becalmed, however, this type of buoyancy regulation, which requires a relatively long period of adjustment, takes the filaments to the surface where they are exposed to excessive amounts of radiation (especially ultraviolet) and death results (Reynolds 1971, Horne 2002). These are merely speculations on mechanisms, however; additional research would be required to demonstrate which ones apply.

Regardless of the mechanism of algal death, it is clear that death of a substantial population of *Aphanizomenon* in Upper Klamath Lake would reduce the potential oxygen supply (by cutting off a portion of the photosynthetic capability of the water column) and would simultaneously generate a large amount of labile organic matter (as a result of the lysis of algal cells), which would raise the oxygen demand of the water column through the respiratory activities of bacteria whose growth would be stimulated by the presence of the organic matter (Figure 3-8). The extended nature of oxygen depletion suggests that it takes many days for the excess organic matter to be consumed, for the photosynthetic capacity of the lake to be regenerated, or both. In the meantime, substantial harm can occur to endangered suckers because oxygen concentrations remain low. An important practical question is whether the episodes of low dissolved oxygen throughout the water column are related to water level. Empirical evidence indicates that no such association exists, as shown Figure 3-8 (other locations and percentiles also lack a pattern). If stabilization of the water column is ultimately a danger to the fish through the induction of high algal mortality

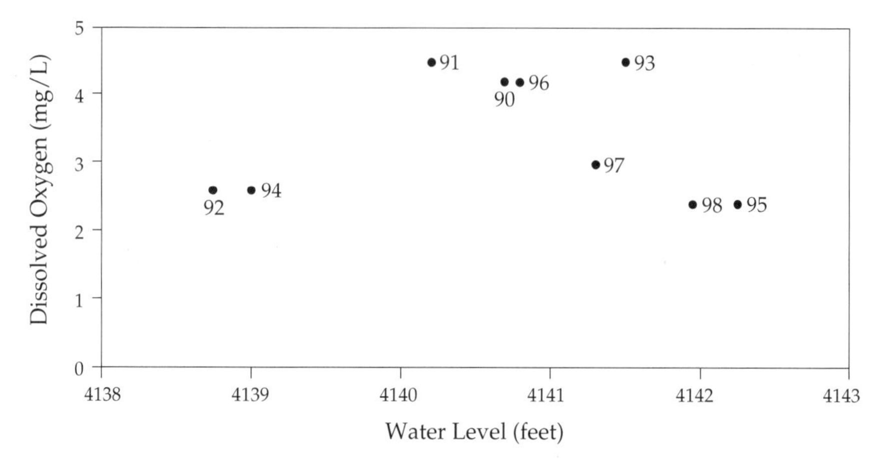

FIGURE 3-8 Relationship between water level (median, July and August) and dissolved oxygen in the water column of Upper Klamath Lake. Oxygen data are given as 75th percentile of minimums for all sampling dates in a given year at three sampling sites in the northern part of lake, which is considered to be especially important as habitat for large suckers. Source: Data from Welch and Burke 2001.

followed by loss of considerable oxygen from the entire water column, conditions leading to high stability would be least favorable to fish (Figure 3-9). Other factors being equal, deeper water columns are more stable, as acknowledged by Welch and Burke (2001); that is, one might expect higher water levels to produce greater mortality than lower water levels. However, given the complicating influence of numerous factors, including weather, associations between depth and extremes of oxygen concentrations may be too variable to detect. At any rate, there is no evidence based on oxygen that favors higher water levels over lower water levels as judged from information collected during the 1990s.

Highly productive lakes may show depletion of oxygen under ice during winter. Photosynthesis typically is weak in winter because of low irradiance and the effects of ice cover and snow on light transmission. Under winter conditions, even though respiration rates are suppressed by low temperature, dissolved oxygen can be completely depleted, and this can lead to the death of fish (winterkill). If all other factors are equal, a shallower lake is more likely to show winterkill than a deeper lake because a deeper lake has larger oxygen reserves and less respiration per unit volume than a shallower lake. Other factors are also important, however, including especially the duration of the period of ice cover and the presence of refugia, such as springs or tributaries, that move oxygen to selected locations where fish may find oxygen.

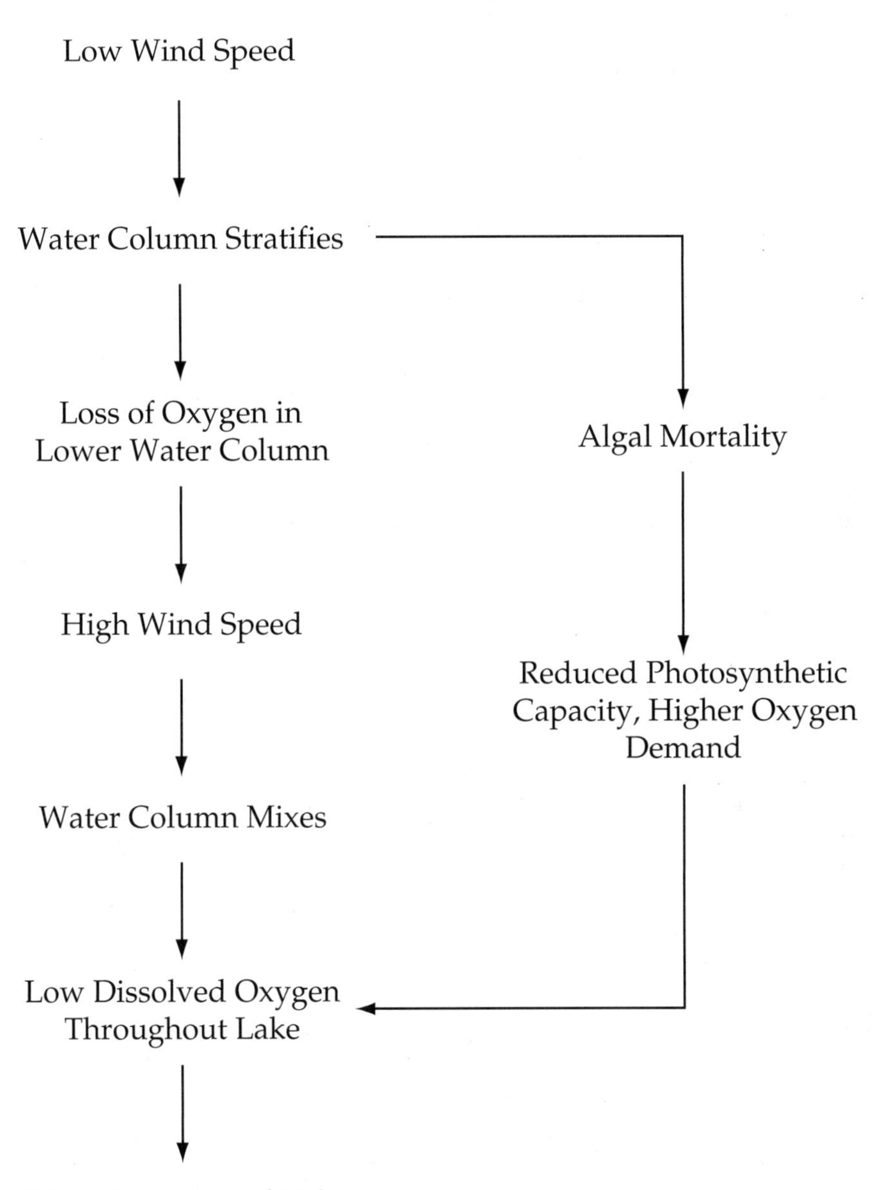

FIGURE 3-9 Probable cause of low dissolved oxygen throughout the water column of Upper Klamath Lake during the growing season leading to mass mortality of fish.

Welch and Burke (2001) and USFWS (2002) have noted risk to the endangered suckers through increased potential for winterkill when the lake is severely drawn down, as it is in dry and critical dry years. No winter mortality has been observed, however, even though the period of observation includes 2 yr that have shown more severe drawdown than any other years in the last 40 yr of record. Sparse data on oxygen under ice do not indicate depletion (USFWS 2002), but much more information is needed. Analogies that Welch and Burke (2001) have shown with studies done elsewhere may be unreliable because of differences in the duration of ice cover and other factors that make comparisons problematic. On a hypothetical basis, winter fish kill seems more likely when the lake is drawn down than when it is not, but winter fish kill may not occur at all, in which case water level is not an issue within the operating ranges of the 1990s. Measurements of oxygen concentrations under ice cover would shed additional light on this issue.

Overview of Water Quality in Upper Klamath Lake

Poor water quality causes the mass mortality of the two endangered sucker species of Upper Klamath Lake and may also cause other, more subtle kinds of harm. The diagnosis and remediation of mechanisms leading to mass mortality or stress of fish require knowledge of the causal connections between human activity and poor water quality. Researchers working on both fish and water quality in the upper Klamath basin have worked out some causal connections (Table 3-2) but in other cases have not yet succeeded in establishing cause-effect relationships. There are two critical sets of causal connections related to water quality: (1) connection of human activity with high phytoplankton biomass and dominance of *Aphanizomenon* in Upper Klamath Lake, and (2) connection of high phytoplankton abundance with chemical conditions that could harm fish.

High phytoplankton biomass has, according to the hypothesis (external phosphorus-loading hypothesis) underlying the TMDL analysis of Boyd et al. (2001), occurred through augmentation of phosphorus loading of Upper Klamath Lake, mostly by nonpoint sources or through weakening of natural interception processes that occur in wetlands or riparian zones. There are, however, two major problems with this hypothesis (Figure 3-10). First, the anthropogenic augmentation of external loading is sufficient to account for only about 40% of the total load; the main factor accounting for very high phosphorus concentrations at present is internal loading rather than external loading. The pH-internal loading hypothesis proposes, however, a mechanism by which a 40% increase in external load could have produced a much larger increase in internal load. According to this line of thinking, the increase in external load raised the maximum algal abundances enough

TABLE 3-2 Status of Various Hypotheses Related to Water Quality of Upper Klamath Lake

Hypothesis	Status
Algal abundance as measured by chlorophyll is positively related to total phosphorus in the water column	Well supported
Algal biomass as measured by chlorophyll is positively related to daytime pH	Well supported
Rate of early-spring development of biomass is positively related to rate of warming in the water column	Well supported
Rate of early-spring phytoplankton growth is inversely related to lake volume	Relationship weak or absent
Mean growing-season average algal biomass is inversely related to lake depth	Inconsistent with field data
Peak algal abundance is inversely related to lake depth	Inconsistent with field data
A large amount of phosphorus in the water column during the growing season originates in sediments (internal loading)	Well supported
pH is the main control on internal loading of phosphorus	Not resolved yet
Interception of anthropogenic phosphorus from the watershed will reduce algal abundance in the lake	Uncertain; unlikely
Lake water level is inversely related to pH	Inconsistent with field data

to increase the maximum pH during the growing season, which in turn greatly augmented internal loading by facilitating the desorption of phosphate from iron hydroxide floc on the sediment surface. It is also possible, however, that internal loading, which can occur by several mechanisms, always has been large enough to saturate algal demand, as suggested by the steady nature of internal loading beginning early in the growing season, before pH reaches its peak. A second weakness in the external phosphorus-loading hypothesis is that it fails to explain why *Aphanizomenon* has become dominant. Nutritional conditions seem to have been favorable for *Aphanizomenon* (or other nitrogen fixers) before land-use changes in the watershed because of an inherently low nitrogen:phosphorus ratio in the lake.

Because of the two major unresolved issues for the external phosphorus-loading hypothesis, alternate hypotheses are still worthy of consideration. One, shown in Figure 3-10, is based not on phosphorus enrichment, but rather on changes in the limnohumic acid content of the lake, which is likely to have been quite high in waters emanating from the extensive wetlands around Upper Klamath Lake. The hypothesis proposes that the basic cause of change in water quality of the lake is reduction in the supply of limnohumic acids to the lake, with a consequent increase in transparency or possibly even a decrease in inhibitory effects (toxicity of the acids to

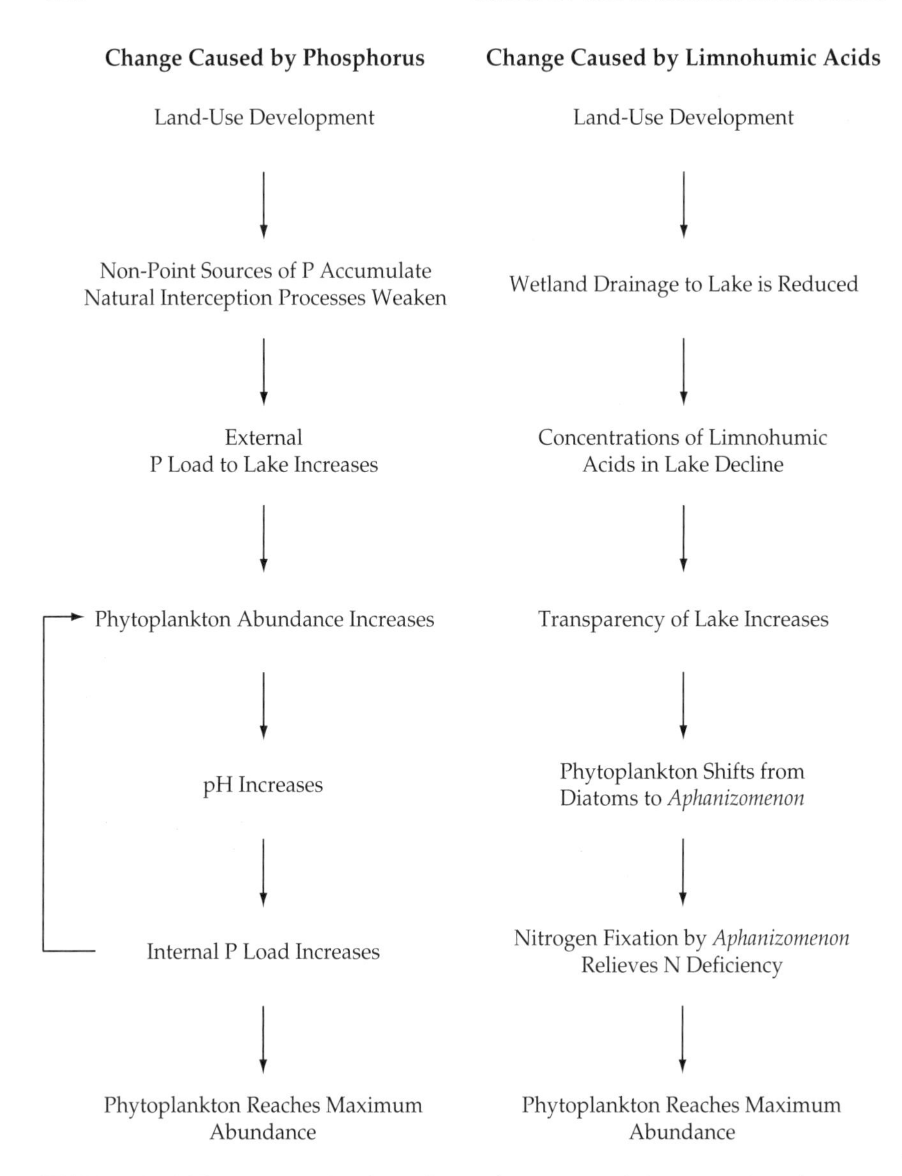

FIGURE 3-10 Two contrasting hypotheses that may explain connections between human activity and high abundances of phytoplankton in Upper Klamath Lake.

algae). Released from suppression by weak light availability or chemical inhibition, *Aphanizomenon* became more abundant in the lake. Unlike the diatoms that preceded it, *Aphanizomenon* was able to offset the low nitrogen:phosphorus ratio of the lake by nitrogen fixation, thus allowing algal growth for the first time to take full advantage of the abundant phosphorus supplies and produce the very high algal abundances that are now characteristic of the lake. The advantage of this hypothesis is that it accounts simultaneously for the change in community composition of phytoplankton and for an increase in biomass. The key factor causing major changes in the lake was, according to this hypothesis, drainage or hydrologic alteration of wetlands, rather than increase in external phosphorus loading.

Figure 3-11 shows causes leading from high algal abundance to water-quality conditions potentially harmful to fish. High abundance of phytoplankton produces high pH, which can be directly harmful to fish. Although the connection of phytoplankton abundance to high pH is well verified, the amount of harm to fish that it causes is still a matter of speculation. A second factor is episodic stratification of the water column, which leads to oxygen deficits in the bottom portion of the water column and appears to cause algal mortality. Mixing caused by windy weather brings oxygen-poor water to the surface, along with ammonia. The importance of ammonia in mass mortality is probably not great, but it could be harmful in more subtle ways to fish. Low oxygen that results from mixing probably is prolonged by algal death and probably is the main reason for mass mortality of fish.

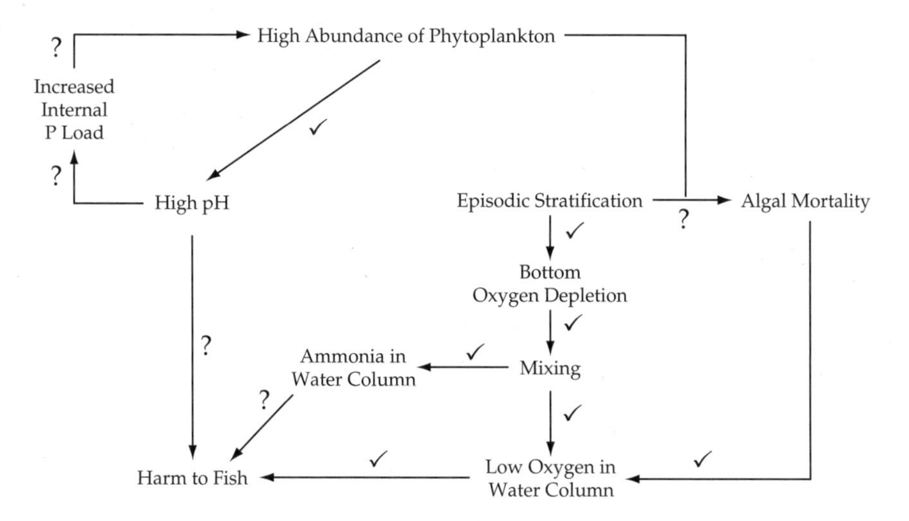

FIGURE 3-11 Potential (?) and demonstrated (√) causal connections between high abundance of phytoplankton and harm to fish through poor water-quality conditions.

Potential for Improvement of Water Quality in Upper Klamath Lake

Two proposals have been made for actions that would improve the water quality of Upper Klamath Lake. Both presume, with substantial scientific support, that an improvement of water quality in Upper Klamath Lake will require suppression of algal abundance. The first proposal, which could be implemented immediately, is for maintenance of water levels in Upper Klamath Lake exceeding levels that have been characteristic of the recent historical past. The second proposal, which deals more with long-term improvement, is for reduction in the anthropogenic component of external phosphorus loading of Upper Klamath Lake.

Higher water levels have been proposed in recent biological opinions for operation of Upper Klamath Lake (USFWS 2001, 2002). USFWS makes a number of kinds of arguments for higher water levels, while noting that empirical evidence of a connection between lake level and water quality is "weak" (USFWS 2001, p. 51). One of the arguments is that higher water levels will improve water quality in Upper Klamath Lake. As shown by the preceding review of available evidence, there is no scientific support for the proposition that higher water levels correspond to better water quality in Upper Klamath Lake. For example, mean and maximum abundances of algae, which are the driving force behind poor water quality, show no indication of a relationship with water level. USFWS acknowledges that no relationship has yet been demonstrated, but it argues that a complex, multivariate relationship may exist but not yet be evident. For example, as noted by USFWS and others (Welch and Burke 2001), an effect of water level on water quality could be contingent on water-column stability, which in turn is under the influence of weather. Other multivariate relationships could be proposed that involve water level as one of several factors explaining the water-quality conditions in a given year. This line of argument leads to the conclusion that water-quality conditions may be explained in the future (after further study) by a suite of variables that include water level, but it also suggests that the influence of water level is too weak to be discerned without consideration of other variables.

The potential usefulness to management of a complex mechanism involving water level as a covariate would be low even if it could be demonstrated. Furthermore, the mode of influence of water level as one of a suite of variables affecting water quality would not necessarily work in the direction of water-quality improvement at higher water levels. For example, as indicated in the foregoing text, higher water level promotes water-column stability, which appears to be the principal means by which water-quality conditions for mass mortality develop in Upper Klamath Lake. All things considered, water level cannot now be managed with confidence for control of water quality.

The second proposal, which is for long-term improvement of water quality through reduction in external phosphorus loading, has been favored by many and is the main recommendation related to water quality of Upper Klamath Lake in a recent TMDL analysis (Boyd et al. 2001). The proposal has three weaknesses: one related to the feasibility of intercepting substantial load, a second related to the internal-load effects of reducing external load, and a third involving the role of increased phosphorus loading in sustaining large algal populations under current conditions.

The TMDL proposal is for reduction of external phosphorus by about 40%. Because the current anthropogenic load is about 40% of the total, the proposal is to return the external phosphorus loading of Upper Klamath Lake to background conditions. Only about 1% of the anthropogenic loading is from point sources (wastewater treatment plants: Boyd et al. 2001). Interception of point-source loads is technically feasible, but interception of nonpoint-source loads, although approachable through best-management practices, is more problematic in that it would require major changes in agricultural practice and other types of land use. Even a reduction of 20% would be ambitious and potentially infeasible in view of the association between non-point sources and privately held lands.

Even a reduction of 40% in total external phosphorus loading would probably be ineffectual without suppression of internal phosphorus loading, given that internal phosphorus loading is very large for Upper Klamath Lake. The authors of the TMDL study have anticipated this problem. Invoking the pH-internal loading hypothesis as described above, they anticipate that a reduction in external loading will result in lower extremes of pH, which in turn will reduce internal loading, thus providing magnified benefits. This is a highly speculative proposition, however. Because soluble phosphorus is available in quantity even at the end of the growing season, it appears that internal loading is sufficient to supersaturate the needs of algae for phosphorus. Furthermore, a pH reduction, if it did occur, might or might not be sufficient to shut off internal loading related to high pH. Finally, high pH is only one mechanism by which phosphorus is mobilized from sediments; other mechanisms would remain as they are and could easily be sufficient to provide the internal loads necessary to generate the high phytoplankton biomass observed in the lake. Thus, reduction of external load as proposed in the TMDL document has results that are quite uncertain.

A third problem with the phosphorus-reduction strategy is that the high abundances of phytoplankton in Upper Klamath Lake may have not become established because of external phosphorus loading, but rather because of other changes in the lake. A drastic decrease in mobilization of limnohumic acid alteration of wetlands and hydrology, for example, fits historical observations more satisfactorily than a phosphorus-based hypothesis, given that Upper Klamath Lake apparently has always had the very low nitrogen:phosphorus ratios that set the stage for dominance by a nitrogen fixer, such as *Aphanizomenon*. The data suggest that other factors were holding back the nitrogen fixers and that human activity reversed or modified one of them, producing the current dominance of *Aphanizomenon*. *Aphanizomenon*, once established, could generate higher abundances than nonfixing algae because of its ability to offset nitrogen deficiency in the water. Thus, the key to improving water quality may be to suppress *Aphanizomenon*.

Restoration of limnohumic acids to the lake would be the most obvious way of restoring any beneficial effects that limnohumic acids might have had before land-use development of the upper Klamath basin watershed. Restoration of wetlands is under way and could increase transport of limnohumic acids to the lake. Although justified in large part by an attempt to intercept nutrients, these programs could have beneficial effects on limnohumic acid supply. One discouraging aspect of restoring limnohumic acid transport to the lake, however, is that many of the wetland sediments surrounding the lake that would have been perhaps the richest source of limnohumic acids have disappeared through oxidation after dewatering. Furthermore, restoration of limnohumic acid supply would require not just restoration of wetlands but also extensive rerouting of water through wetlands, with attendant loss of water through evapotranspiration. Nevertheless, this option is virtually unstudied and deserves more attention. It could be compatible with nutrient-removal strategies justified by improvements in water quality of streams.

Current proposals for improvement of water quality in Upper Klamath Lake, even if implemented fully, cannot be counted on to achieve the desired improvements in water quality. Thus, it would be unjustified to rely heavily on future improvements in the water quality of Upper Klamath Lake as a means of increasing the viability of the sucker populations.

Oxygenation as a Management Tool

The possibility that oxygenation of deep water could be used as a means of reducing mass mortality of endangered suckers in Upper Klamath Lake has been mentioned by USFWS in its biological opinion (2002; see also Martin 1997). An engineering study of the possibility is already available (Horne 2002). Because of the size of Upper Klamath Lake and the speed with which it can become anoxic toward the bottom of the water column during episodes of stratification, it is unlikely that oxygenation could be used in preventing low concentrations of dissolved oxygen from developing in the lower water column during stagnation or in restoring oxygen when the water column mixes at depressed oxygen concentrations.

Even so, it is conceivable that oxygenation could be used in such a way as to provide specific refuge zones to which the endangered suckers would be attracted when they experience stress due to low dissolved oxygen. Of particular interest would be the adult suckers, which cluster in specific locations (USFWS 2002).

It is doubtful that the potential for aeration to reduce mass mortality of large suckers can be developed entirely from calculations and estimations. Pilot testing for proof of concept seems well justified for the near future. Potential success of this approach is uncertain, however, in that use of oxygenation specifically to create refugia in large lakes apparently does not appear in the literature on oxygenation.

CLEAR LAKE

Clear Lake was created in 1910 at the location of a smaller natural lake and associated marsh on the Lost River (Figure 3-12). One purpose for the creation of the reservoir was to allow storage of runoff for irrigation of lands below the dam. An additional purpose was to promote evaporative loss of water that otherwise would flow to Tule Lake and Lower Klamath Lake, which were intended for dewatering to allow agricultural development. In addition to high evaporative losses associated with its low mean depth, Clear Lake has extensive seepage losses.

Clear Lake is divided into east and west lobes that are separated by a ridge; the dam is on the east lobe. Willow Creek, a tributary of Clear Lake, is critical to the sucker populations, which appear to rely primarily or even exclusively on this tributary for spawning. The lands surrounding Clear Lake are not under intensive agricultural use. The area surrounding the reservoir consists primarily of Clear Lake National Wildlife Refuge, and the watershed above the lake is largely encompassed by the Modoc National Forest.

Although Clear Lake would store as much as 527,000 acre-ft at its maximum height, which corresponds to a lake area of 25,000 acres (USBR 2000a), its average area has been close to 21,000 acres, which corresponds to a storage of about 167,000 acre-ft and a mean depth of 8 ft (USBR 2002a, USBR 1994); it has never reached maximum storage. Average annual inflow is 117,000 acre-ft, which suggests a mean hydraulic residence time of 1–2 yr (computed from input and volume). Clear Lake is similar to Upper Klamath Lake in being shallow in relation to its area. It differs from Upper Klamath Lake in its considerably longer hydraulic residence time and its very low output of water relative to input. One other important feature, which has to do with water management, is the high interannual and interseasonal variation in storage volume of Clear Lake, which corresponds to great variations in area and mean depth (USBR 1994).

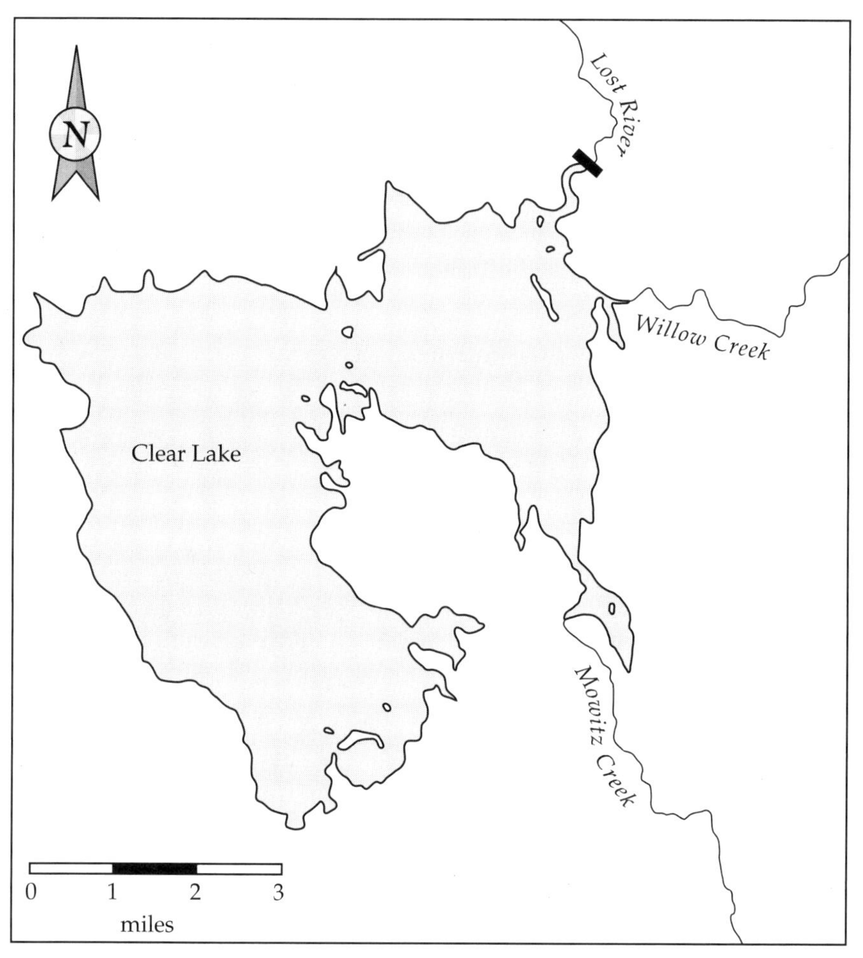

FIGURE 3-12 Map of Clear Lake.

Clear Lake contains both shortnose and Lost River suckers (USFWS 2002). Both species show evidence of stability and ecological success in Clear Lake, as indicated by diverse age structure and high abundance (USFWS 2002, USBR 1994; Chapter 5). Interannual variations in the welfare of the populations have been scrutinized, however, because of questions related to the maximum permissible drawdown of the reservoir in a dry year or in a succession of dry years. Monitoring of water quality and condition of fish in 1991–1995 provided a good opportunity to evaluate extreme drawdown because the water level in 1992 declined to its lowest point since the drought of the 1930s.

Although water-quality records collected in 1991–1995 (USBR 1994, Hicks 2002) are useful, the breadth of information that is available for Clear Lake is much narrower than that of Upper Klamath Lake. Apparently, there has been no sampling for phytoplankton or for nutrients that would allow comparisons with Upper Klamath Lake. Observations suggest that Clear Lake has far lower population densities of phytoplankton than Upper Klamath Lake; there is no evidence of massive blooms of bluegreen algae, for example. Aquatic macrophytic vegetation like that found in Upper Klamath Lake is virtually absent from Clear Lake because of its wide range of water levels.

The water column of Clear Lake typically has a turbid appearance suggestive of fine inorganic particulate material that is continually suspended by wind-generated currents (USBR 1994). The transparency of the lake has been measured only sporadically. During 1992, when water levels were exceptionally low, Secchi depths ranged from 0.1 to 0.4 m, which indicated extremely low penetration of irradiance in the lake (M. Buettner, USBR, personal communication, 23 January 2003). In more typical years, transparency is low but not nearly at the extreme of 1992 (for example, June 1989, 0.4–1.5 m across 24 stations; M. Buettner, USBR, personal communication, 23 January 2003). Although Clear Lake is generally characterized as allowing less light penetration than Upper Klamath Lake, the scanty data on light penetration that are available suggest that the transparencies may fall within the same range for the two lakes (for example, see Kann 1998 for data on Upper Klamath Lake). Because transparency may be related to the welfare of sucker larvae through predation, which may be more pronounced in transparent waters, further study of this subject seems warranted.

In 1991–1995, recording sensors were used for measuring temperature, specific conductance, pH, and dissolved oxygen; vertical profiles also were taken for these variables. Although interpretation of the records is complicated by occasional malfunction of the sensors, which is characteristic of this type of data collection, the overall results are useful. The temperature record indicates that the lake is unstratified; if it does stratify, it does so only sporadically over the deepest water (near the dam). The pH varies seasonally but does not reach the extremes observed in Upper Klamath Lake, presumably because high rates of algal photosynthesis, the driving force behind extreme pH in Upper Klamath Lake, are not characteristic of Clear Lake (USBR 1994). The oxygen data indicate that the lake does not show episodes of strong oxygen depletion like those in Upper Klamath Lake. One incident of oxygen concentration as low as 1 mg/L near the dam apparently was associated with drainage of the east lobe of the reservoir during 1992 as the lake was drawn down to the extremes of that year. Monitoring under ice showed concentrations of oxygen near saturation, even during an interval of especially long ice cover during 1992, a year of very low water level (USBR 1994).

Mass mortality of suckers in Clear Lake is unknown. Loss of fish occurs through the dam but does not appear to be seriously decreasing the populations. The populations were studied for signs of stress during the dry year of 1992. Although mortality was not observed, there were several indicators of stress, including higher rates of parasitism and poor body condition. These indicators disappeared quickly as water levels climbed in 1993 at the end of the drought (USBR 1994). The indications of stress associated with water levels of 1992 have served as a basis of proposed thresholds of drawdown in Clear Lake (USFWS 2002).

The potential of Clear Lake to provide information about Upper Klamath Lake has not been well exploited. The agencies have invoked Clear Lake for comparative purposes in several instances, but the background information on the reservoir is not sufficiently broad and does not extend over sufficient intervals of time to allow good comparisons. Comparative population and environmental studies in the two lakes could open up new possibilities for diagnosing mechanisms that are adversely affecting endangered suckers in Upper Klamath Lake.

GERBER RESERVOIR

Gerber Reservoir was established on Miller Creek, a tributary of the Lost River, in 1925 (Figure 1-1). The lake can store as much as 94,000 acre-ft of water but often is substantially drawn down and shows considerable interannual and intraannual variability in volume, mean depth, and area (USBR 1994, 2002b). Nevertheless, characteristic depths of Gerber Reservoir probably are substantially greater than those of Upper Klamath Lake or Clear Lake. Statistics are not readily available, but the sampling record (USBR 2002b) suggests that in most years a substantial area of the lake would have water deeper than 15 ft. Extreme drawdown occurred in 1992, when the lake was reduced to less than 1% of its maximum volume (USBR 2002b). Even under those conditions, the water near the Gerber Reservoir dam was 15 ft deep.

As might be expected, given that it is smaller and deeper than Clear Lake or Upper Klamath Lake, Gerber Reservoir shows a tendency toward stability of thermal stratification, as indicated by loss of oxygen near the bottom during summer. Stability may be interrupted by mixing, and entrainment of water through the outlet may lead to a replacement of bottom waters, which could produce changes (oxygenation, warming) similar to those expected as a consequence of mixing.

Little information is available on the water quality of Gerber Reservoir. The lake appears to have less inorganic turbidity than Clear Lake, presumably because it is deeper and smaller. *Aphanizomenon flos-aquae* probably is present and apparently creates blooms but not to the same degree of Upper Klamath Lake (USFWS 2002). *Aphanizomenon* probably fares better in this reservoir than in Clear Lake because the latter has more suspended inorganic turbidity, which shades the water column.

Information on temperature, specific conductance, pH, and dissolved oxygen was collected for the first half of the 1990s by automated monitoring and occasional vertical profiles (USBR 2002b), as was the case for Clear Lake. The pH reaches higher extremes than in Clear Lake but is less extreme than in Upper Klamath Lake. This probably reflects a gradient of algal photosynthesis across the three lakes. Dissolved oxygen in Gerber Reservoir is substantially depleted in deep water both in summer and in winter, but without any obvious effect on fish. No episodes of mass mortality of the shortnose sucker, which occupies Gerber Reservoir, have been reported. During 1992, when drawdown of the lake was severe, the lake was aerated (USFWS 2002); sampling indicated that the fish had reached suboptimal body condition during the drought. Under other circumstances, the population appears to have been stable in that it has shown no indication of stress, has preserved a diversified age structure, and has been abundant. For reasons primarily having to do with water quality, the low water levels of 1992 serve as a guideline for setting thresholds to protect the fish from stress.

LOWER KLAMATH LAKE

Lower Klamath Lake has been reduced to a marshy remnant by dewatering. It has occasional connection to the Klamath River through which it appears to receive some recruitment of young suckers, but there is no adult population. Water quality apparently has not been studied in any systematic way. Development of an adult population is unlikely unless the depth of water can be increased, which would involve incursion of the boundaries of the lake onto lands that are used for agriculture. If the lake were deepened, water quality might be adequate for support of suckers.

TULE LAKE

Tule Lake historically was very large and capable of supporting, in conjunction with the Lost River, large populations of the shortnose and Lost River suckers (Chapter 5). It has been reduced to remnants as a means of allowing agricultural use of the surrounding lands. Water reaches Tule Lake from Upper Klamath Lake or from the Lost River drainage via irrigated lands or from Clear Lake or Gerber Reservoir. Water is removed from Tule Lake (now appropriately called Tule Lake Sumps) by Pump Station D (USBR 2000a).

There are two operational sumps at Tule Lake now: 1A and 1B. In the recent past, Sump 1B has been much less likely to hold adult suckers than Sump 1A; it is shallower and has shown a higher rate of sedimentation than Sump 1A. It also appears to have worse water quality than Sump 1A. Sump 1B is being manipulated by USFWS for increase of marshland in the Tule Lake basin.

Some water-quality information is available on Tule Lake through monitoring during the 1990s (USBR 2001a) and fish have been sampled (Chapters 5 and 6). It appears that the sucker population consists of a few hundred individuals, including shortnose and Lost River suckers, and that these favor specific portions of Sump 1A (the "doughnut hole" or a location in the northwest corner) that presumably provide more favorable conditions than the surrounding area. Monitoring of Sump 1A has not shown any incidence of strongly decreased oxygen concentrations or extremely high pH, as would be the case in Upper Klamath Lake (USBR 2001a). These adverse conditions may occur in Sump 1B, however. The fish of Tule Lake, although not very abundant, appear to be in excellent body condition, and this suggests they are not experiencing stress.

Suckers migrate from Tule Lake Sumps; migration terminates on the Lost River at the Anderson Rose Diversion Dam (USFWS 2002, Appendix C), in the vicinity of which spawning is known to occur. Water-quality conditions there for spawning appear to be acceptable (USBR 2001a). Larvae are produced but apparently are not passing into the subadult and adult stages.

From the water-quality perspective, it appears that the Tule Lake population is potentially closer to survival conditions than the Upper Klamath Lake population. An unresolved mystery, however, is the fate of larvae. It is not clear whether water quality prevents the larvae from maturing, or if other factors are responsible for their loss.

Sedimentation threatens the apparently good conditions for adults in Sump 1A. Without aggressive management, the favorable portion of Sump 1A may become progressively less favorable in the future.

RESERVOIRS OF THE MAIN STEM

There are five main-stem reservoirs (Table 3-1); because Copco 2 is extremely small, it generally does not receive independent consideration. The composite residence time of water in the main-stem reservoirs, which extend about 64 mi from Link River Dam to Iron Gate Dam, averages about 1 mo. At moderately low flow (for example, 1,000 cfs), hydraulic residence time is close to 2 mo; and at moderately high flow (such as 6,000 cfs), it is close to 10 days. Thus, some of the processes that would make these lakes distinctive from each other and from their source waters are not expressed because of the relatively rapid movement of water through the system.

The main source of water for the main-stem reservoirs is Upper Klamath Lake, but it is not the only source. Agricultural returns and drainage water enter the system upstream of the Keno Dam (Figure 1-2) by way of the Klamath Strait Drain (about 400 cfs, summer) and the Lost River Channel (about 200–1,500 cfs, fall and winter). In addition, cold springs provide about 225 cfs all year at a point just below the J.C. Boyle Dam; and two tributaries, Spencer Creek and Shovel Creek, provide 30–300 cfs to J.C. Boyle Reservoir and Copco Reservoir. Fall Creek and Jenny Creek provide 60–600 cfs to Iron Gate Reservoir. During the wet months, sources other than the Link River, which brings water from Upper Klamath Lake, provide about one-third of the total flow reaching Iron Gate Dam; in midsummer, these sources may account for up to 50% of the total water reaching Iron Gate Dam (PacifiCorp 2000, Figure 2-7). Thus, source waters of diverse quality influence the quality of water in the reservoirs. The waters of Upper Klamath Lake often bring large amounts of algal biomass to the upper end of the system, along with large amounts of soluble and total phosphorus. When Upper Klamath Lake is experiencing senescence of its algal population, the entering waters also may have low concentrations of dissolved oxygen and an abundance of decomposing organic matter. Irrigation tailwater and other drainage would carry abundant nutrients and could carry organic matter but would probably lack substantial amounts of algae. Spring waters and tributary waters would be the coolest and cleanest of the water sources.

The reservoirs differ physically in several ways that are likely to influence water quality. Keno Reservoir and J.C. Boyle Reservoir are shallow and have the lowest hydraulic residence times. Physically, they resemble rivers more than lakes. In each, the water is pooled at the lower end and may run swiftly at the upper end, thus potentially benefiting from reaeration (gas exchange). The two lower reservoirs are much deeper and have hydraulic residence times that are short on an absolute scale but much longer than those of the two upper reservoirs.

None of the reservoirs has very deep withdrawal. Thus, for the two reservoirs that support stable stratification (Copco and Iron Gate), withdrawals reflect the characteristics mostly of epilimnetic (surface) water, although their withdrawal cone may extend a short distance into the hypolimnetic (deep) zone at times (Deas 2000). For example, the temperature of water leaving Iron Gate Dam during midsummer, when the hypolimnion has a temperature of about 6°C, reaches 22–23°C (PacifiCorp 2000, Figure 4-5; Deas 2000, Figure 6.5) because the powerhouse withdrawal is at about

12 m depth when the lake is full. For Copco, withdrawal is at about 6 m when the lake is full. Thus, cold hypolimnetic water of the two deepest reservoirs tends to be much more static hydraulically than the upper water column during the stratification season, as would be the case in a natural lake of similar depth; the main withdrawal occurs by way of the epilimnion. A small withdrawal (about 50 cfs) for the Iron Gate Hatchery does occur from the hypolimnion at Iron Gate Reservoir, however.

The quality of water in the reservoirs and leaving the reservoir system has been studied many times by numerous parties dating back to the 1970s. PacifiCorp has sponsored a number of studies in conjunction with its Federal Energy Regulatory Commission (FERC) licensing and other regulatory requirements, and USBR has sponsored studies of water quality because of its oversight responsibilities for the Klamath Project. The city of Klamath Falls has also studied water quality, particularly in the upper end of the system, and the Oregon Department of Environmental Quality has studied and analyzed water quality from the viewpoint of fisheries. Other information is available from the U.S. Geological Survey, the U.S. Army Corps of Engineers, and the North Coast Regional Water Quality Control Board. In its consultation document on FERC relicensing, PacifiCorp (2000) provides an overview of the monitoring programs.

Monitoring to date provides useful information but shows several deficiencies. Most of the monitoring has been limited to water-quality variables that can be measured with meters (temperature, pH, specific conductance, and dissolved oxygen). There is much less information on nutrients, total phytoplankton abundance, phytoplankton composition, total organic matter, and other important variables. Thus, interpretations are necessarily limited in scope. Also, there have been few efforts to synthesize and interpret the data, most of which exist merely as archives. Hanna and Campbell (2000) have modeled temperature and dissolved oxygen in the reservoirs. The temperature modeling is useful for planning, but the oxygen modeling fails to incorporate primary production, which could be important. Deas (2000) has done extensive modeling for Iron Gate Reservoir that is especially useful for temperature and dissolved oxygen. A full, system-level understanding of the reservoirs is not yet available, however.

During the cool months (October or November through May), all the lakes are isothermal and appear to mix with sufficient vigor to remain almost uniform chemically (see, for example, Figure 3-13). During the warm season, there may be substantial differences in temperature and water quality with depth. Keno Reservoir and J.C. Boyle Reservoir are not deep enough to sustain thermal stratification during summer. They may stabilize briefly, however, in which case oxygen may be depleted from deep water (see Figure 3-14), but such depletions probably are interrupted by episodes of mixing. Copco and Iron Gate reservoirs, in contrast, stratify

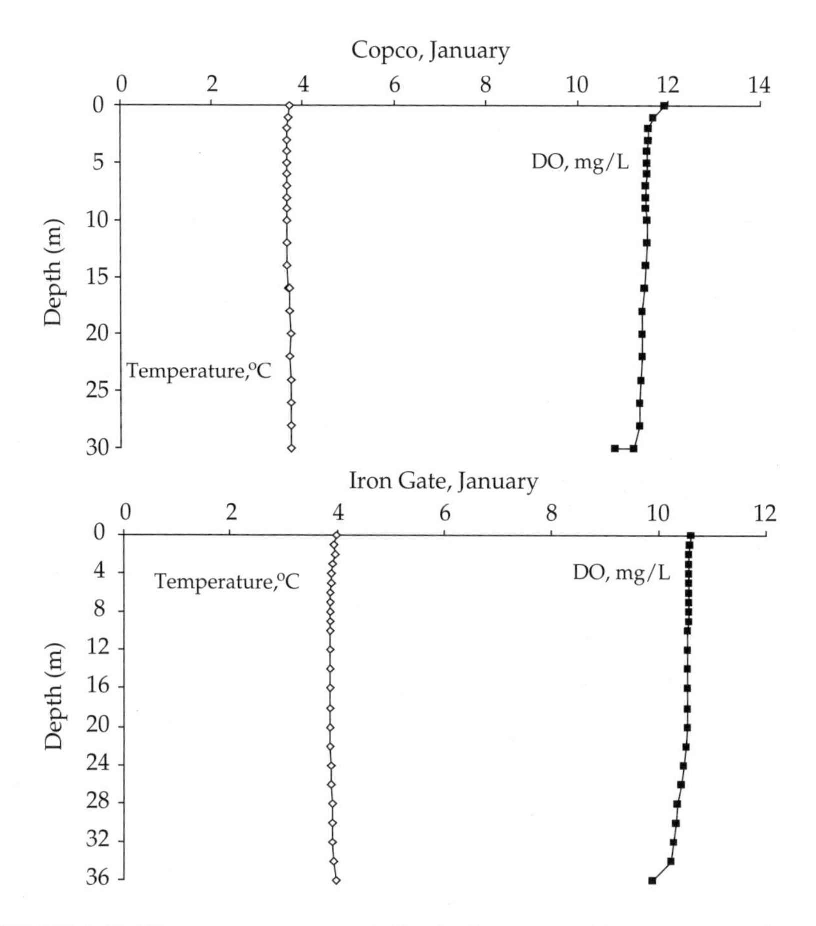

FIGURE 3-13 Water temperature and dissolved oxygen (DO) in Copco and Iron Gate Reservoirs, January 2000. Source: Data from USBR 2003.

stably on a seasonal basis. Thus, the water near the bottom of these two reservoirs can be classified as hypolimnetic and has a much lower temperature than that of the upper water column. As expected, oxygen is depleted in the hypolimnion of both lakes. Although the rate of oxygen depletion varies across years (Deas 2000), both reservoirs apparently have an anoxic hypolimnion for as much as 4 or 5 mo beginning in the last half of summer.

Periodic episodes of severe oxygen depletion may occur in the upper two reservoirs. One such event appears to have occurred in 2001, when the entire water column of Keno Reservoir became hypoxic or anoxic (Figure 3-15). It is not known how often such an event occurs. Because no mass mortality of fish in the reservoirs have been recorded, it is possible that the

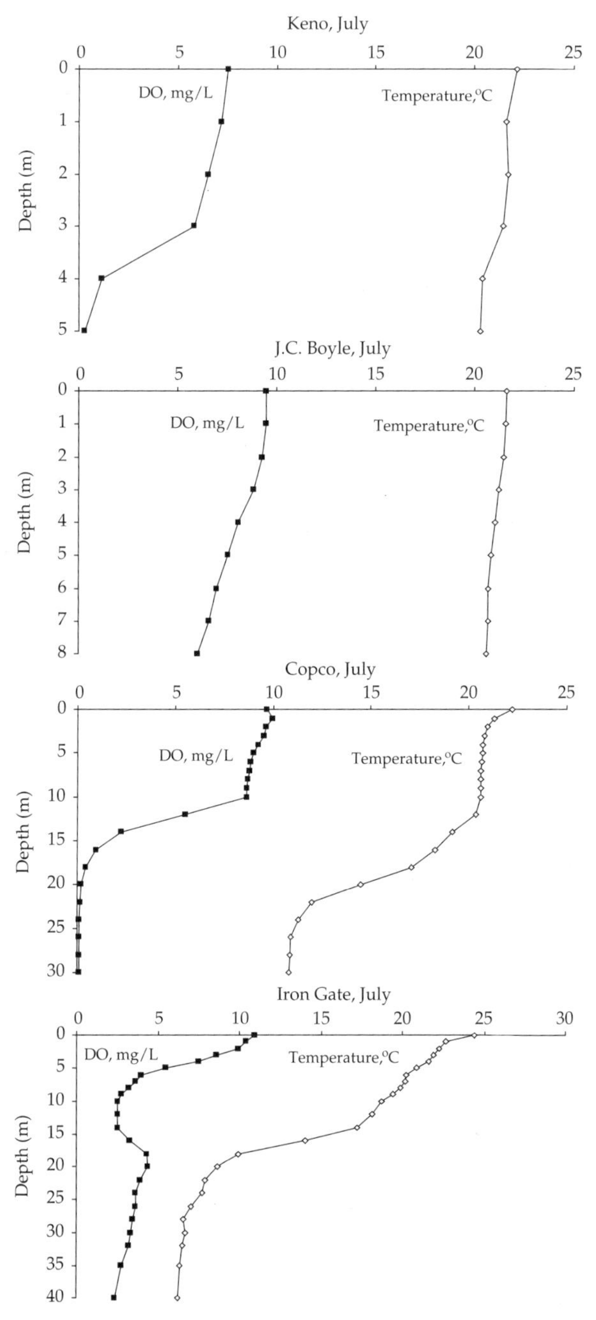

FIGURE 3-14 Water temperature and dissolved oxygen (DO) in all main-stem reservoirs, July 2000. Source: Data from USBR 2003.

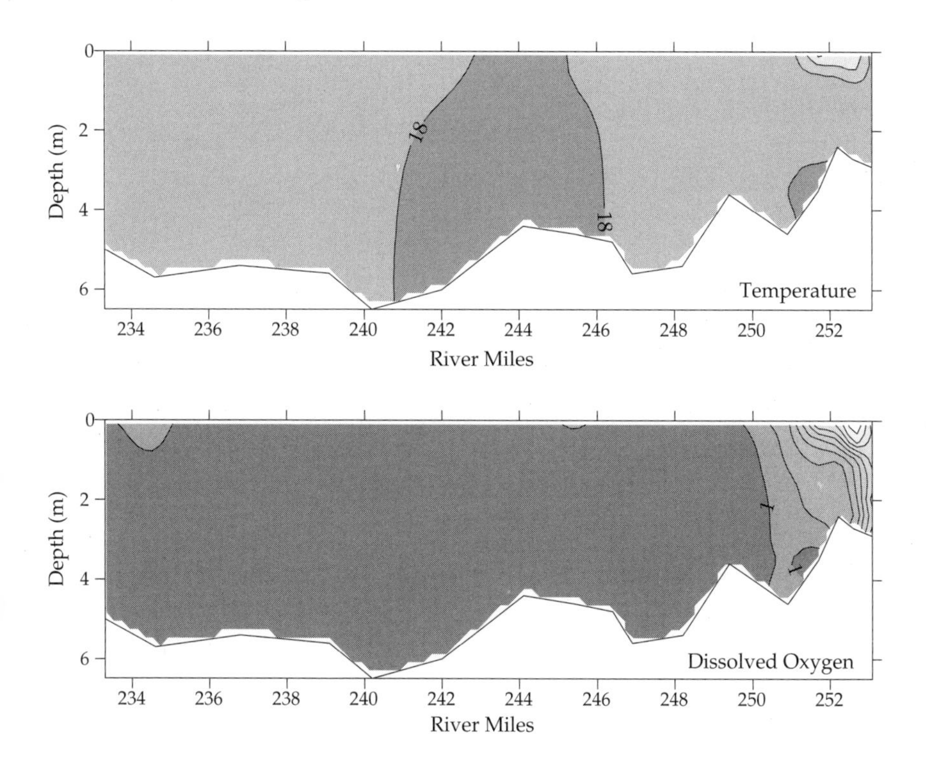

FIGURE 3-15 Longitudinal transect data on Keno Reservoir (Lake Ewauna), 13–14 August 2001. Isolines indicate temperature at 1°C intervals (top panel, increasing from 18°C) and dissolved oxygen at intervals of 1 mg/L (bottom panel, increasing from 1 mg/L). Darker tones indicate lower temperature or lower dissolved oxygen; the darkest zone on the bottom panel indicates concentrations of dissolved oxygen below 1 mg/L (i.e., without or almost without oxygen). Source: Data from USBR.

fish under these circumstances seek inflowing water of high oxygen concentration to sustain them until the episode dissipates.

Although the reservoirs receive abundant supplies of algae from Upper Klamath Lake, they do not appear to sustain such high rates of algal growth as Upper Klamath Lake, as indicated by comparisons of pH. Upper Klamath Lake shows extremes of pH extending above 10, but such extremes are not characteristic of the reservoirs. For example, monitoring of Copco, Iron Gate, and J.C. Boyle reservoirs in 1996–1998 by PacifiCorp showed the highest pH to be about 10.0, and even this was quite unusual (PacifiCorp 2000, Figure 4-10). More recent data are similar in this respect (Table 3-3).

Concentrations of phosphorus (means) tend to be about the same in the main-stem reservoirs as in Upper Klamath Lake. There is ample phos-

TABLE 3-3 Summary of Grab-Sample Data for Surface Waters in the Main-Stem Reservoir System, 2001[a]

			Concentration, µg/L					
							Chlorophyll *a*	
Location		pH	NH_4^+–N	NO_3^-–N	SRP	Total P[b]	2001	2000
Keno (1 m)	Mean	7.50	1,080	80	160	390	62	—
	Max	8.82	1,220	90	240	730	—	—
J.C. Boyle (1 m)	Mean	7.31	190	1,120	250	260	50	5
	Max	7.86	260	1,760	290	450	—	20
Copco (1 m)	Mean	7.91	90	620	150	280	5	10
	Max	8.90	130	880	220	560	—	31
Iron Gate (1 m)	Mean	8.28	260	370	160	180	5	11
	Max	9.45	260	630	280	410	—	46
Below Iron	Mean	7.87	80	980	170	190	4	—
Gate (0.5 m)	Max	8.68	90	1,710	210	360	—	—

[a]N = 4 in most cases (monthly, June–September); N = 1 for chlorophyll in 2001 (July); additional chlorophyll data for 2000 (N = 6) are shown for three of the reservoirs. Chlorophyll shown at concentrations below about 20 µg/L is only a rough approximation because of limitations on analytical sensitivity.

[b]Total P less than SRP (soluble reactive P) for some dates.

Sources: USBR 2003; PacifiCorp, unpublished data, 2001.

phorus in available form for stimulation of phytoplankton growth, but it is not clear whether a net accumulation or a net loss of phytoplankton biomass occurs in the reservoirs because information on phytoplankton biomass shows some internal inconsistencies. Observations from the field suggest that substantial blooms of *Aphanizomenon* occur in both Copco and Iron Gate reservoirs (USFWS 2002). This would not be surprising, given the strong seeding of these reservoirs with *Aphanizomenon* from Upper Klamath Lake and the presence of large amounts of nutrients. Although the residence times for the two large reservoirs are not great enough to allow the establishment of large populations of algae starting from a very small inoculum, a large inoculum could double several times over the duration of residence in the two reservoirs and thus generate a bloom. Alternatively, a bloom could simply be transferred from Upper Klamath Lake. The difficulty with the observations, however, is that they are not confirmed by monitoring data in 2000 and 2001. For both of those years, analysis of chlorophyll *a* showed abundances of algae ranging from low to high but not extreme in the sense of Upper Klamath Lake (Table 3-3). There are several possible explanations. Field reports might be biased by appearance of some algae at the surface while underlying populations are not extraordinarily high. There could be something wrong with the chlorophyll analyses, or perhaps large blooms occur very seldom. These matters are unresolved.

One special concern with respect to coho salmon and other salmonids in the Klamath River main stem is the condition of water as it leaves Iron Gate Dam. Oxygen concentrations below Iron Gate Dam are seasonally below saturation but generally exceed 75% of saturation (Deas 2000), have a temperature that reflects surface waters in the lakes, and have lower concentrations of nutrients and algae than would be typical of Upper Klamath Lake. Because there is some question about the consistency of data on algae, however, no firm conclusions are possible about the export of *Aphanizomenon* to the main stem via Iron Gate Dam.

It appears that the upper two reservoirs have the poorest water quality, as judged from concentrations of nutrients and dissolved oxygen. The two lower reservoirs, although they develop anoxia in deep waters during summer, maintain better water quality than the upper two reservoirs in their surface waters. The major question of *Aphanizomenon* blooms in the system seems unresolved because of internal inconsistencies in the data. In general, the water-quality environment seems to be comparable with or slightly better than that of Upper Klamath Lake in the two upper reservoirs, which may have very low dissolved oxygen but do not seem to have the pH extremes that Upper Klamath Lake does. The two lower reservoirs appear to have better quality overall than the two upper reservoirs, although their deep waters are essentially uninhabitable for fish during the summer months because of their lack of oxygen. More synthetic work on the reservoirs is needed.

CONCLUSIONS

1. Water-quality conditions in Upper Klamath Lake are harmful to the endangered suckers. Mass mortality of large fish is caused by episodes of low dissolved oxygen throughout the water column. Very high pH and high concentrations of ammonia, although more transitory than the episodes of low dissolved oxygen, may be important agents of stress that affect the health and body condition of the fish.

2. Poor water quality in Upper Klamath Lake is caused by very high abundances of phytoplankton, which is dominated by *Aphanizomenon flos-aquae*, a nitrogen fixer. Suppression of the abundance of *Aphanizomenon* is essential to the improvement of water quality.

3. Very high abundance of *Aphanizomenon* in Upper Klamath Lake is almost certainly caused by human activities, but mechanisms are not clear. One hypothesis is that increased algal abundance has occurred because of an increase in phosphorus loading in the lake. An alternative hypothesis, which is more consistent with the shift in dominance to *Aphanizomenon*

and with the naturally rich nutrient supply of phosphorus to the lake, is that loss of wetlands and hydrologic alterations have greatly reduced the supply of limnohumic acids to the lake. According to this untested hypothesis, loss of limnohumic acids greatly increased the transparency and may also have reduced inhibitory effects caused by the limnohumic acids. These changes allowed *Aphanizomenon* to replace diatoms as dominants in the phytoplankton. Total phytoplankton abundance then increased because of the ability of *Aphanizomenon* to offset nitrogen depletion by nitrogen fixation, which diatoms could not do.

4. Substantial evidence indicates that adverse water-quality conditions are not related to water level. Further study extended over many years may ultimately show multivariate relationships that involve water level. Control of water quality in Upper Klamath Lake by management of water level, within the range of lake levels observed during the 1990s, has no scientific basis at present.

5. Suppression of algal abundance in Upper Klamath Lake could involve drastic reduction in external phosphorus load or reintroduction of a substantial limnohumic acid supply, depending on the mechanism by which *Aphanizomenon* has become dominant. Both of these remedial actions, if undertaken on a scale sufficient to suppress the abundance of *Aphanizomenon,* could be achieved only over a period of many years and could prove to be entirely infeasible.

6. Because remediation of water quality in the near term seems very unlikely, recovery plans for the endangered suckers in the near term must take into account the potential for continued mass mortality of suckers.

7. Use of compressed air or oxygen to offset oxygen depletion near the bottom of Upper Klamath Lake has been suggested as a means of moderating mass mortality of adult suckers. Such a technique cannot be expected to offset oxygen depletion throughout the lake, but it has some potential to provide refuge zones. The endangered suckers may be particularly well suited for this type of treatment because the large suckers, which are susceptible to mass mortality, congregate in known locations.

8. Researchers have provided a great deal of useful information related to water quality of Upper Klamath Lake. Needs for additional information include studies designed to show the mechanism for *Aphanizomenon* death; physical studies, including continuous monitoring of temperature and oxygen and associated analytical and modeling work, that demonstrate more definitively the mechanisms that promote alternation of stratification and destratification during the growing season for the lake; studies of the effects of limnohumic acids on *Aphanizomenon* and of the former limnohumic acid supply to the lake; studies of diel pH cycling in the lake; and studies of water quality under ice.

9. Clear Lake and Gerber Reservoir lack extremes of pH, oxygen depletion, and algal blooms that occur in Upper Klamath Lake. Better water quality, in combination with other favorable factors given in more detail in Chapters 5 and 6, appear to explain steady recruitment, diverse age structure, and good body condition of these populations. Deterioration of body condition of the listed suckers at a time of extreme drawdown provide a rationale for the lower allowable thresholds of water level in these lakes. The lakes and their tributary spawning areas have exceptional value for protection against loss of the two endangered sucker species. Additional studies of limnological variables (and those of fish populations) have special value for use in comparison with water quality and population characteristics of suckers in Upper Klamath Lake.

10. Tule Lake, which supports suckers in good body condition but does not show evidence of successful recruitment, may have water quality that would allow recovery of this subpopulation if problems involving spawning habitat and larval survival were resolved. Lower Klamath Lake, which now lacks adult suckers, might well support a sucker population if water levels were raised.

11. Of the four major main-stem reservoirs, Keno and J.C. Boyle appear to have the poorest water quality because they are shallow, have the strongest influence from Upper Klamath Lake, and show the least benefit of dilution by waters entering from other sources. Copco and Iron Gate reservoirs have better water quality but develop anoxia in hypolimnetic waters during summer. Water released to the Klamath main stem from Iron Gate Dam often is below 100% saturation with oxygen but seldom less than 75% of saturation and may be excessively warm in summer for salmonids because it is drawn mostly from the epilimnion. Algal populations in Iron Gate Reservoir appear not to reach the extremes that are typical of Upper Klamath Lake.

4

Current and Historical Status of River and Stream Ecosystems

Aquatic ecosystems of the Klamath River basin have been extensively modified by human activities that have changed hydrology and channel morphology, increased fluxes of nutrients, increased erosion, introduced exotic species, and changed water temperatures. Efforts at restoration of declining native species need to recognize the unique characteristics of various portions of the basin in the current context of land use and human activities. This chapter considers the major streams and rivers of the Klamath basin and analyzes anthropogenic changes in conditions that affect especially the coho salmon and endangered suckers but also other fishes and aquatic life generally. Each section of this chapter considers either a specific section of the main-stem Klamath River or of its tributaries; locations are designated in river mi (RM) from the ocean.

TRIBUTARIES TO UPPER KLAMATH LAKE (RM 337-270)

Streams and rivers above Upper Klamath Lake are a source of nutrients to the lake and provide spawning and larval habitat for endangered suckers. The main sources of surface water for Upper Klamath Lake are the Williamson, Sprague, and Wood rivers (Kann and Walker 2001; Chapter 2). Groundwater and direct precipitation account for most of the balance of inflow.

For Upper Klamath Lake, external loading of phosphorus, a key nutrient that promotes algal blooms (Chapter 3), comes primarily from the Williamson, Sprague, and Wood drainages. Geologic features of this region

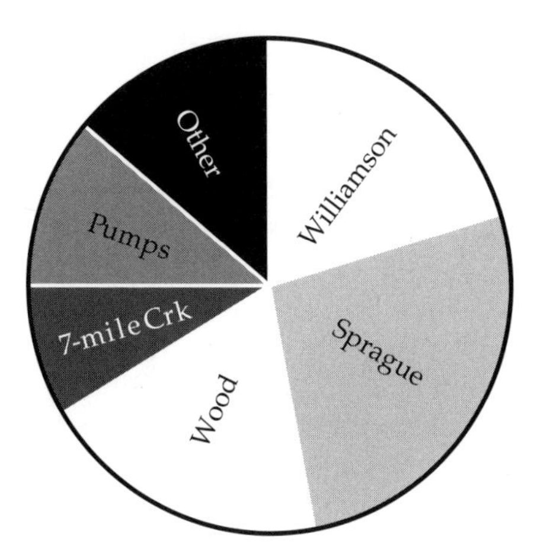

FIGURE 4-1 Relative external phosphorus loading from tributaries and other sources to Upper Klamath Lake. Source: Data from Kann and Walker 2001.

cause its streams and rivers to carry naturally high phosphorus loads (Chapter 3). Background concentrations of phosphorus, however, are augmented by human activity related to land use and river modifications. The Williamson and Sprague watersheds contribute 86 metric tons of phosphorus to Upper Klamath Lake per year (Kann and Walker 2001). The Williamson accounts for 21% of the total load, and the Sprague accounts for 27% (Figure 4-1).

Recent changes in hydrology may have affected total nutrient loading of Upper Klamath Lake. Annual runoff from the Williamson and Sprague drainages increased from the period 1922–1950 to the period 1951–1996 (Risley and Laenen 1999). The cause of the change is uncertain, but it is independent of climatic variability and probably is related to a combination of river channelization, reduction in area of wetlands, timber harvest, and other factors that reduce evapotranspiration in the watershed (Risley and Laenen 1999). Increased flows from the Williamson and Sprague drainages, coupled with current land-use practices, probably have increased phosphorus transport within the basin through greater erosion that leads to higher transport of suspended sediments, which carry phosphorus. Estimates of sedimentation rates from cores taken in Upper Klamath Lake support the hypothesis that transport of sediments from the watershed has increased in recent decades (Eilers et al. 2001).

Although its watershed is much smaller than that of the Williamson River, the Wood River is an important phosphorus source and has a high export of phosphorus per unit area of watershed (Figure 4-1). The balance of the phosphorus load to Upper Klamath Lake comes from Seven Mile Creek, agricultural pumps, and miscellaneous sources. Virtually all of this phosphorus is from nonpoint sources, including both natural and anthropogenic components.

Rivers and streams above Upper Klamath Lake support populations of cold-water fishes, including Klamath redband and bull trout (Chapter 5). During summer, temperatures can be undesirably high for these fishes in many stream reaches. For example, one threshold temperature that is used by government agencies to assess suitable rearing habitat for cold-water fishes is 17.8°C. The Williamson and especially the Sprague during late summer exceed this temperature (Boyd et al. 2001). In addition, concentrations of dissolved oxygen in the main stem of the Sprague River (mouth to junction of the North and South Forks) fall below Environmental Protection Agency water-quality targets (Boyd et al. 2002). Modeling indicates that restoration of riparian vegetation potentially could reduce temperatures in the Sprague through shading (Boyd et al. 2002), and also could have a beneficial effect on oxygen concentrations because water holds more dissolved oxygen at low temperatures than at high temperatures. In addition, shading could reduce the accumulation of algae and rooted aquatic plants on the sides and beds of tributaries. Plants produce oxygen through photosynthesis and thereby potentially increased concentrations of dissolved oxygen during the day, but nocturnal respiration and the degradation of accumulations of nonliving organic matter that they produce can cause oxygen depletion. Hence, temperature management via restoration of shading may help to alleviate a number of water-quality problems. Water-quality problems in the streams are less likely to affect endangered suckers than some of the other native fishes, however (Chapter 5).

Efforts are under way to restore wetlands associated with the Williamson, Wood, and Sprague rivers. The rationale for the projects is to restore wetland-river connections that promote such processes as nutrient trapping and sediment retention, to provide habitat for young fish, and to damp variations in river flow. Wetlands are sources of dissolved organic matter and tend to enrich water with complex humic compounds that may be related to changes in the composition of phytoplankton blooms observed in Upper Klamath Lake (Chapter 3).

THE LOST RIVER

The Lost River main stem (Figure 1-3) was an important spawning site for suckers and supported a major fishery, but few suckers use the river now (Chapter 5). Water that historically would have entered the Lost River from October to April is held back by Gerber and Clear Lake dams; summer flows are reduced by withdrawals and are dominated by irrigation tailwater. Free interchange of water and fish with the Klamath main stem is blocked in various ways. Not surprisingly, water quality of the Lost River is poor throughout the year, as indicated by low oxygen concentrations and high concentrations of suspended solids (Shively et al. 2000a, USFWS 2001), and physical habitat is greatly changed from its original state. The Lost River is now so degraded that restoration of conditions suitable for sucker spawning seems unlikely unless land-use or water-management practices change.

THE MAIN-STEM KLAMATH: IRON GATE DAM TO ORLEANS (RM 192-60)

Below Iron Gate Dam, the Klamath River runs unobstructed to the ocean. Alterations in flow and high temperatures make conditions in the main-stem Klamath less suitable than was the case historically for salmonids that use the river for spawning, rearing, and migration (Chapter 7). Four major tributaries (the Shasta, Scott, Salmon, and Trinity rivers) enter the Klamath main stem below Iron Gate Dam. These are considered in detail below.

The effect of management on the annual cycle of water flow has been the subject of considerable research on historical flows in the main stem. Before the creation of the Klamath Project and other modifications of flow, the Klamath River had a relatively smooth annual hydrograph with high flows in winter and spring that declined gradually during summer and recovered in fall. This pattern reflects the seasonal cycle of winter rainfall and spring rainfall and snowmelt in the basin (Risley and Laenen 1999). There is still an annual cycle, but its magnitude and seasonal dynamics have changed (Hardy and Addley 2001).

Figure 4-2 illustrates hydrologic change on the basis of a comparison of mean monthly flows for the periods 1905–1912 (pre-project) and 1961–1996 (post-project). Data on the earlier period are estimates based on measured discharges at the Keno gaging site extrapolated to discharges for the Iron Gate Dam site; data on the later period are based on direct measurements at the Iron Gate Dam (for methods, see USGS, Fort Collins, CO, unpublished material, 1995; Balance Hydrologics 1996; Hardy and Addley 2001). Flows over the period 1905–1912 have been adjusted to correct for the above-average precipitation that occurred then.

Post-project flows exhibit a shift in peak annual runoff from a mean maximum centered on April to a mean maximum centered on March (Figure 4-2). The later recession in spring flows extends to mean minimum flows lower than the historical minimums. Low-flow conditions during summer are more prolonged than they were before the project was built. The same analyses indicate that post-project flows during fall are slightly

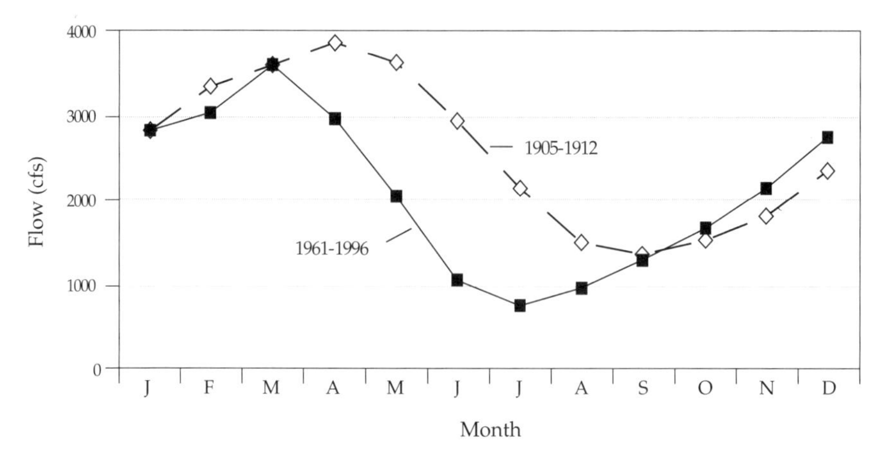

FIGURE 4-2 Mean monthly flows at Iron Gate Dam in 1961–1996 compared with reconstructed flows for 1905–1912. Source: Data from Hardy and Addley 2001.

higher than pre-project flows. The annual volume of flow from the upper Klamath basin is probably reduced. Estimated average annual runoff at the Iron Gate Dam site has declined by about 370,000 acre-ft since the construction of the Klamath Project (Balance Hydrologics 1996), as might be expected in view of the amount of water that is used for irrigation above Iron Gate Dam (Table 1-1). The magnitude of the change in water yield is a matter of dispute among groups concerned with water use in the upper basin. Nevertheless, there is no doubt that changes in seasonality of flow and at least some change in water yield have occurred since the full development of the Klamath Project.

As noted by the U.S. Geological Survey (USGS, Fort Collins, Colorado, unpublished material 1995) in its review of the hydrology of the Klamath River, the changes in flow below Iron Gate Dam are attributable to water-management practices in the upper and lower Klamath basin. The shift toward an earlier peak in annual runoff appears to be associated with increased flows in the Klamath River from the Lost River diversions and the loss of seasonal hydrologic buffering that originally was associated with overflow into Lower Klamath Lake and Tule Lake. The persistent low-flow conditions that occur in summer below Iron Gate Dam reflect irrigation demand in the Klamath Project and other parts of the upper Klamath basin and irrigation diversions on the Scott and Shasta rivers and other tributaries (discussion below).

Release of water from Iron Gate Dam has both direct and indirect effects on water temperature in the Klamath River. The magnitude of these effects depends on three principal factors: the temperature of the water as it is released from the dam, the volume of the release, and the meteorological conditions. The temperature of water released from Iron Gate Dam varies seasonally; a peak at about 22°C (+/– 2°C) occurs in August (Figure 4-3). In summer, the volume of flow exerts substantial control over the rate of daytime warming and nocturnal cooling. Low flows have long transit times and thus show greater change per unit distance. For example, a 500-cfs release takes 2.5 days to reach Seiad Valley, a distance of about 60 river mi, whereas a 1,000-cfs release moves the same distance in 2 days and a 3,000-

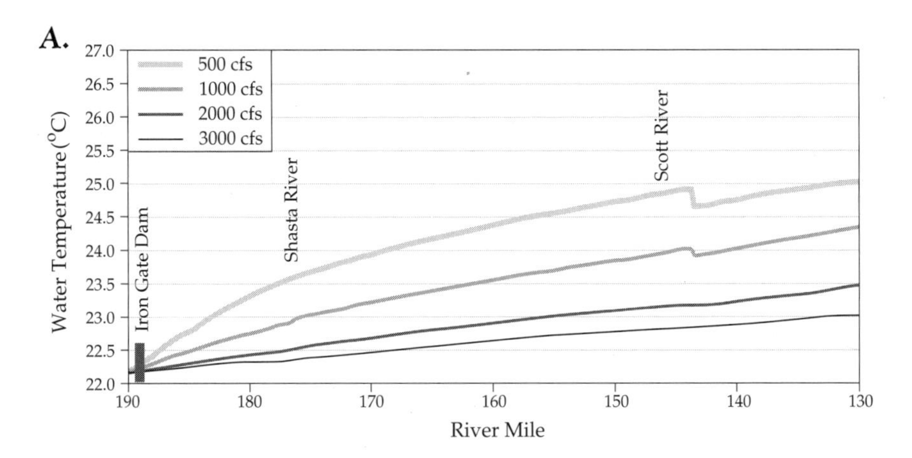

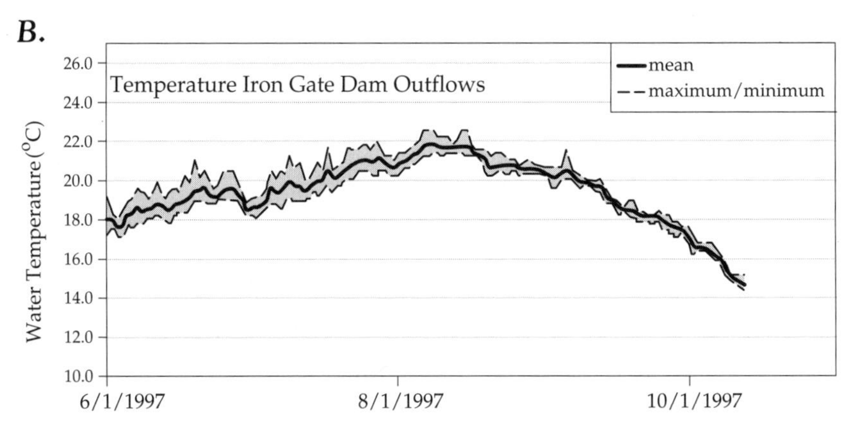

FIGURE 4-3 Simulated and measured temperature in the Klamath River below Iron Gate Dam. A) Simulated daily mean temperatures from Iron Gate Dam to Seiad Valley for flows of 500–3,000 cfs for conditions in August. B) Measured temperature of releases from Iron Gate Dam, June–October 1997. Note the minor diel change in temperature during the warmest summer releases. Source: Deas 2000. Reprinted with permission from the author; copyright 2000, University of California Press.

cfs release does so in 1.25 days (Deas 2000). Warming and cooling per unit distance are reduced by short transit time and by greater depth. Higher flows extend the reach of river below Iron Gate Dam that supports lower mean water temperatures (Figure 4-4), but also may result in higher daily minimum temperatures over a portion of the reach below Iron Gate Dam (see below).

Increased releases from Iron Gate Dam may benefit coho salmon (Hardy and Addley 2001, NMFS 2001). The potential benefit from the releases is

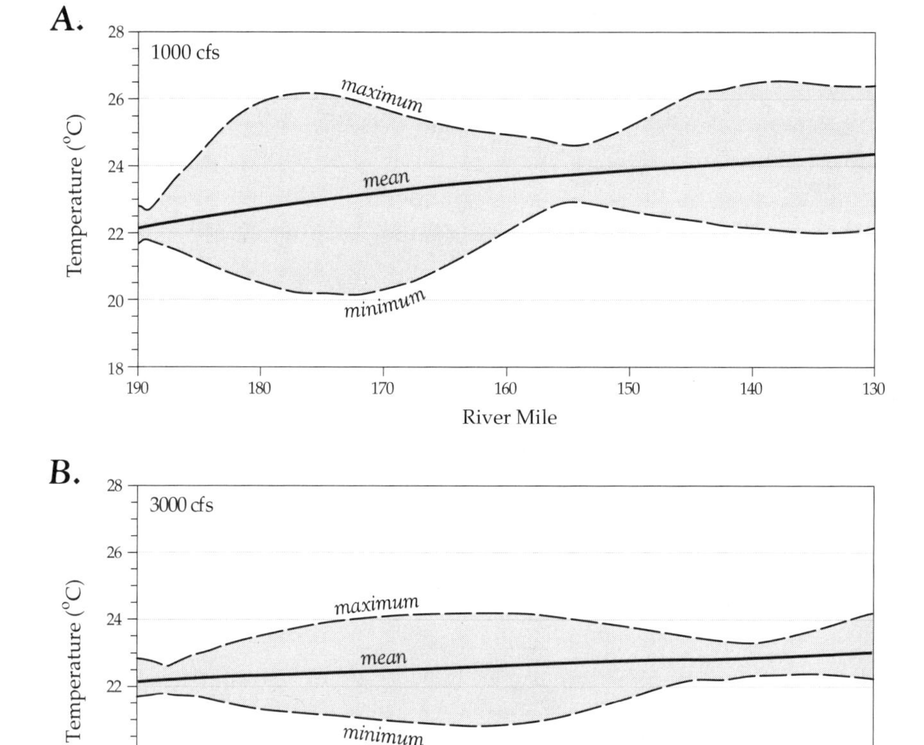

FIGURE 4-4 Simulated daily maximum, mean, and minimum water temperatures on the Klamath River from Iron Gate Dam to Seiad Valley for Iron Gate Dam releases of 1,000 cfs (A) and 3,000 cfs (B) under meteorological conditions of August 14, 1996. Source: Deas 2000. Reprinted with permission from the author; copyright 2000, University of California Press.

confounded, however, by relationships between minimum, mean, and maximum temperatures. For example, water released from Iron Gate Dam in August has a mean temperature near 22°C, which is well above the acute tolerance threshold for coho (Chapters 7 and 8). Field-calibrated models developed by Deas (2000) and models presented by Hardy and Addley (2001) show a considerable increase in the daily mean water temperature with distance downstream for flows that are typical of August. As noted in Chapters 7 and 8, however, bioenergetics of salmonids depend not only on the mean temperature but also on the diel range of temperature; low minimum temperatures are especially important for coho salmon.

Simulations conducted by Deas (2000) provide insight into the thermal response of the Klamath River to increases in flow during late summer (Figure 4-4). Under moderate flow conditions in mid-August (1,000 cfs), with typical accretions from tributaries, maximum daily temperatures increase rapidly downstream of Iron Gate Dam to a peak of 26°C within 15 mi. Daily minimum temperatures caused by nocturnal cooling reach a minimum of 20°C within about the same distance. By the time this water reaches Seiad Valley (RM 130), maximums are greater than 26°C, and minimums are 22°C; the average gain from Iron Gate Dam is 2°C. Tripling the flow from Iron Gate Dam (Figure 4-4B) provides modest reduction in mean and maximum daily temperatures, particularly in the first 20 mi of the river downstream from the dam. The increased volume of water and shorter transit time, however, reduce the effect of nocturnal cooling in the reach between Iron Gate Dam and Seiad, and raise minimum temperatures for about two-thirds of the reach. Although increased flows reduce mean and maximum temperatures, the increase in minimum temperatures may adversely affect fish that are at their limits of thermal tolerance (Chapters 7 and 8).

Two additional complications arise from increased releases from Iron Gate Dam. First, during low-flow conditions, tributaries can influence main-stem temperatures. Temperatures in the Klamath River at 1,000 cfs are affected substantially by the Scott River and minimally by the Shasta River. Modification of flow and temperature regimes in these tributaries through better water management could improve main-stem temperatures. Increase in flow to 3,000 cfs, however, eliminates any thermal benefit from the tributaries (Deas 2000).

In regulated rivers such as the Klamath, there often is a node of minimum diel temperature variation about 1 day's travel time from a dam (Lowney 2000) and an antinode of maximum variation at half this distance. The muted minimums and maximums of the thermal node reflect a single diel cycle of roughly equal heating and cooling during 1 day's travel time. Conversely, the large variation in temperatures at the antinode reflects only half a diel heating or cooling cycle. Reduction in maximum temperature is one of the benefits of the thermal nodes. These nodes, however, also exhibit greatly increased minimum temperatures. In the Klamath River under flow and meteorological conditions typical of August, the highest minimum daily temperatures will occur at the node and may be points of greatest thermal stress for salmonids. Increases in flow will cause the node to shift downstream because of decreased transit times (Figure 4-4), thus increasing the amount of river that is subjected to increased temperature minimums.

The main-stem Klamath—like the lakes, reservoirs, and rivers of the upper basin—has concentrations of nitrogen and phosphorus that are quite high relative to many aquatic systems (Campbell 2001; Figure 4-5); they indicate eutrophic conditions. In addition, much of the nitrogen and phosphorus is readily available for plant uptake (for example, the forms nitrate and soluble reactive phosphorus). As a consequence of high nutrient concentrations, the river has the potential to support high rates of primary production. Even when nutrient concentrations are high, however, blooms of phytoplankton, such as those in Upper Klamath Lake, do not occur in streams or rivers of moderate to high velocity because flow limits the accumulation of suspended algae. Conditions may be favorable in the main stem for the growth of phytoplankton during low flow, when the water is moving slowly, and growth of attached algae and aquatic vascular plants also can be stimulated by nutrients. Stimulation of any kind of plant growth can affect oxygen concentrations.

During summer, oxygen concentrations in the Klamath River often fall below 7 mg/L and, for brief periods, below 5.5 mg/L (Campbell 2001). For

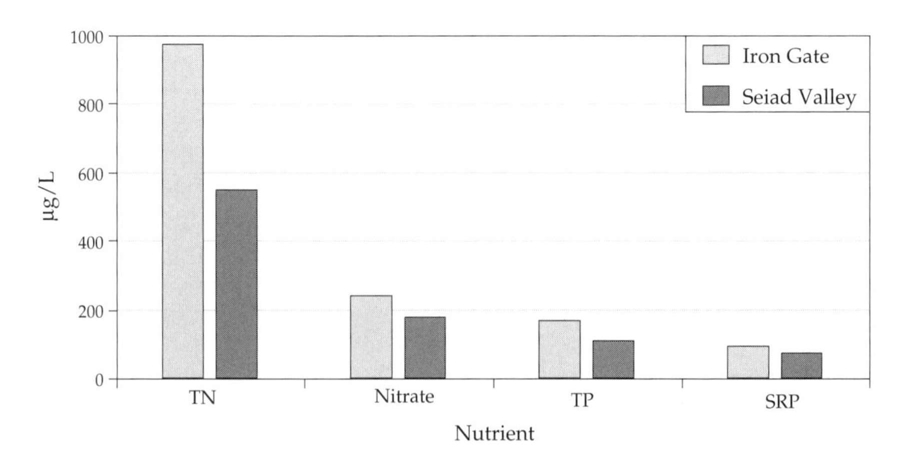

FIGURE 4-5 Mean annual concentrations of total nitrogen (TN) and total phosphorus (TP), nitrate (NO_3^- expressed as N), and soluble reactive phosphorus (SRP) at two stations on the Klamath River. Source: Data from Campbell 2001.

example, average concentrations were below 7 mg/L on 36 days at the Seiad Valley monitoring station in 1998. More severe and extended periods of low oxygen concentrations occur at Iron Gate Dam because of degradable organic matter (such as dead phytoplankton) originating in reservoirs. Low oxygen concentrations, especially below 5.5 mg/L, are unfavorable to salmonids (Chapter 7).

THE SHASTA RIVER (RM 177)

Flow of the Shasta River is dominated by discharge from numerous cool-water springs and not by surface runoff. The stable, cool flows and high fertility of the Shasta historically created a highly productive, thermally optimal habitat for salmonids.

The Shasta River maintains about 35 mi of fall-run Chinook habitat, 38 mi of coho habitat, and 55 mi of steelhead habitat (West et al. 1990). The amount of habitat has not declined since 1955 but is substantially smaller than the original amount. Use of remaining habitat is contingent on flow and water quality, both of which may be inadequate in dry years. Mean annual runoff from the Shasta River is 136,000 acre-ft, which is less than 1% of the runoff of the Klamath River at Orleans. Runoff within the basin peaks during winter, when daily flow is near 200 cfs (Figure 4-6).

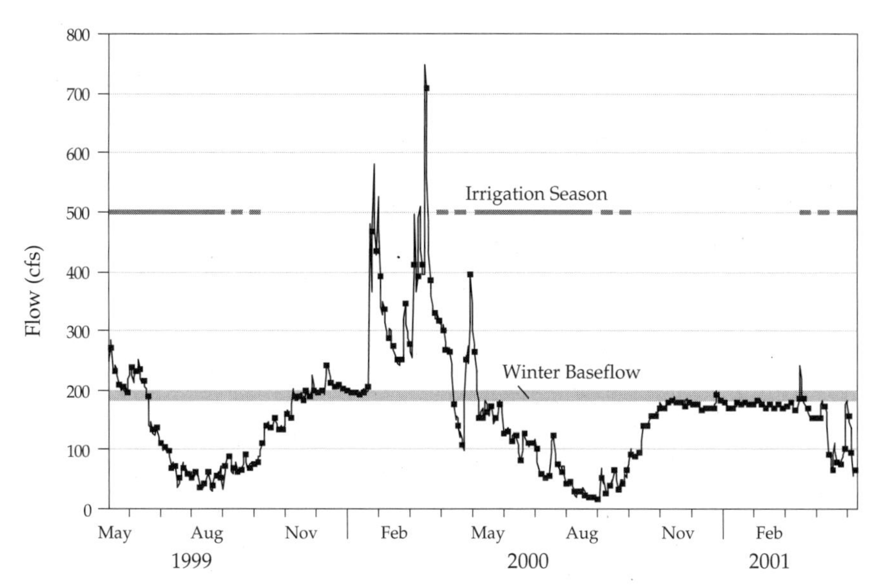

FIGURE 4-6 Annual hydrograph for the lower Shasta River (at Yreka, California), from May 1999 to May 2001. Note base-flow recovery during fall and sustained base flow throughout the winter of 2001.

Peaks are associated with rain at times when there are no irrigation diversions (note that peaks did not occur in 2001, a year of drought). Flow declines rapidly with the onset of irrigation in late March. Flow minimums typically averaging less than 30 cfs occur during summer. Flow increases rapidly in the fall when irrigation ends. Winter base-flow conditions typically are 180–200 cfs, regardless of precipitation.

The hydrology of the Shasta River is affected by surface-water diversions, alluvial pumping, and the Dwinnell Dam (Figure 4-7). Historically, springs and seeps dominated the hydrograph of the Shasta River. Mack (1960) reported that one small tributary, Big Springs (Figure 4-7), supplied a consistent 103 cfs to the Shasta River before water development. Flow from the springs and numerous small accretions in the reach above them would have supplied flows close to or exceeding today's bankfull condition, even during summer months. Flows of that magnitude would have had very short transit times (less than 1 day to the Klamath River), thus maintaining cool water throughout summer for the entire river. Consistency of flow and cool summer water were the principal reasons that the Shasta River was historically highly productive of salmonids. During summer, the Shasta River may also have cooled the main-stem Klamath near the confluence of the Shasta and the main stem.

Since 1932, surface-water resources in the Shasta valley have been under statutory adjudication (Decree 7035). Three of the four major irrigation districts have a cumulative appropriative right to divert more than 110 cfs from the Shasta River from April 1 to October 1 (Gwynne 1993). Dwinnell Dam is used by the fourth major irrigation district to store winter flows of the Shasta River and Parks Creek. Dwinnell Dam, constructed in 1928, has a capacity of 50,000 acre-ft. The California Department of Water Resources Watermaster Service has been apportioning water within the basin since 1934. Riparian water rights below Dwinnell Dam are not adjudicated and are not regulated by the watermaster, and the 1932 adjudication did not address groundwater, which is critical for support of base flow.

Seven major diversion dams and numerous smaller dams or weirs are on the Shasta River and its tributaries below Dwinnell Dam (Figure 4-7). When the diversions are in operation, they substantially and rapidly reduce flows in the main stem (Figure 4-6). During the drought of 1992, flows in the Shasta dropped from 105 cfs on March 31 to 21 cfs by April 5. The numerous diversions on the Little Shasta River now routinely lead to complete dewatering of its channel in late summer. Although surface diversions play an important role in causing the low flows of the Shasta, there is little quantitative information on the relative role of each diversion, and records either have not been kept or are not available from the watermaster service that apportions flows.

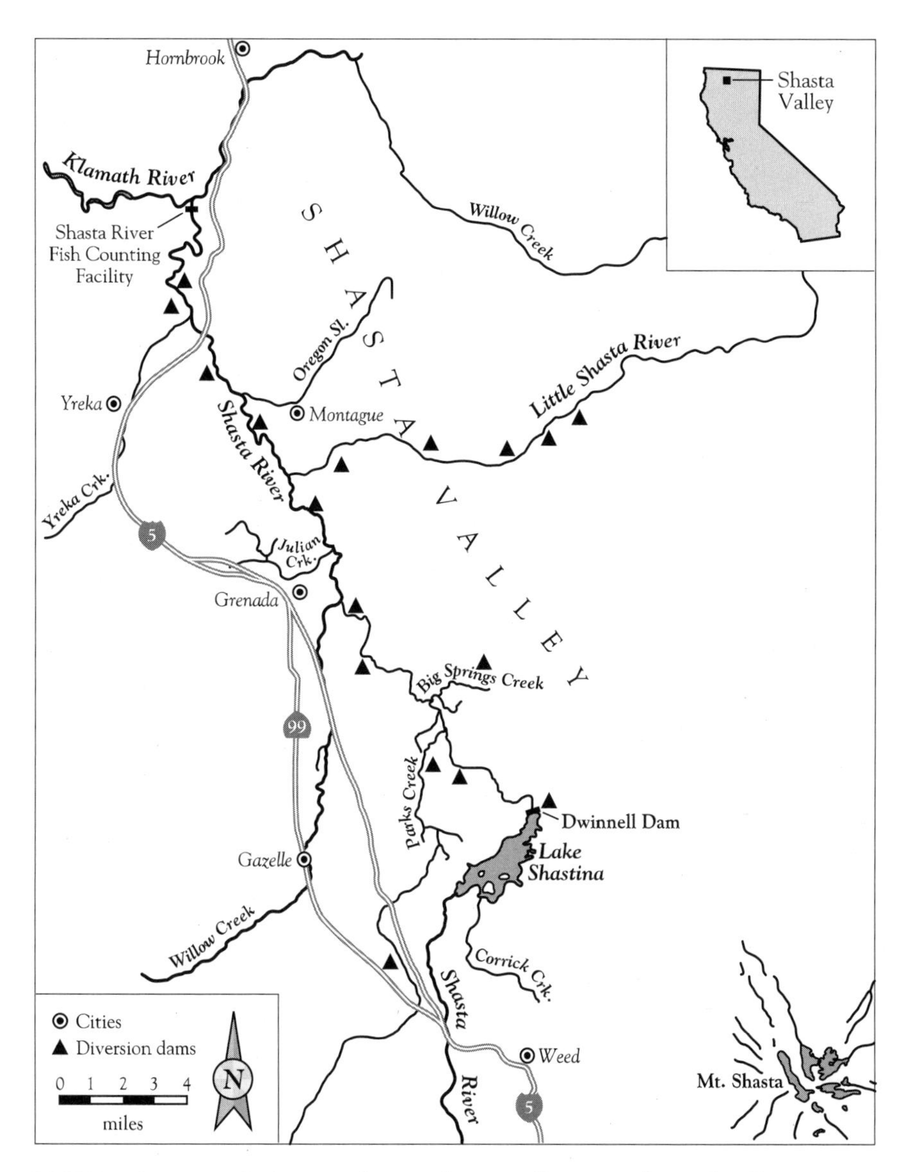

FIGURE 4-7 Map depicting substantial water diversions from the Shasta River below Dwinnell Dam. Note that the Shasta River flows north and drains into the Klamath River. Source: Modified from Gwynne 1993.

Dwinnell Dam affects the hydropattern of the Shasta River. Peak winter flows associated with large precipitation events have been strongly suppressed. Absence of flushing flows reduces sediment transport and reduces the availability of spawning gravels downstream of the dam (Ricker 1997). With the exception of above-average water years, when Lake Shastina is full, no flow is released from Dwinnell Dam except for small amounts to specific water users downstream. Water in Parks Creek is diverted into Lake Shastina, thus decreasing winter flows in the creek. In addition, seepage losses from Lake Shastina are large; they exceed the total amount of water supplied to irrigators (Dong et al. 1974).

Groundwater is not part of the adjudication of water rights in the Shasta basin, and little is known about its influence on surface flows. The exceptionally high specific capacity of the aquifers and the large recharge area make groundwater one of the most important and resilient resources in the valley. Well records of the California Department of Water Resources (CDWR) indicate a great increase in the number of irrigation wells in the valley since the 1970s. The shift toward groundwater production from use of surface diversions may have had a measurable effect on surface flows and may have exacerbated low-flow conditions. For example, the Big Springs Irrigation District ceased using surface diversions and switched to groundwater wells in the 1980s to meet its water needs; these highly productive wells may have contributed to the reported dewatering of the springs that historically fed Big Springs Creek.

Recent surveys have shown that channel conditions in the main stem of the Shasta River and its most important tributary, Parks Creek, generally are poor and may limit salmonid production. Replicate habitat surveys summarized by Ricker (1997) and Jong (1997) focus on Chinook spawning gravels and indicate that the percentage of fines in gravels is high throughout the main stem and Parks Creek. The fines, which are detrimental to egg survival and emergence of fry, are associated with accelerated erosion and lack of flushing flows that maintain and recruit coarse gravels.

In some reaches, particularly in the lower canyon and the reach below the Dwinnell Dam, limited recruitment of coarse gravels is contributing to a decline in abundance of spawning gravels (Buer 1981). The causes of the decline in gravels include gravel trapping by Dwinnell Dam and other diversions, bank-stabilization efforts, and historical gravel mining in the channel.

Loss of vegetation in the riparian corridor poses a widespread and important threat to salmonid habitat. In the lowermost reach of the Shasta River, the loss is explained principally by mining. In the valley above the lower Shasta, grazing has been responsible for most of the loss. Where intense unfenced grazing has occurred, trampling and removal of vegetation have commonly led to accelerated bank erosion, loss of shading, reduced accumulation of local woody debris, loss of pool habitat to sedimentation, loss of channel complexity and cover, and degradation of water quality. Riparian fencing programs and construction of stock-water access points are under way in the Shasta valley, but efforts to date are modest (Kier Associates 1999).

The Shasta River contains seven major diversion dams and multiple smaller dams or weirs (Figure 4-7). Dwinnell Dam eliminated access to about 22% of habitat historically available to salmon and steelhead in the watershed (Wales 1951). The reach between Big Springs and Dwinnell Dam, which has the potential to support a range of salmonids, receives minimal flows from the dam.

Although Dwinnell Dam is the most important diversion structure on the Shasta River, numerous other diversions have an important but unquantified effect. Many of the structures create low-water migration barriers and during summer create water-quality problems by acting as thermal and nutrient traps. Unscreened diversions have been identified as a serious problem for salmonid spawners, outmigrants, and juveniles (Chesney 2000).

Surface diversions and groundwater withdrawals have eliminated or substantially degraded flows on the Shasta River and its tributaries. The alterations are most evident during late spring through early fall, when increasing air temperatures and low flow coincide with poor water quality. The low flows also reduce habitat for salmonids and increase the adverse effects of diversion structures on migration.

Substantial reduction of flows by water withdrawal and the associated poor water quality probably are principal causes of decline in salmonid production in the Shasta watershed. The 1932 adjudication of surface waters in the basin, as currently administered, is insufficient to supply the quantity and quality of water necessary to sustain salmonid populations in the basin.

A major bottleneck for salmonid production in the Shasta River watershed is high water temperature (Figure 4-8). Daily minimum temperatures in the lower main stem in summer are typically greater than 20°C, and daily maximums often exceeding 25°C. Salmonids, especially coho salmon, rarely persist under such conditions. McCullough (1999) found that salmonids are typically absent from waters in which daily maximum temperatures regularly exceed 22–24°C for extended periods, although bioenergetic considerations or presence of thermal refugia may push distribution limits into slightly warmer water (see Chapter 7). Growth and survival are usually highest when temperatures stay within an optimal temperature range; this range differs among species and life-history stages, but for juvenile salmonids in the Klamath system, optimal temperatures are 12–18°C (Moyle 2002); bioenergetic considerations also alter optimal temperatures for growth and survival (McCullough 1999). The Shasta River becomes progressively cooler as elevation and flows increase, but temperatures remain largely suboptimal for salmonids for most of its length from late June through early September (Figure 4-8). Higher temperatures also are associated with reduced amounts of dissolved oxygen (DO) in the water. DO concentrations below saturation are apparently uncommon in the Shasta River, but where they occur, they coincide with high temperatures and low flows (Campbell 1995, Gwynne 1993). The causes of high temperatures include chronic low flow due to agricultural diversions, lack of riparian shading, and addition of warm irrigation tailwater. Temperature simulations for the Shasta River conducted by Abbott (2002) demonstrate the importance of flow (Figure 4-9) and riparian vegetation to river temperatures. Low flows with long transit times typical of those now occurring in the summer on the Shasta River cause rapid equilibration of water with air

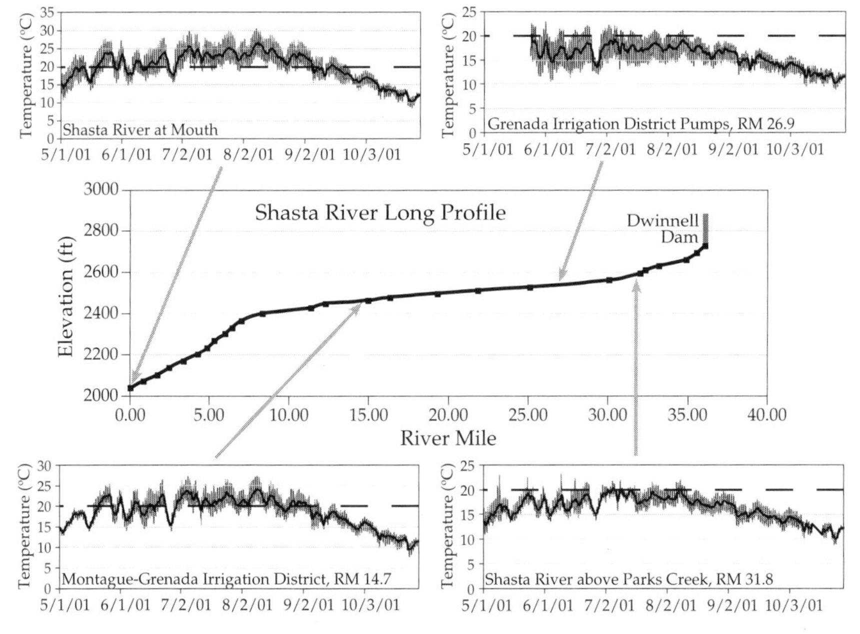

FIGURE 4-8 Temperature (thin line) and daily average temperature (wide line) within the Shasta River below Dwinnell Dam during the summer of 2001. The dashed line at 20°C is for comparison between plots. Note that the generally cool, spring-fed upper reaches of the river have temperatures suitable for salmon. Low flow, warm tailwater return flows, and lack of riparian cover on the lower main stem lead to high temperatures unsuitable for salmonids. Source: Abbott 2002. Reprinted with permission from the author; copyright 2002, University of California Press.

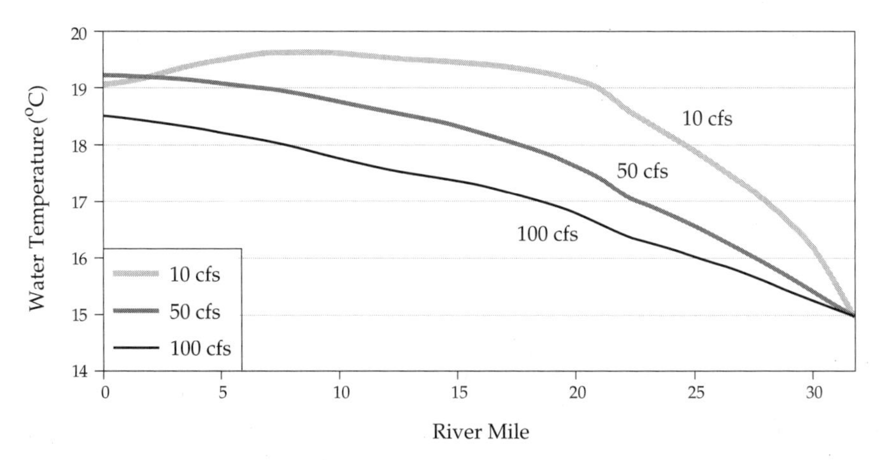

FIGURE 4-9 Simulation of daily mean water temperatures in the Shasta River at three flows for August 2001 conditions. Simulations assume no significant shading. Source: Abbott 2002. Reprinted with permission from the author; copyright 2002, University of California Press.

temperatures, which produces water temperatures exceeding acute and chronic thresholds for salmonids well above the mouth of the river. Small increases in flow could reduce transit time substantially and thus increase the area of the river that maintains tolerable temperatures. Increases in riparian vegetation also could help to sustain lower water temperatures. Unlike other large tributaries, the Shasta River has a relatively narrow channel that could be strongly affected by riparian shading. Simulations of the effect of mature riparian forests for weather conditions of August 2001, and in drought conditions, showed lowering of daily mean water temperature at the mouth of the river from 21.4°C to 17.1°C and lowering of average maximum temperatures from 31.2°C to 24.2°C (Abbott 2002).

THE SCOTT RIVER (RM 143)

The watershed of the Scott River historically has provided important spawning and rearing habitat for coho salmon and, on the basis of records of spawning runs as recent as winter 2001–2002 (USFWS 2002), remains one of the most important tributary watersheds for coho in the lower Klamath basin.

The hydrology and water budget of the Scott River watershed are poorly documented. One USGS gage at Fort Jones provides the longest continuous record of flows (1942–2002). The gage is 16 mi upstream of the Klamath River and does not take into account accretions from the tributaries to the Scott River Canyon. Mean annual runoff within the basin is 489,800 acre-ft (range 54,200–1,083,000 acre-ft). Flows within the tributaries are poorly documented.

The hydrograph of the Scott River, like that of the Salmon River, shows two seasonal pulses (Figure 4-10) that are unaffected by any large impoundments. The winter pulse is caused by high precipitation from mid-December through early March and is highly important geomorphically because it accounts for most of the annual sediment transport (Sommerstram et al. 1990, Mount 1995). The second pulse is caused by the spring snowmelt, which begins in late March and in wet years continues through June (Figure 4-10).

From late June through November, flows in the Scott River and its tributaries are low (Figure 4-10). During average to dry years, the tributaries with large alluvial fans are disconnected from the Scott River except through subsurface flow (Mack 1958, CSWRCB 1975). The loss of flow is caused by high seepage in the alluvial fans and diversions for irrigation. Along the main stem of the Scott River, surface flow ceases in several reaches during August and September of average and dry years. Discontinuous flow occurs into the fall. During average and wet years, continuity of

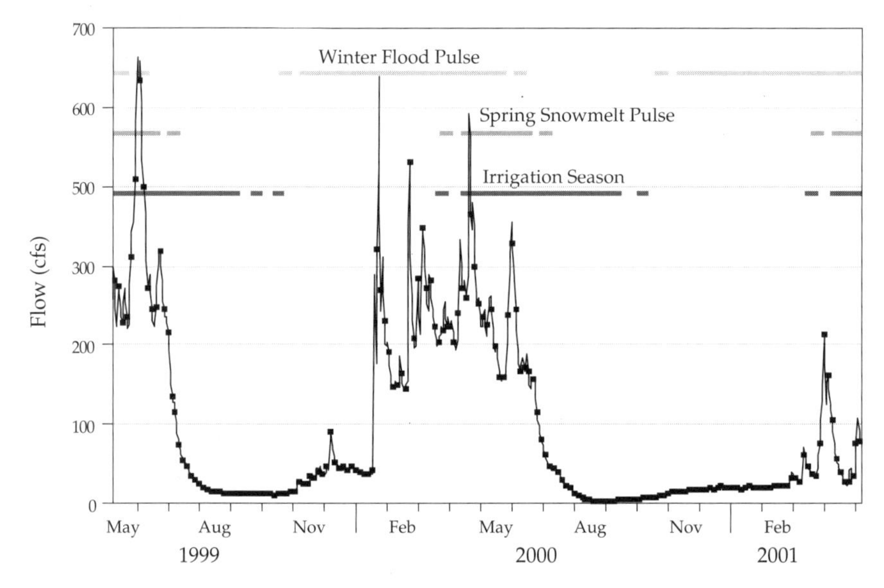

FIGURE 4-10 Annual hydrograph of Scott River at Fort Jones, California, May 1999 through May 2001. Note the significant decline in flows at the start of the irrigation season and weak recovery of flows during the dry winter of 2000–2001.

flow is restored between late October and early November as evapotranspiration declines and irrigation decreases. During dry years, low-flow conditions persist until substantial rainfall occurs. Unlike the Shasta River, the Scott River shows lack of significant recovery of base flow during late fall and winter in years of low rainfall, indicating lack of resiliency in the groundwater reservoirs.

Because low base flow during summer and early fall is a natural element of the Scott River hydropattern, dry conditions in some reaches of the river may have occurred at some times before water management. Water management has decreased fall flows and has increased the frequency and duration of negligible flow.

The main groundwater source for irrigation and domestic water use in the Scott valley is the extensive alluvium under the river (Mack 1958, CDWR 1965). High rates of recharge to the valley aquifer, whose volume exceeds 400,000 acre-ft, are a byproduct of the fan heads of west-side tributaries, which receive seepage through the river bed; direct recharge from seepage of precipitation; infiltration losses from irrigation ditches; and deep percolation of irrigation water.

Groundwater levels in the valley aquifer reflect drawdown during the irrigation season and recharge during the wet season. The combination of high specific capacity of the shallow alluvial aquifers of the basin and high hydraulic gradients produces rapid seasonal changes in groundwater levels. Where subsurface water-bearing sediments are hydraulically connected in the Scott Valley, groundwater pumping can cause serious losses in channel flow (Mack 1958, CSWRCB 1975). Thus, pumping may be an important contributor to low-flow and no-flow conditions. There has been no comprehensive analysis of the water budget of the Scott River.

The Scott River and most of its tributaries are adjudicated under California water law. Adjudication and enforcement play key roles in the water budget of the Scott River. The Scott Valley Irrigation District initiated adjudication proceedings by petition to the California State Water Resources Control Board (CSWRCB) in 1970. Investigations cited above revealed the hydraulic connections between shallow groundwater and surface flows, indicating that adjudication should include both surface-flow rights and pumping rights adjacent to the river. At the time, this type of adjudication was not allowed under California statutes. Special legislation was developed for the innovative adjudication of the Scott River. Most of the shallow groundwater in the valley probably is linked to the surface flows. Recognizing this, the CSWRCB staff arbitrarily chose an adjudicated zone extending about 1,000 ft from the main-stem channel of the Scott River (CSWRCB 1975).

In 1980, the Siskiyou County courts decreed the Scott River adjudication, recognizing 680 diversions capable of diverting up to 894 cfs from the river and its tributaries above the USGS gage at Fort Jones (CH2M HILL 1985). Adjudications had been completed earlier on Shackleford and Mill creeks and on French Creek. Since 1989, the Scott River and its tributaries French, Kidder, Shackleford, and Mill creeks have been considered fully appropriated by CSWRCB.

The CDWR has provided a watermaster service to minimize litigation over water rights. Although a watermaster oversees 102 decreed water rights on several tributaries in the basin, no watermaster service has been requested for the main stem.

During the adjudication process, the state and federal governments both failed to negotiate successfully for water that would favor robust populations of fish. There are now no adjudicated rights for fish upstream of the USGS gage in Fort Jones. Below the Fort Jones gage, the U.S. Forest Service (USFS) was allotted flow of 30 cfs during August and September, 40 cfs during October, and 200 cfs from November through March to protect fish. With no watermaster service, USFS, a junior appropriator, commonly does not receive its adjudicated flows during late summer and fall.

Assessments of limiting factors for coho salmon have been summarized by Siskiyou County Resource Conservation District (Scott River Watershed CRMP Council 1997, West et al. 1990) and are given in Chapter 8. The limiting factors can be grouped into two classes: those associated with tributary flows and conditions, and those associated with the main stem of the Scott River.

Tributaries that drain the west side of the watershed and the East and South Forks of the Scott have substantial habitat for coho and other salmonids. Juvenile salmon occupy the uppermost reaches of the tributaries, where they benefit from the consistently low water temperatures and perennial flows (West et al. 1990). West-side tributary reaches that are above the major diversions maintain high water quality and favorable temperatures throughout the year, including August and early September (SRCD 2001). Maximum weekly average temperatures range from 15 to 17°C, and diel fluctuations are less than 3°C.

The principal limiting factor in the upper tributary reaches is excessive sediment derived from logging, particularly in tributaries with granitic soils (CH2M HILL 1985, Lewis 1992). Highly erodible decomposed granite has led to a serious loss in volume and number of pools in tributaries and associated degradation of spawning and rearing habitat. Logging over the past 50 yr has taken place on a mix of USFS land and land held by a few large private timber companies. Historical logging practices have been poor, particularly on private land, and have left a legacy of degraded hillslope and stream conditions.

Within the lower reaches of the west side, where tributaries contain surface diversions or large alluvial fans, low or negligible flow may be a limiting factor for coho and other salmonids. The loss of base flow in these tributaries may have occurred historically during dry years, particularly where there were large alluvial fans. Diversions and groundwater withdrawal, however, probably have increased the frequency and length of dry conditions, particularly in Etna, Patterson, Kidder, Mill, and Shackleford creeks (Mack 1958). The dewatering of these tributaries eliminates potential rearing habitat for coho and causes loss of connectivity and reduction of base flow in the main stem. Dry conditions in these creeks can persist into fall, thus blocking tributary access for spawning coho, steelhead, and Chinook.

West et al. (1990) documented 128 mi of potential spawning and rearing habitat for coho in the Scott River, mostly on the main stem. Degradation of habitat, however, is considerable; less than 30% is rated good to fair (SRCD 2001). California Department of Fish and Game (1999) rated the holdover of adults before spawning as fair, spawning habitat as fair, and juvenile rearing habitat as poor. The decline in salmonid habitat conditions on the main stem of the Scott is caused by channel alterations, low flow, and poor water quality.

The main-stem channel of the Scott River has been extensively altered over the last 150 yr by placer and hydraulic mining, logging, grazing in the riparian corridor, unscreened irrigation and stockwater diversions, elimination of wetlands, and flood-management or bank-stabilization efforts. These activities have cumulatively degraded salmonid habitat on most reaches of the main stem above the canyon. The most important limitations appear to arise from loss of optimal channel complexity and depth, loss of riparian vegetation, and unscreened diversions. There are 153 registered diversions in the Scott Valley, of which 127 are listed as active by SRCD. Fish screens have been installed on 65 of these diversions; another 38 are funded but not yet built (SRCD 2001).

Seasonal low flows are consistently recognized as one of the most important limiting factors for all salmonids that use the main stem of the Scott River (CH2M HILL 1985, West et al. 1990, SRCD 2001). Low flows and dry conditions contribute to the decline in spawning and rearing habitat in the river and exacerbate poor water quality during summer and early fall. During years when seasonal rains arrive late, low-flow conditions can persist into the fall, and limit access of salmon to spawning sites in tributary streams.

Low-flow and dry conditions are a natural aspect of the main-stem Scott in dry years, but the adjudication of the Scott River and its tributaries offers little protection for stream flow and related temperature requirements of salmonids in the watershed even during normal years. The adjudicated water rights are sufficient to allow removal of all flow from the river during the summer and early fall. The shift from surface diversions, which are naturally self-limiting, to groundwater wells, has exacerbated the apparent overappropriation of water in the watershed. That problem is compounded by a limited watermaster service in the basin and insufficient records, so it is not known whether diverters are adhering to their appropriative rights. The net result is that limited management and overappropriated water have seriously affected flows in the river.

The frequency and duration of low-flow conditions has increased since the 1970s (summary in Drake et al. 2000); the most important effects occur in September (Figure 4-11A), as confirmed by analysis of double-mass

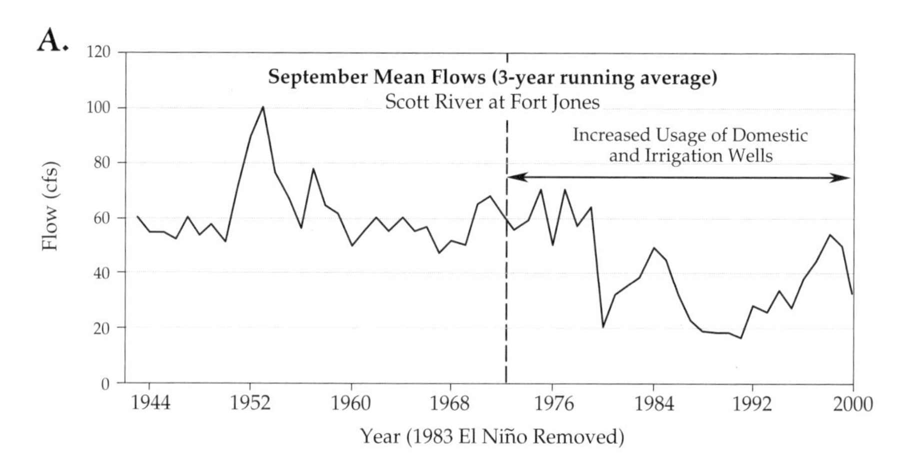

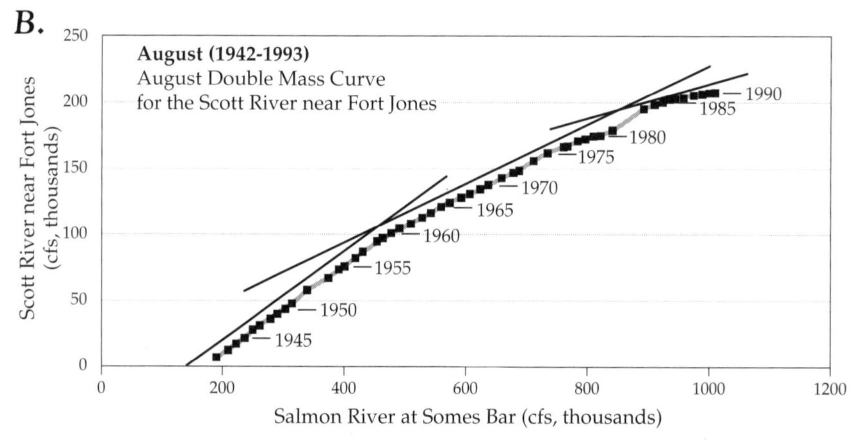

FIGURE 4-11 Declines in late summer and early fall flows on the Scott River. A) 3-yr running average of September mean flows, 1942–2002. Note the shift in low-flow conditions in late 1970s. B) Double-mass curve of August flow volumes on the Scott vs the Salmon River showing decline in August volume in the Scott relative to the Salmon during last 50 yr. Source: Bartholow 1995.

curves that compare runoff between the Scott River and the nearby Salmon River, which is not subject to diversion (Figure 4-11B). The decline in late summer and fall runoff is a considerable challenge to restoration of salmonid holding, spawning, and rearing conditions in the Scott River. In the absence of credible information and hydrologic models, there has been widespread speculation about the causes of declining flows in the Scott River. For example, Drake et al. (2000) postulated that the principal cause of declining late summer and fall flows in the Scott River is climate change. Drake et al. analyzed the relationship between precipitation in the Scott River watershed and fall runoff. Their work demonstrated a modest statistical correlation between declining precipitation in April at two snowpillows (snow-accumulation sensors) in the western edge of the watershed and declining runoff in September. On the basis of that correlation, Drake et al. (2000) ascribed the fall runoff shifts to declines in the water content of the April snowpack caused by climate change. They concluded that changes in land-use practices and water use were not responsible for declining flows.

The analysis by Drake et al. correlated fall flows only to two snow gages that showed declines in April snowpack. Five other gages in the basin showed no long-term changes in precipitation. As Power (2001) noted, the two stations that Drake et al. used are also invalid for comparative purposes because encroachment of forest vegetation has progressively reduced the catch of the snowpillows since their installation. Thus, it remains likely that the decline in fall flows can be attributed to changes in land cover and water-management practices in the watershed.

Cropping patterns in the Scott River valley have changed during the last 50 years (Figure 4-12A). In 1953, there were 15,000 acres of irrigated agriculture and 15,000 acres of natural subirrigation in the Scott valley (Mack 1958). Land surveys (CDWR 1965; CDWR, Red Bluff, CA, unpublished material, 1993) show that the amount of irrigated land has not changed substantially since 1953, but land use has. Grain declined from 7,000 acres in 1953 to 2,000 acres in 1991; alfalfa increased by 40% from 10,000 acres to more than 14,000 acres. Alfalfa has evapotranspiration rates that are several times greater than those of grain. Increased cultivation of alfalfa, including a tendency to seek four cuttings per year (SRCD records) rather than the traditional three, may have caused a decline in fall flows.

The change in cropping patterns is mirrored by a shift from surface diversions to irrigation wells (Figure 4-12B). CDWR records of well drilling in the Scott valley indicate a large increase in irrigation and domestic wells during the 1970s and 1990s. During the 1950s, there were about 60 domestic wells and six irrigation wells in the valley. During the 1970s, more than 300 domestic wells and 100 irrigation wells were drilled in the valley. That shift from surface diversions to wells increased the amount and reliability of water for irrigation. Because of the high specific capacity of shallow aqui-

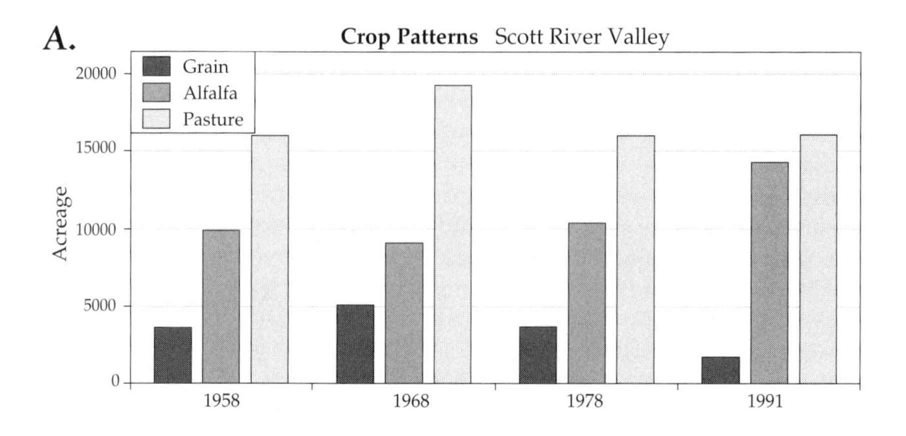

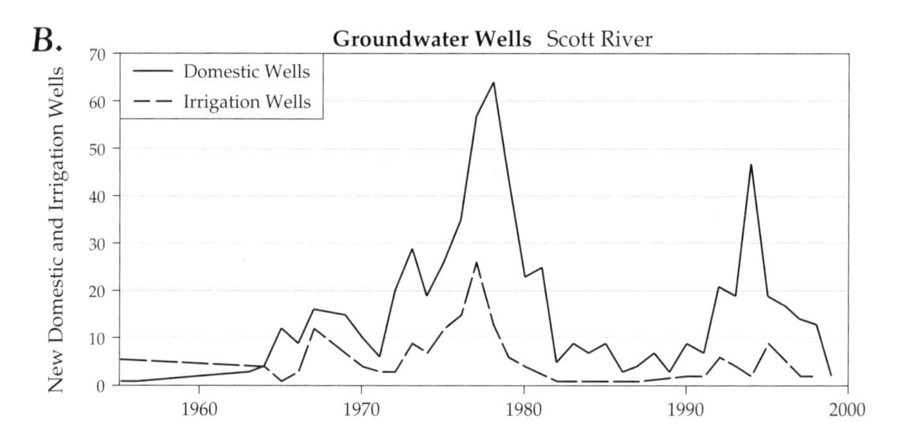

FIGURE 4-12 Changes in cropping and water wells in the Scott Valley. A) Increase in alfalfa production from 1958 to 1991. Sources: citations in text. B) New domestic and irrigation wells recorded in the Scott Valley from 1954 to 1999, showing increase in well-drilling activity in the 1970s. Source: Data from CDWR records, provided by K. Maurer.

fers in the Scott basin, pumping also decreased the contribution of shallow groundwater to base flow in the Scott River.

Water temperatures of the Scott River in July through September exceed thresholds for chronic and acute stress of coho and other salmonids (Figure 4-13). Ambient air temperature is the primary control on maximum weekly average temperature (MWAT)—warmest 7-day period for 1995–2000—of the main stem during summer and early fall (SRCD 2001).

MWAT increases downstream along the main stem of the Scott River because of the long hydraulic residence time of summer flow (Figure 4-13). Local cooling of main-stem temperatures is associated with augmentation

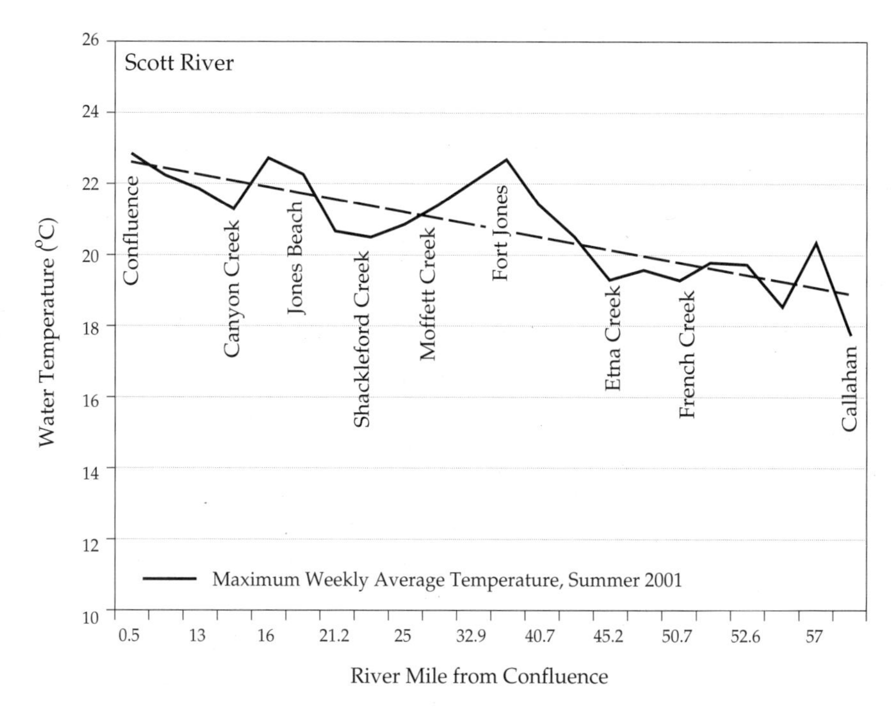

FIGURE 4-13 Plot of downstream changes in maximum weekly average water temperature on the main stem of the Scott River during summer. Note the irregular pattern of change in temperature, presumably associated with accretions from groundwater and the effects of irrigation return flows. Source: Modified from SRCD 2001.

of baseflow by shallow groundwater. Local warming of the Scott is associated with reaches of the river where water loss and tailwater return flows occur, but the current monitoring program is not capable of resolving heat flux.

Dissolved oxygen of the Scott River has been monitored sporadically. Dissolved-oxygen data are available from 1967 to 1979 at Ft. Jones (Earthinfo, Inc. 1995) and from 1961 to 1967 and 1984 (CDWR 1986). The lowest concentrations of oxygen occur during late August and early September, when flows are low and temperatures are high. The data suggest that problems with low concentrations of dissolved oxygen, if any, are limited temporally and spatially.

Extensive, locally driven efforts are under way in the Scott Valley to address the decline in water quality, and in salmonid spawning and rearing habitat. These efforts are led by the SRCD and the local Watershed Council, with cooperation from state and federal agencies, and have been well funded through aggressive grant acquisitions. Only a handful of these efforts have monitoring programs that allow assessment of their effectiveness, and there appears to be no independent review of the restoration and monitoring programs. More importantly, these efforts have yet to address comprehensively water budgets and water uses, including the contribution of groundwater to surface flows and water quality. Until a comprehensive water budget is developed, significant progress at restoring coho and other salmonids is unlikely to occur.

THE SALMON RIVER (RM 62)

Within the lower Klamath watershed, the Salmon River remains the most pristine tributary; it has a natural, unregulated hydrograph, no significant diversions, and limited agricultural activity. Although it is not well documented, runs of all the remaining anadromous fishes in the Klamath watershed (Chapter 7, Table 7-1) occur in the Salmon River (Moyle et al. 1995, Moyle 2002).

The Salmon River's unique characteristics stem from its mountainous terrain and public ownership of land. At 750 mi^2, the Salmon River is the smallest of the four major tributary watersheds in the Klamath basin. Even so, the annual runoff from the Salmon is twice that of the Scott and 10 times as great as that of the Shasta River. High runoff reflects the steep slopes and high annual precipitation (50 in) of the watershed. Runoff in the basin is dominated by a winter pulse associated with high rainfall and a spring snowmelt pulse from April through June (Figure 4-14). During summer and late fall, low-flow conditions predominate, particularly in smaller tributaries. Unlike the Scott and Shasta, the Salmon River watershed is almost entirely federally owned (Chapter 2).

The Salmon River watershed supports about 140 mi of fall-run Chinook spawning and rearing habitat and 100 mi of coho and steelhead habitat (CDFG 1979a). Logging roads, road crossings, and frequent fires in the basin appear to contribute to high sediment yields. Historical and continuing placer mining has reduced riparian cover and disturbed spawning and holding sites in the basin as well. Increased water temperatures have been noted in the Salmon River during late-summer low-flow periods, but their cause is unclear; they may be natural or may be in part a byproduct of logging and fires. The high summer temperatures may also be in part a function of the orientation of the watershed and naturally low base flow during late summer (Kier Associates 1998).

THE TRINITY RIVER (RM 43)

The Trinity River has the largest tributary watershed in the lower Klamath basin (2,900 mi^2). The watershed extends up to 9,000 ft in the

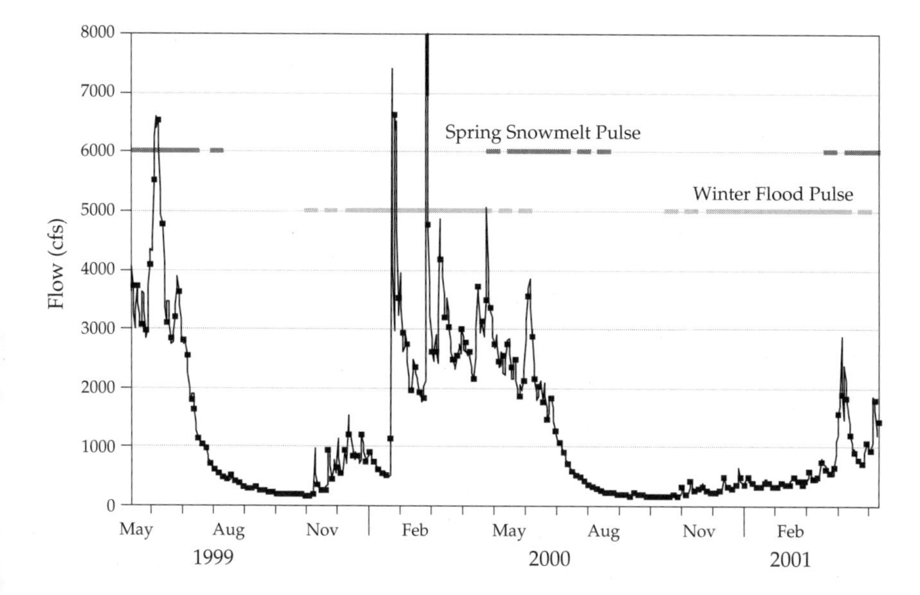

FIGURE 4-14 Annual hydrograph of the Salmon River at Somes Bar, California, May 1999–May 2001.

Trinity Alps and the Coast Ranges and flows more than 127 mi to its confluence with the Klamath at 230 ft asl, 43 mi above the Klamath River mouth (Figure 4-15). It is the largest contributor of tributary flow to the main-stem Klamath. Prior to construction of the Trinity River Diversion (TRD), the Trinity River accounted for close to one-third of the average total runoff from the Klamath watershed (based on USGS gaging records)—more than twice the runoff from the entire upper basin.

Hydrologically, the Trinity watershed is broadly similar to the Scott and Salmon watersheds. Prior to construction of the Trinity River Diversion (TRD) project in 1963 (discussed below), runoff averaged close to 4.5 MAF annually. The bulk of this runoff was concentrated into two seasonal pulses (Figure 4-16)—winter floods associated with mixed rain-snow events that typically occur between mid-December and mid-March, and a spring snowmelt pulse that begins in late March–early April and, depending upon snowpack conditions, ceases in July. The summer and fall are dominated by baseflow conditions. Historically, late summer and early fall flows on the Trinity were quite low, indicating limited natural baseflow support. During years of below-average moisture, tributaries to the Trinity commonly dry up.

Precipitation patterns and associated runoff vary considerably throughout the Trinity watershed. Precipitation averages 57 in. annually, but approaches nearly 85 in. in the Hoopa Mountains and the Trinity Alps. In the

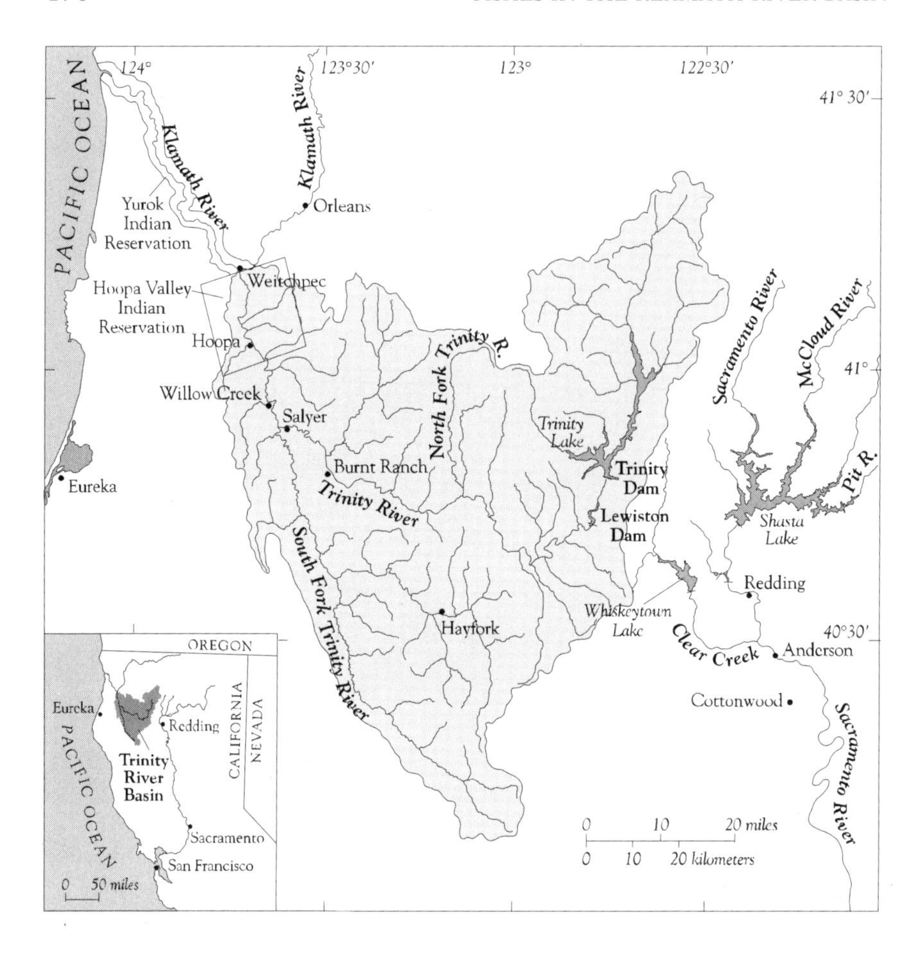

FIGURE 4-15 Index map of the Trinity River watershed. Source: Modified from USFWS/HVT 1999.

high-altitude, northeastern portions of the watershed, the annual hydrograph is dominated by snowmelt runoff during the spring and early summer. In contrast, the lower-elevation watersheds, such as the South Fork and North Fork, are dominated by winter rainfall flood pulses.

As noted in Chapter 2, the tectonic, geologic, and climatic setting of the Trinity River has amplified the influence of land-use activities on fish. Highly unstable rock types, which are associated with the Coast Range Geologic Province on the west and the Klamath Mountains Geologic Province on the east, coupled with high rates of uplift, lead to naturally high erosion rates (Mount 1995). Like the western portions of the Scott watershed, the eastern portions of the Trinity watershed contain deeply weath-

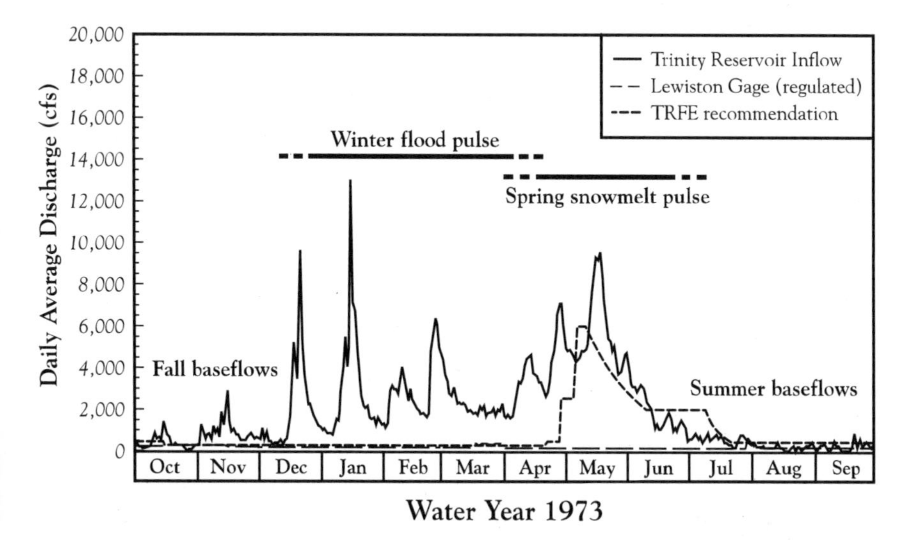

FIGURE 4-16 Example of regulated (dotted line, current recommended outflow) and unimpaired (solid line, inflow to Trinity Diversion Project) flows on the Upper Trinity River for water year 1973, a normal water year (40–60% exceedance probability for annual flow volume). Source: Modified from USFWS/HVT 1999.

ered granitic rocks that yield highly erodible soils dominated by decomposed granite. In both the eastern and western portions of the watershed, highly unstable metamorphic rock units are associated with numerous and widespread slope failures. Landslides play a dominant role in hillslope evolution on the South Fork Trinity and in canyon reaches of the main stem.

Approximately 80% of the Trinity watershed is federally owned and is managed by USBR and USFS. The remainder is a mix of private ownership and lands within the Hoopa Valley and Yurok Indian reservations. Land-use practices on public and private land within the Trinity watershed have played a central role in the precipitous decline of salmon runs in the latter half of the 20th century.

As with most tributary watersheds of the Klamath system, logging, mining, and grazing have reduced the quantity and quality of salmon habitat in the Trinity watershed. The greatest effects have occurred in the South Fork of the Trinity and on the main stem below Lewiston Dam and above the confluence of the main stem with the North Fork.

The South Fork is the largest tributary of the Trinity River, and was historically a significant producer of Chinook and coho salmon and steelhead trout (Pacific Watershed Associates 1994). The South Fork and its main tributary, Hayfork Creek, comprise 31% of the Trinity watershed and 6% of the total Klamath watershed. The South Fork, which is undammed, is the largest unregulated watershed in California. Currently, more than 56 mi of the river are protected under the California Wild and Scenic Rivers Act.

The South Fork has high background sedimentation rates, but intense logging in the 1960s on highly unstable soils, coupled with a large storm in 1964, produced sedimentation rates significantly above background levels. Adverse effects of sediment on aquatic life caused EPA to require a total maximum daily load (TMDL) study for sediment in the South Fork (EPA 1998). Loss of riparian cover and deep pools also appears to have affected water temperature.

Most regional and national attention has been focused on the main stem of the Trinity River. Mining, logging, and grazing practices within this portion of the watershed contributed high volumes of sediment to the main stem and degraded habitat prior to creation of the TRD (EPA 2001). Logging on sensitive soils produced high loads of fine sediment in the mainstem Trinity. Prior to TRD operations, however, seasonal high flows associated with the winter and spring flood pulses appear to have maintained habitat of reasonable quality, thus preventing a significant decline in steelhead and salmon (McBain and Trush 1997).

In 1955 Congress authorized construction of the TRD project to divert water from the upper Trinity River into the Sacramento River as part of the Central Valley Project (CVP). The primary beneficiaries of these diversions are farms of the San Joaquin Valley serviced by the Westlands Water District. The TRD consists of two dams: the Trinity Dam, which has an impoundment capacity of 2.4 MAF, and Lewiston Dam, which impounds Lewiston Reservoir and provides the diversion for the CVP.

The closure of Lewiston Dam in 1963 led to loss of access to spawning sites and degradation of habitat. Located at Trinity RM 112, Lewiston Dam currently blocks access to more than 109 mi of potential spawning habitat in the upper watershed (USFWS 1994). Additionally, the Trinity and Lewiston Dams trap all coarse sediment that would normally be supplied by the upper watershed.

When completed, the TRD diverted more than 88% of the annual runoff from the upper watershed to the CVP. After 1979, these diversions were decreased to 70% of the annual runoff. The magnitude of the diversions and associated flow release schedules eliminated winter and spring flood pulses in the main stem of the Trinity (Figure 4-16). The effects of these manipulations are most acute between Lewiston Dam and the North Fork Trinity (RM 112-72). Below the North Fork, tributary flow and sediment supply reduce the adverse effects of upstream water management (USFWS/HVT 1999).

Changes in hydrology on the Trinity River, loss of sources of coarse sediment, and continued influx of fine sediment from hillslope erosion have created significant changes in habitat conditions downstream of the TRD. Channel response to changes in flow regime included reductions in cross section, reduction in lateral migration, establishment of riparian vegetation on channel berms, loss of backwater habitat, and loss of spawning gravel. The new channels have been static, reduced in size, and deficient in suitable habitat.

In 1981 the Secretary of the Interior authorized a Trinity River Flow Evaluation (TRFE) study of ways to restore the fishery resources of the Trinity River (USFWS/HVT 1999). The final TRFE report recommends releases from TRD based on five water-yr types: extremely wet, wet, normal, dry, and critically dry. The hydrographs consistent with these recommendations still allow for delivery of water to the CVP, but shape the hydrographs so that they support the life-history needs of salmonids, including reintroducing disturbance to control establishment and growth of riparian vegetation, coarse sediment transport to establish pools and riffles and to clean spawning gravels, and sufficient flows to reduce water temperatures for rearing. The TRFE also contained an adaptive management approach that calls for assessment of the effect of changes in flow regime and adjustments as necessary to improve the success of the program.

The TRFE and the associated federal environmental impact statement (EIS) and environmental impact report (EIR) were the product of multiple years of collaborative effort on the part of agencies and stakeholder groups. This program was subjected to rigorous external peer review, which led to numerous, substantive revisions in proposed remediation measures. The TRFE was used in the Department of the Interior's Record of Decision (ROD; Trinity River Mainstem Fishery Restoration, USFWS 2000). A lawsuit filed by the Westlands Water District in 2001 contended, however, that the underlying studies did not adequately address the economic impacts of the CVP water on users and electricity consumers, and failed to account for the effects of changes in flow on ecosystems of the Sacramento-San Joaquin Delta. In 2001, U.S. District Court Judge Oliver Wanger ruled against the Department of the Interior (DOI) and ordered it to complete a supplemental EIS, which is still in preparation. Consequently, the recommended TRFE flow releases have not occurred. In response to the lower Klamath fish kill of September 2002, the presiding judge was asked by the Hoopa Valley Tribe to allow some operational flexibility in order to help avoid fish kills in September 2003. The judge allowed 50,000 acre-ft to be set aside for emergency increases in flow to reduce the chances of a fish kill. In August 2003, the Trinity Management Council requested that DOI allow a sustained flow release in September 2003 due to low-flow conditions and predictions of a large salmon run. As of September 2003, these modifications in flow were under way.

Given the size of the Trinity River watershed and its large amount of runoff, the operations of the TRD must affect the quality of habitat in the lowermost Klamath River and its estuary. There is little published information, however, on the effects of the Trinity on the lowermost Klamath and the estuary. Information provided here is principally derived from an analysis of USGS gaging data (1951–2002) from the Trinity and the Klamath, and from the Trinity River Flow Evaluation study (USFWS/HVT 1999).

Following construction of the TRD, the contribution of the Trinity to the total flow of the Klamath River declined from 32% to approximately 26% (Figure 4-17). This decline is not equally distributed throughout the year. The largest effect of the TRD occurs in the spring, during filling of the Trinity Reservoir. Prior to construction of the TRD, snowmelt runoff from the Trinity provided approximately 290,000 acre-ft, or approximately one-third of the inflow to the estuary, to the Klamath River in June. Following construction of the TRD, the average contribution of the Trinity in June declined to 160,000 acre-ft; during this same period, inflow to the Klamath estuary declined by approximately 200,000 acre-ft per yr.

During the late summer and early fall the Trinity, prior to construction of the TRD, contributed a relatively small amount to the total flow of the Klamath River (less than 15% in September). In the period following con-

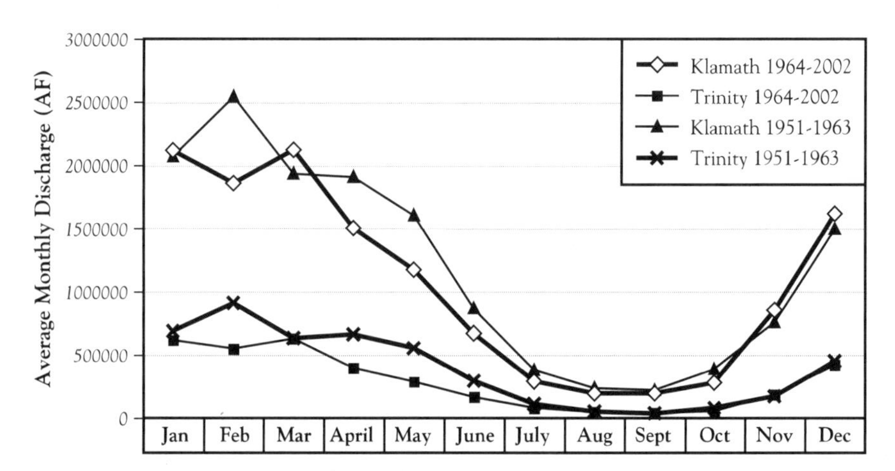

FIGURE 4-17 Average monthly discharge of the Klamath River at Klamath (USGS 11530500) and the Trinity River at Hoopa (USGS 11530000) for the period 1951–2002. The Trinity River Diversion project was constructed in 1963. Note the reduction in spring flows associated with operation of the TRD.

struction of the TRD, there was a decline of 11% in average September flow of the Klamath main stem above the Trinity. Because of minimum flow requirements for the TRD, however, average flows from the Trinity increased during this period, partially offsetting the declines in flow from Iron Gate Dam and boosting the Trinity's relative contribution to 20%.

Spring and early summer water temperatures are of concern in the lower Klamath and Trinity due to their effect on outmigrating steelhead and salmon smolts. Field and modeling studies conducted in 1992–1994 at the confluence of the Klamath and Trinity demonstrate the relative importance of flow to water temperatures (Appendix L in USFWS/HVT 1999). Although temperature differences between the Klamath and the Trinity River can be considerable (up to 5°C or more), temperature regimes usually are quite similar at the confluence because of the long distances of travel (> 100 mi) for water released from both Iron Gate Dam and Lewiston Dam, and the broadly similar release schedules of the two reservoirs. Differences between the two rivers become pronounced only when there are large disparities in flow volumes. For example, when the Trinity flow releases are very large (by a factor of 2 to 3) compared to flow within the Klamath main stem, the Trinity cools the Klamath because its waters reach the confluence more quickly than at low flow.

The Trinity River Mainstem Fishery Restoration program (USFWS 2000) is, by necessity, focused principally on restoring spawning and rearing habitat within the main-stem Trinity River. Thus, from the viewpoint of coho recovery, the EIS process cannot be expected to result in the improvements of tributary habitat that coho require. Also, the program does not appear to have invested significant effort in evaluating its beneficial effects on the lower Klamath and its estuary. With the exception of the participation of the Hoopa Valley and Yurok Tribes, there also appears to be only minimal effort to coordinate management of the Trinity watershed with efforts to manage the rest of the Klamath watershed. The proposed flow release schedule contained within the 2001 ROD, which is currently held up in litigation, may, however, provide substantial benefit downstream of the Trinity, thereby increasing the welfare of salmon and steelhead throughout the Klamath watershed.

MINOR TRIBUTARIES TO THE LOWER KLAMATH MAIN STEM (RM 192-0)

Many small tributaries enter the main-stem Klamath between Iron Gate Dam and the mouth of the river. They drain mountainous, largely forested watersheds, but most are creeks affected to some degree by logging, past mining, grazing, and agriculture. In many of the tributaries along the stream corridors, water withdrawal leads to reductions in summer base flows.

Water quality has not been extensively studied, but the tributaries may be particularly important in providing cold-water habitats for salmonids (Chapter 7). Of these creeks, 47 are known to have coho populations (NMFS 2002), but little is known about the specific conditions of these populations in relation to habitat and changing conditions in the basin.

In the more mountainous sections of the basin, slopes are steep, soils are unstable, and streams are affected by erosion that is exacerbated by roads and disturbance in the riparian zone. Large floods that have occurred about once per decade also have led to erosion, debris jams, and aggradation of sediments where tributaries enter the Klamath. In some cases, the bars, which consist of aggraded sediments, block flow during low-flow conditions, thus preventing fish passage, but many of the blockages have been removed in recent years (Anglin 1994).

MAIN-STEM KLAMATH TO THE PACIFIC (RM 60-0)

Over its final 60 mi the Klamath flows first southwest from Orleans to Weitchipee, where the fourth major tributary, the Trinity River, enters at RM 43. The Klamath then flows northwest to the ocean. The estuarine portion of the Klamath River is relatively short in relation to the watershed. Because intrusion of salt water varies seasonally, the length of the estuary is variable. The greatest intrusions occur at low flow, but brackish water (15–30 ppt) extends only a few mi upriver even at low flow (Wallace and Collins 1997). Tidal amplitudes in the estuary vary up to 2 m.

Flows in the lowermost Klamath are driven by a seasonally varying mixture of main-stem flow and accretions of water from tributaries. For example, water reaching the river via the Iron Gate Dam contributes less than 20% of the flow at Orleans in May and June (1962–1991). The other 80% of the flow is derived primarily from tributaries. The percentage of flow that comes from Iron Gate Dam increases over the summer. In September, over 60% of the flow originates from Iron Gate Dam (Hydrosphere Data Products, Inc. 1993). As noted above, the Trinity River and operations of the TRD exert substantial influence over hydrologic conditions of the lower Klamath and its estuary. Changes in release, even under the new ROD, have led to declines in late winter through early summer flows at the mouth of the Klamath. Fall flows, on the other hand, are augmented by increased flows from the Trinity.

Although alteration of hydrographs in a number of headwaters and tributaries has been quite substantial (e.g., Lost River, Shasta River), the overall effect of water development on total annual flow of the downstream reaches of the Klamath River is surprisingly small. Runoff from the upper Klamath basin has been reduced from approximately 1.8 million acre-ft to 1.5 million acre-ft in a year of average moisture (USGS 1995, Hardy and

Addley 2001, Balance Hydrologics 1996), and irrigation has depleted the mean annual flow at Orleans (above the Trinity), where the flow is approximately 6 million acre-ft, by less than 10%. There has been a noticeable shift in the timing of runoff, however. Peak annual runoff occurs in March instead of April and the flows of late spring and early summer tend to be lower than they were historically. In late summer, water temperatures at Orleans exceed 15°C typically from June into September (Figure 4-18). River temperatures in excess of 20°C occur on most dates in July and August and in many years, high temperatures extend into fall. For example, temperatures over 18°C have been observed in late October. Temperatures in the Klamath may have always been high (over 15°C) in summer and fall, but it is likely that the loss of cold water from tributaries has resulted in a net increase in temperatures over the annual cycle, particularly during summer under either normal or low-flow conditions.

Even though hydrologic change in the lowermost Klamath main stem seems too small to have caused large changes in the estuary, significant impairment of the estuary could have occurred through warming of the river water and through increased organic loading caused by eutrophication and alteration of flow regimes in headwaters. The estuary could show adverse chemical conditions as a result of these changes, and coho in the estuary thus could be affected. The extent of these changes and their potential effect on coho have not been well documented, however. Information on water quality of the lowermost Klamath River is sparse.

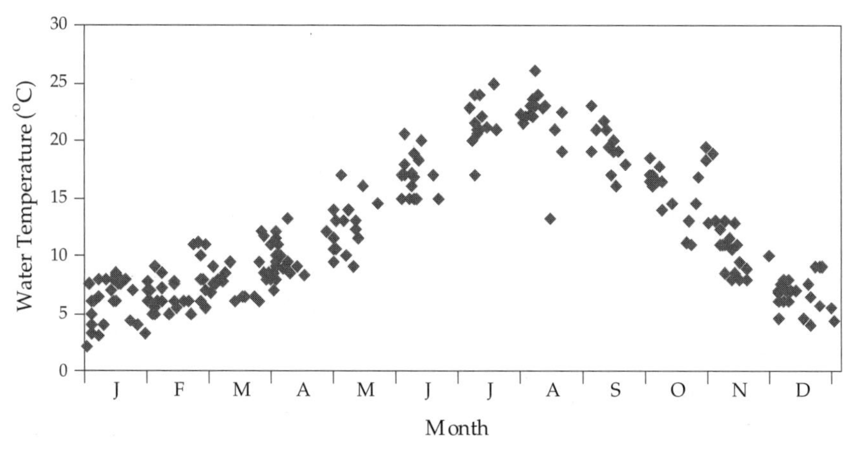

FIGURE 4-18 Water temperature (instantaneous daytime values) of the Klamath River at Orleans based on observations at USGS station 18010209, 1957–1980, plotted on a single annual time span.

CONCLUSIONS

Most flowing waters of the Klamath basin show substantial environmental degradation involving loss of coarse gravels, excessive suspended sediment, impaired channel morphology, loss of woody riparian vegetation, major alteration of natural hydrographic features, and excessive warmth. These changes affect not only the main stems of the Klamath River and major tributaries, but also small tributaries where salmon are or could be present. While to some extent historical, degradation continues through a variety of water-management and land-use practices including irrigation, grazing, mining, and timber management. Documentation is poor for some locations, and especially so for small tributaries.

In the upper basin, the tributaries that drain into Upper Klamath Lake are poorly understood except in regard to nutrient transport. Knowledge of basic hydrology and water use is sparse, as are conditions relevant to spawning of listed suckers and refugia for sucker fry. Topics of special interest include substrate and channel quality, sediment load, and status of riparian vegetation. In the lower basin, research has documented extensive modifications of riparian habitats, especially along the Scott and Shasta rivers. Adverse changes in stream-channel structure, sediment transport, flow, and temperature are commonplace even on federal lands.

Nutrients, dissolved oxygen, temperature, flows, and physical habitat of the main stem of the Klamath River have been extensively studied. Still, additional research that would clarify the interactions between hydrology and temperature, especially as affected by water-management strategies, is needed. Considerable research on this topic is in progress, but field investigations have focused primarily on the river between Iron Gate Dam and Orleans. Conditions in the lowermost reaches of the Klamath River, including the estuary, have received less attention but are important to salmonids, as shown by the mass mortality of salmonids in 2002 (Chapter 7).

The Klamath system as a whole is nutrient-rich and productive. High concentrations of phosphorus, a key nutrient, are typical of Klamath waters because of natural sources. Anthropogenic sources may be important in some cases as well. Water-quality conditions, except temperature, are within satisfactory bounds in most cases for flowing waters. The greatest impairments involve physical features, including temperature for salmonids.

5

Fishes of the Upper Klamath Basin

The upper Klamath basin is an ancient, isolated, and unusual environment for fish. Most, or possibly all, of the native species that live in the upper basin are endemic to it. The distinctiveness of the upper basin and its fishes has been recognized since the first ichthyologists explored it in the late 19th century (Cope 1879, Gilbert 1898), but the application of new kinds of genetic analysis to the fishes is revealing even more diversity and complexity than was previously known (e.g., Docker et al. 1999, Tranah 2001).

Since the shortnose and Lost River suckers were listed as endangered species in 1988, a great deal of attention has been paid to their biology, especially in Upper Klamath Lake, whereas the rest of the species and the rest of the basin have received comparatively little attention. The other endemic fishes, some of which may be considered for listing in the future, interact with the endangered suckers and thus complicate management practices intended to benefit them. In addition, nonnative fishes, which are abundant in the basin, affect the endangered suckers. Overall, the upper basin's land and water should be managed through an ecosystem-based approach with all native fishes in mind under the assumption that management favoring native fishes is likely to have positive effects on other ecosystem components. Failure to do this is likely to result in listing of additional species as threatened or endangered. The purposes of this chapter are to describe the factors that led to the high endemism of fishes in the upper Klamath basin and to its invasion by nonnative fishes, to give a brief summary of the biology and welfare of each of the native fishes with special attention to the listed suckers, an overview of the nonnative fishes, and to identify gaps in knowledge about all the fishes.

NATIVE FISHES

The fishes of the upper Klamath basin originated when a large river draining the western interior of North America flowed through the Klamath region on its way to the ocean (Minckley et al. 1986, Moyle 2002). Uplift and erosion have since caused the water in the region to flow at different times into the Great Basin to the east, into the Columbia River via the Snake River to the northwest, into the Sacramento River via the Pit River to the south, and into the lower Klamath River to the west. As the connections to large drainage basins shifted back and forth, fishes from each of the basins entered the upper Klamath basin (Minckley et al. 1986, Moyle 2002). Species that persisted through periods of change, which included drought and volcanism, evolved into the endemic fauna of the upper Klamath basin (Table 5-1). These fishes are adapted to the shallow lakes, meandering rivers, and climatic extremes of the upper Klamath basin. The closest relatives of modern fishes of the upper Klamath basin are now found in the Great Basin, Columbia River, Pit River, and lower Klamath River.

The present connection of the upper Klamath basin to the lower Klamath basin probably is fairly recent (Pleistocene, less than 1.8 million years BP), but the connection formed and was blocked more than once. Connection of the upper and lower basins led to colonization of the upper basin by anadromous Chinook salmon, steelhead, and Pacific lampreys. Repeated isolation of anadromous fishes, which occurred when the connection between the two parts of the Klamath basin was broken, left behind resident populations that now differ from parent stocks (such as redband trout and Klamath River lamprey).

The lower basin contains mainly fast-flowing, cool-water rivers and streams that are ideally suited for anadromous fishes but inhospitable to fishes of the upper basin, which are adapted to lakes or warmer streams and rivers of lower gradient. Thus, the two basins have remarkably different fishes. The absence of major physical barriers to movement of fish before installation of dams explains the former use of the upper basin by anadromous fishes and the apparent occasional entry into the upper basin of the Klamath smallscale sucker, which is abundant in the lower basin.

Only five families of fishes—Petromyzontidae, Cyprinidae, Catostomidae, Salmonidae, and Cottidae—are native to the upper basin, and the species in these families have many unusual adaptations to the environment of the basin. The lampreys and suckers of the upper basin show some interbreeding (hybridization) among species.

TABLE 5-1 Native Fishes of the Upper Klamath Basin

Species	Adult Habitat[a]	Status[b]	Comments
Pacific lamprey, *Lampetra tridentata*	R, L	C?	Same species as in lower river but land-locked and probably distinct
Klamath River lamprey, *L. similis*	R	C	Also in lower river
Miller Lake lamprey, *L. milleri*	R, L, W	U	Once thought extinct
Pit-Klamath brook lamprey, *L. lethophaga*	W, C	C?	Shared with Pit River
Klamath tui chub, *Siphatales bicolor bicolor*	L, R, W	A	Abundant and widespread
Blue chub, *Gila coerulea*	R, W	C	Special concern species in California
Klamath speckled dace, *Rhinichthys osculus klamathensis*	W, C, R, L	C?	May be more than one form
Shortnose sucker, *Chasmistes brevirostris*	L, R	L	Listed as endangered
Lost River sucker, *Deltistes luxatus*	R, L	L	Listed as endangered
Klamath largescale sucker, *Catostomus snyderi*	R, L, W	C?	May be more than one form; declining?
Klamath smallscale sucker, *C. rimiculus*	R, W, C	R	Common in lower basin
Klamath redband trout, *Oncorhynchus mykiss* subsp.	R, L	C?	Fishery; may be more than one form: lake and stream
Coastal steelhead, *O. mykiss irideus*	R, C	E	Anadromous, common in lower basin
Chinook salmon, *O. tshawytscha*	R	E	Anadromous, common in lower basin
Bull trout, *Salvelinus confluentus*	C	L	Threatened species
Upper Klamath marbled sculpin, *Cottus klamathensis klamathensis*	C, W, R	C	Widespread in basin
Klamath Lake sculpin, *Cottus princeps*	L, R	A	Abundant in Upper Klamath Lake
Slender sculpin, *Cottus tenuis*	L, R	R	Gone from much of former range

[a]Adult habitat: L, lakes; R, river; W, warm-water creeks; C, cold-water creeks.
[b]Status in upper basin: A, abundant; C, common; E, extirpated; L, listed under federal Endangered Species Act; R, rare; U, unknown.

Petromyzontidae: Lampreys

The lampreys of the upper Klamath basin are all derived from anadromous Pacific lampreys that became land-locked, perhaps multiple times over millions of years. Their evolution and ecology are poorly understood. Four species are recognized (Docker et al. 1999, Lorion et al. 2000, Moyle 2002), but additional species may be uncovered as genetic studies proceed. There are two basic life cycles among lampreys: one with predatory adults and one with nonpredatory adults. Both types spend the first 3–7 yr of their lives living in mud and sand as eyeless, wormlike larvae (ammocoetes) that feed on algae and organic matter. The ammocoetes metamorphose into silvery, eyed adults. Adults of the predatory forms attach to other fish with their sucking-disc mouths, through which they remove blood and body fluids. Typically the prey survives the attack of a predatory lamprey, but the attack may impair growth and survival (Moyle 2002). Predatory lampreys engage for about a year in this feeding behavior, which enables them to grow to produce a larger number of gametes than do nonpredatory lampreys. The adults of the nonpredatory form do not feed; they live only long enough to reproduce (Moyle 2002).

The Pacific lamprey is regarded as a land-locked version of the predatory anadromous species, but the form native to the upper Klamath basin probably should be a separate taxon. Nothing is known about its ecological differences from the slightly smaller (14–27 cm) Klamath River lamprey. The Klamath River lamprey is a nonmigratory predatory species that is widespread in the upper and lower basins. Little is known about its biology except that it preys on native suckers and cyprinids, especially in reservoirs (Moyle 2002). The Miller Lake lamprey is the smallest (less than 15 cm) predatory lamprey known anywhere in the world; it occurs mainly in the Sycan and Williamson rivers, where resident prey species are abundant (Lorion et al. 2000). The Miller Lake lamprey is closely related to the nonpredatory Pit-Klamath brook lamprey, which is abundant and widespread in small streams in the upper Klamath and Pit River basins. Because of the long (about 1 million years) separation of the Pit and Klamath basins, genetic studies will probably show that the two populations belong in different taxa. The exact distribution of the four species in the watershed is not known, because most collections are of ammocoetes, which are difficult to identify in the field.

Cyprinidae: Minnows

The Klamath tui chub is widespread in the interior basins of the western United States and is divided into a number of subspecies (Moyle 2002). Some, including the Klamath tui chub, may eventually be recognized at the species level. Tui chubs are chunky, omnivorous minnows that can become large (about 30 cm) and have high longevity (20–35 yr), especially in large lakes. In the Klamath basin, they are the most abundant and widely distributed native fish. They occur in streams, rivers, reservoirs, and lakes (Simon and Markle 1997a,b; Buettner and Scoppettone 1991) and in a wide array of habitats (Bond et al. 1988). They are tolerant of high temperature (over 30°C), low dissolved oxygen (below 1 mg/L), and high pH (10–11; Falter and Cech 1991, Castleberry and Cech 1992, Moyle 2002). Despite those tolerances, they typically are among the most abundant species in fish kills of Upper Klamath Lake (Perkins et al. 2000b), although counts of dead chubs usually do not distinguish tui chubs from blue chubs. In the last 30 years, the tui chub has declined in abundance in the Lost River, where it has changed from a dominant member to a minor component of the fish fauna (Shively et al. 2000a).

In contrast with tui chub, the blue chub is confined largely to the Klamath basin and a few adjacent basins into which it may have been introduced (Moyle 2002). It is especially abundant in lakes, reservoirs, and other warm, still habitats (Bond et al. 1988, Buettner and Scoppettone 1991). It may be the most abundant native fish in Upper Klamath Lake, although it may also be in decline there, along with most other native fishes (Simon and Markle 1997b, Moyle 2002). It clearly is in decline elsewhere in the upper Klamath basin. For example, Contreras (1973) found that the blue chub was the most abundant species in the upper part of the Lost River but that the tui chub was the most abundant in the lower half of the river. More recent sampling indicates that both species have been largely replaced by fathead minnows, brown bullheads, and other nonnative species that tolerate poor water quality (Shively et al. 2000a). Not much is known about the biology of the blue chub except that it is omnivorous, schools, and reaches a length of about 25 cm. It is somewhat less tolerant of high temperatures and low dissolved oxygen than the tui chub (Castleberry and Cech 1992) and is common in fish kills of Upper Klamath Lake (Perkins et al. 2000b).

The speckled dace is even more widespread than the tui chub in western North America and probably consists of a complex of species (Moyle 2002). Dace from both the upper and lower Klamath basins are recognized as just one subspecies, but the two forms probably are distinct, and the upper basin probably supports more than one taxon (M. E. Pfrender, Utah State University, personal communication, 2002). The speckled dace is common in the upper basin but is most abundant in cool streams associated with rocks and gravel (Buettner and Scoppettone 1991, Bond et al. 1988). Even so, Castleberry and Cech (1992) found that it could withstand high temperatures (28–34°C) and low concentrations of dissolved oxygen (1–3 mg/L). The status of the speckled dace in the basin is not known, because collections are biased toward the larger fishes. It apparently has become very uncommon in the Lost River, however (Shively et al. 2000b).

Catostomidae: Suckers

The four species of suckers in the Klamath basin (Table 5-1) have an interesting and long evolutionary history (Moyle 2002) and probably once were the most abundant fishes, in terms of biomass, in the lakes and large rivers. The listed shortnose sucker and Lost River sucker, which have been the focus of most fish studies in the upper basin, will be treated in the last part of this chapter. The Klamath smallscale sucker is rare (or perhaps absent since the construction of Copco Dam) in the upper basin, although it is found in upper Jenny Creek, a tributary to Copco Reservoir. It is abundant in the lower basin (see Chapter 7). The Klamath largescale sucker is resident in the upper basin. All four species show some evidence of hybridization with each other (Tranah 2001).

The Klamath largescale sucker, which becomes large (about 50 cm) and has a long lifespan (31 yr or more), as do the shortnose and Lost River suckers, is one of the least understood fish in the basin (Moyle 2002). It appears to be mainly a resident of large rivers, although a small population exists in Upper Klamath Lake, and it is rare or absent in the Lost River (Koch et al. 1975, Buettner and Scoppettone 1991, Shively et al. 2000a). It apparently is common and widely distributed in the Williamson, Sprague, and Wood rivers (Reiser et al. 2001). In Upper Klamath Lake, the Klamath largescale sucker is found mainly near inflowing streams; this suggests a low tolerance for lake conditions, but it has been found at temperatures near 32°C in environments of dissolved oxygen at 1 mg/L and pH over 10 (Moyle 2002). Lake populations of largescale suckers migrate for spawning in March and April; peak spawning activity occurs a month or so earlier than that of shortnose and Lost River suckers. Radio-tagged fish have migrated as far as 128 km upstream, presumably to find gravel for spawning (Reiser et al. 2001). The Klamath largescale sucker hybridizes with the shortnose and Lost River suckers. Genetic studies by Tranah (2001) suggest that the largescale suckers in the Sprague River belong to a different taxon from other largescale suckers in the basin.

The status of the Klamath largescale sucker is poorly understood. The lake populations probably are similar to those of the shortnose and Lost River suckers in having declined in abundance. The status of stream populations is not known, although they are assumed to be widespread and abundant (Reiser et al. 2001).

Salmonidae: Salmon and Trout

The bull trout is a predatory char that is widely distributed in the northwestern United States but is considered a relict species in the Klamath basin. It apparently entered the Klamath basin when it was connected to the

Snake River but then became isolated. Genetic evidence reflects isolation and suggests that the bull trout of the upper Klamath basin could be assigned to a distinct taxon or evolutionarily significant unit (Ratliff and Howell 1992). The bull trout is known from only 10 creeks in the upper Klamath basin—four tributaries to the Sprague River, four to the Sycan River, and two to Upper Klamath Lake (Ratliff and Howell 1992, Buchanan et al. 1997)—although it has been extirpated or is at risk of extirpation in most of these creeks. An important characteristic of streams containing bull trout is high water quality; temperatures do not exceed 18°C in these streams (Moyle 2002). The bull trout tends to disappear from streams with degraded water quality even if the streams can support other kinds of trout. The bull trout also declines when the brook trout invades its habitat. Hybridization between the bull trout and the brook trout has taken place in some Klamath basin streams (Markle 1992). Threats to the existence of the bull trout are not peculiar to the Klamath basin; they occur throughout its range. Thus, the bull trout of the upper Klamath basin was listed by the U.S. Fish and Wildlife Service (USFWS) in 1998 as threatened.

The bull trout, like the endangered suckers of the upper basin, demands special attention in the future. Unlike the suckers, however, the bull trout is spatially separated from the Klamath Project and most other water management because its distribution is restricted primarily to headwaters that are remote from Upper Klamath Lake or the lower reaches of tributaries that are so important to suckers. At present a good deal of attention is being given to the welfare of bull trout, but much work remains to be done.

The redband trout is a resident rainbow trout whose ancestors entered the upper Klamath basin when it was connected to the Columbia Basin via the Snake River (Behnke 1992). Coastal rainbow trout (steelhead) later entered the upper basin, but the redband trout derived from the Columbia Basin maintained its identity and is recognizable by its morphology and color. Behnke (1992) indicates that there are two types of redband trout in the basin: a small form resident in isolated streams and the form present in Upper Klamath Lake; he suggests that the lake form is so distinctive (for example, it has large numbers of gill rakers, an adaptation to life in lakes) that it deserves subspecies designation (as *O. m. newberrii*). The Oregon Department of Fish and Wildlife (ODFW), however, regards all redband trout in the interior basins of Oregon as belonging to one taxon, even though it states that the Klamath Lake redband trout is "unique in terms of life history, meristics, disease resistance, and allozyme variation" (Bowers et al. 1999). The various stream populations in the basin also show genetic evidence of isolation from one another (Reiser et al. 2001). Regardless of taxonomic position, these fish have persisted because of their ability to thrive in lake and stream conditions that would be lethal to most salmonids.

Behnke (1992) wrote of observations he made on Upper Klamath Lake in September 1990 (p. 181): "In clear-water sections influenced by spring flows, hundreds of large, robust trout from about 1 to 5 kg could be readily observed. In shallow (2 m) Pelican Bay, in the midst of a bloom [of bluegreen algae] (I estimated a Secchi disk clarity of about 40 cm), I caught a magnificent trout of 640 mm and 2.3 kg." This is consistent with continuing reports of a strong summer fishery for trout, especially in Pelican Bay (e.g., Hoglund 2003). Water temperatures in Upper Klamath Lake in summer are 20–25°C, occasionally spiking to 27°C, and dissolved oxygen may drop below 4 mg/L for several days (Perkins et al. 2000b). Springs and the mouths of streams in Pelican Bay, which apparently have higher water quality than the lake, may serve as refuges for the trout, especially during episodes of very poor water quality in the lake. Trout have been reported in the lake's summer fish kills, but the only example of mass mortality was in 1997, when about 100 large trout were found dead (Perkins et al. 2000b).

The lake population of redband trout is adfluvial; it migrates up into the Wood, Williamson, and Sprague rivers for spawning during spring. The rivers also support resident populations of these trout, as does the river below Upper Klamath Lake, mostly above Boyle Dam (Bowers et al. 1999). Isolated populations, which are genetically distinct from the Klamath Lake and river populations, exist in the upper Williamson and Sprague rivers and in Jenny Creek, which flows into Iron Gate Reservoir (Bowers et al. 1999).

Hatchery rainbow trout (coastal stock) in the past have been stocked in Klamath basin streams, and some interbreeding with native redband trout was noted. Stocking now is limited to Spring Creek, which flows into the lower Williamson River. The hatchery fish apparently have poor survival because they are not resistant to endemic disease and are not adapted to high pH (Bowers et al. 1999).

Redband trout are doing surprisingly well in the Klamath basin, considering all the changes that have taken place. The fishery for the lake and river populations is an important recreational resource. The populations of small streams are vulnerable, however, to habitat degradation by roads, grazing, and other activities. The lake and river populations will need protection from adverse water quality and nonnative species and probably would benefit from improved habitat in the rivers and improved access to upstream habitat (Bowers et al. 1999).

Cottidae: Sculpins

The sculpins are a poorly studied group in the Klamath basin despite the presence of at least three endemic species (Klamath Lake sculpin, slender sculpin, and Upper Klamath marbled sculpin). There may be additional taxa in the watershed as well (Bentivoglio 1998).

The Klamath Lake sculpin apparently is the most abundant sculpin in Upper Klamath Lake. It is caught in large numbers in the lake with bottom trawls (D. Markle, Oregon State University, personal communication, 2001) and in smaller numbers with beach seines and trap nets (Simon and Markle 1997b). The abundance of this sculpin is estimated to be in the millions (Simon et al. 1996). It is present only in Upper Klamath and Agency lakes and in springs and creeks that flow into the west side of Upper Klamath Lake (Bentivoglio 1998). The present distribution coincides with the known historical distribution of the species. Little is known about its environmental requirements, but it lives mainly in offshore areas with bottoms of sand and silt and appears to be able to withstand widely varied lake conditions. No Klamath Lake sculpins have been reported in the fish kills of Upper Klamath Lake, but dead fish of this species would not float and so would be easy to overlook. The apparent ability of the Klamath Lake sculpin to live in conditions of poor water quality (especially low dissolved oxygen) is similar to that of prickly sculpin (*Cottus asper*) in Clear Lake of central California which, like Upper Klamath Lake, is subject to massive blooms of cyanobacteria (Moyle 2002).

The slender sculpin apparently once was common in the Williamson, Sprague, Sycan, and Lost rivers and in Upper Klamath Lake (Bentivoglio 1998). Bentivoglio (1998) collected sculpins throughout the upper basin in 1995–1996, however, and found slender sculpins only in the lower Williamson River and a few in Upper Klamath Lake. Simon and Markle (1997b) also recorded small numbers in Upper Klamath Lake. Little is known about the ecology of this fish, although it seems to require coarse substrates and high water quality; it is especially characteristic of cold springs. Its closest relative is the rough sculpin (*C. asperimmus*) of the Fall River in California (Robins and Miller 1957), which requires cold, spring-fed streams (Moyle 2002). It is fairly long-lived for a sculpin (7 yr) but is small (rarely longer than 75 mm; Bentivoglio 1998). Overall, the slender sculpin appears to have disappeared from much of its native range and is uncommon in most areas where it is found today.

The Upper Klamath marbled sculpin is the most widely distributed sculpin in the Klamath basin (A. Bentivoglio, USFWS, personal communication, 2002). It is found in most streams and rivers in the basin in a wide range of conditions, including summer temperatures over 20°C (Bond et al. 1988). It is most abundant among coarse substrates in the larger streams where water velocities are moderate to low (Bond et al. 1988). In the Lost River basin, it is known mainly from riffles in Willow and Boles Creeks (Koch et al. 1975) but has become scarce in recent years (Shively et al. 2000a). It is largely absent from the reservoirs in the basin, at least in California (data in Buettner and Scoppettone 1991), but is fairly common in Upper Klamath Lake (Simon et al. 1996, Simon and Markle 1997b). It occurs mostly on soft bottoms in the lake and apparently enters the water column to feed at night (Markle et al. 1996). It has been recorded in at least one of the fish kills of Upper Klamath Lake (Perkins et al. 2000b). The marbled sculpin, like most stream sculpins, generally hides under or among rocks, where it feeds on benthic invertebrates (Moyle 2002). Females glue their eggs to the bottoms of rocks and logs where developing embryos are tended by males until they hatch. The larvae are benthic and do not move far from their natal site. They become mature in their second summer and live 4–5 yr (Moyle 2002). The details of their ecology and life history in the upper Klamath basin have not been described.

NONNATIVE FISHES

In the last century, the upper Klamath basin has been invaded by 17 nonnative species (Table 5-2), 15 of which were introduced for sport fishing or for bait. Most of the 17 are not particularly common in the basin, but some are abundant and widespread (or are spreading), and their effects on native fishes are poorly understood. One of the most recent invaders is the fathead minnow, which is now one of the most abundant fishes in Upper Klamath and Agency lakes (Simon and Markle 1997a). The Sacramento perch, which was introduced into Clear Lake in the 1960s, has the potential to become very abundant in other lakes of the basin (Moyle 2002). Other introduced species—especially yellow perch, brown bullhead, and pumpkinseed—are locally abundant, especially in reservoirs and sloughs or ponds (Buettner and Scoppettone 1991, Simon and Markle 1997b). Brook trout, brown trout, and nonnative strains of rainbow trout are common in coldwater streams and have replaced native redband trout and bull trout in many areas. One concern is that future changes in water quality in the basin may promote further expansion of nonnative species.

The fathead minnow, which is native to eastern North America, appeared in the Klamath basin in the early 1970s, perhaps as a result of release of fish used in bioassay work (Simon and Markle 1997a). By 1983, it was common in Upper Klamath Lake and by the early 1990s it had spread to the Lost River system (Simon and Markle 1997a, Shively et al. 2000a). It was collected in the lower Klamath River in 2002 (M. Belchik, unpublished memo). Fathead minnows often are the most abundant species at sampling sites. Their effects on other fishes are not well understood, although declines in catches of tui chub and blue chub have been associated with their ascendance.

The Sacramento perch is native to central California, where it has largely disappeared from its native habitats. It survives mainly when introduced into alkaline waters outside its native range (Moyle 2002). It was introduced by the California Department of Fish and Game into Clear Lake in the 1960s and spread throughout the Lost River and into the Klamath

TABLE 5-2 Nonnative Fishes of the Upper Klamath Basin

Species	Adult Habitat[a]	Status[b]	Comments
Goldfish, *Carassius auratus*	L, R, P	U	Locally common
Golden shiner, *Notemigonus chrysoleucas*	L, R, P	R	Bait fish
Fathead minnow, *Pimephales promelas*	L, P	A	Probably still spreading
Brown bullhead, *Ameiurus nebulosus*	P, L, W	A	Widespread
Black bullhead, *A. melas*	P, L	U	Localized populations
Channel catfish, *Ictalurus punctatus*	L, R	?	May not be established
Kokanee, *Oncorhynchus nerka*	L	U?	Localized populations?
Rainbow trout, *O. mykiss*	L, R, C	C	Widely planted, hatchery strains
Brown trout, *Salmo trutta*	C, R	C	—
Brook trout, *Salvelinus fontinalis*	C	U	Localized in headwaters
Sacramento perch, *Archoplites interruptus*	L, P, R, W	C	Spreading
White crappie, *Pomoxis annularis*	L, R	U	Abundant in a few reservoirs
Black crappie, *P. nigromaculatus*	L, P	U	Recorded in Lost River
Green sunfish, *Lepomis cyanellus*	P, W	C	Widespread in reservoirs
Bluegill, *L. macrochirus*	P, W	U	Locally abundant
Pumpkinseed, *L. gibbosus*	L, R, P	C	Widespread
Largemouth bass, *Micropterus salmoides*	P, L, R	C	Common in reservoirs
Yellow perch, *Perca flavescens*	L, R, P	A	Abundant in large reservoirs

[a]Habitats are listed in order of importance for each species: C, cold-water streams; L, lakes; P, ponds and reservoirs; R, rivers; W, warm-water streams.
[b]Status in upper basin: A, abundant; C, common; R, rare; U, uncommon.

River downstream to Iron Gate Reservoir (Buettner and Scoppettone 1991). It is not particularly abundant in most areas where it is present. It has not yet established itself in Upper Klamath Lake. If it does colonize Upper Klamath Lake, it will probably become abundant there, as it has in other shallow lakes (Moyle 2002). It feeds primarily on insect larvae (especially midges), but adults can be piscivorous (Moyle 2002).

ENDANGERED SUCKERS OF THE KLAMATH BASIN

All four native sucker species of the Klamath basin are endemic. The endangered Lost River sucker and shortnose sucker are part of a species group of suckers that are large, long-lived, late-maturing, and live in lakes but spawn primarily in streams; collectively, they are commonly referred to as lake suckers. Lake suckers populated much of the Snake River, Great Basin, and Lahontan Basin region (Miller and Smith 1981, Scoppettone and Vinyard 1991). Present-day species in the genus *Chasmistes* include not only the shortnose sucker (*C. brevirostris)* but also the cui-ui (*C. cujus*) of Pyramid Lake, Nevada; the June sucker (*C. liorus*); and a species that recently became extinct, the Snake River sucker (*C. muriei*) of Wyoming. Lost River suckers and shortnose suckers (Figure 5-1) are closely related to the more speciose and widely distributed sucker genus *Catostomus*; some recent taxonomic treatments place Lost River suckers in this genus (e.g., Moyle 2002).

The lake suckers differ from most other suckers in having terminal or subterminal mouths that open more forward than down, an apparent adaptation for feeding on zooplankton (small swimming animals) rather than suctioning food from the substrate (Scoppettone and Vinyard 1991). Zooplanktivory can also be linked to the affinity of these suckers for lakes, which typically have greater abundances of zooplankton than do flowing waters.

Historically, Lost River suckers and shortnose suckers occurred in the Lost River and upper Klamath River and their tributaries, especially Tule

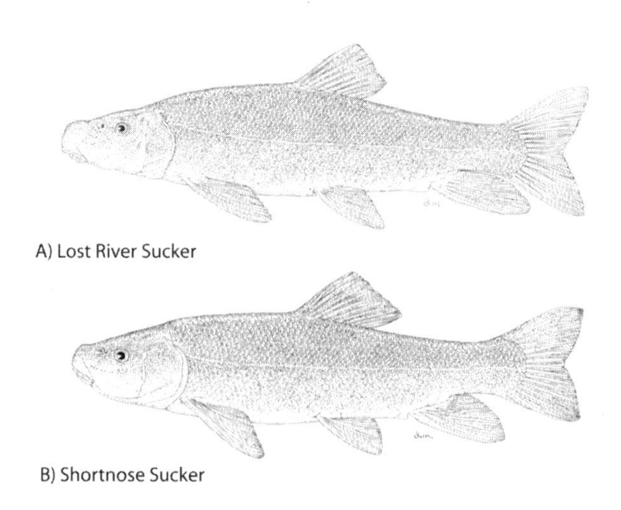

FIGURE 5-1 Endangered suckers of the Klamath River basin. (A) A Lost River sucker from Clear Lake; (B) a shortnose sucker from Clear Lake. Source: Moyle 2002, pp. 199, 203. Drawings by A. Marciochi. Reprinted with permission; copyright 2002, University of California Press.

Lake, Upper Klamath Lake, Lower Klamath Lake, Sheepy Lake, and their tributaries (Moyle 2002; USFWS 2002, Appendix D). Their current distribution (Table 5-3; Figures 5-1 and 5-2) reflects a combination of local extirpations and redistribution through water management. Suckers no longer occur in Lower Klamath Lake or Sheepy Lake, which were extensively drained in the 1920s; the populations in Tule Lake apparently do not reproduce successfully. Juveniles of Lost River and shortnose suckers have been found in much of the Lost River, but they probably originate in Miller Creek (Shively et al. 2000a). An additional population, probably consisting of shortnose suckers, was extirpated from nearby Lake of the Woods, Oregon, in 1952 when government agencies poisoned the lake to remove potential competition with trout (53 Fed. Reg. 27130 [1988]). The endangered suckers also are found in the main-stem reservoirs of the Klamath irrigation project (Chapter 3; Figure 1-4), but these populations appear to be nonreproducing (Desjardins and Markle 2000, USFWS 2002). Reproducing populations exist in Clear Lake and perhaps the Lost River. Shortnose suckers also have a reproducing population in Gerber Reservoir (Moyle 2002, USFWS 2002).

Accounts of sucker distribution often are complicated by difficulties in distinguishing species, especially when the fish are young. Lost River suckers and shortnose suckers are partly distinguished from Klamath largescale suckers and Klamath smallscale suckers by greater maximum size. The Lost River sucker can be 26–40 in. long, the shortnose sucker no longer than 21 in., the Klamath largescale sucker no longer than 18 in., and the Klamath smallscale sucker, a poorly studied species, at least 18 in. The Lost River sucker differs from the shortnose sucker and the Klamath largescale sucker with respect to some anatomical features of the head, mouth, lips, gill rakers, and body shape (Cunningham et al. 2002); it can generally be distinguished by its longer head and narrower, smaller mouth (see Figure 5-1).

The life histories of Lost River suckers and shortnose suckers are in some ways similar to those of anadromous salmon. Salmon spawn in freshwater and live most of their lives at sea before returning to their natal (birth) rivers to spawn and die. Similarly, the adults of the endangered suckers commonly ascend from lakes to rivers to spawn, the eggs hatch in gravel, and the larvae float or swim downstream to lakes, where they grow and mature before returning to rivers or springs to spawn. Unlike salmon, lake suckers spawn repeatedly. It is not known which individuals return consistently to their natal rivers to spawn, but at least 50% do return at least one time to a river in which they have previously spawned (Cunningham et al. 2002). There are many exceptions to these generalizations. For example, some individuals or subpopulations spawn in lakes, whereas others live their entire lives in rivers or streams. The repeated spawning of the

TABLE 5-3 Current and Former Distribution of Adult Lost River Suckers and Shortnose Suckers in the Klamath Basin

Habitats[a]	Map Code	Lost River Suckers	Shortnose Suckers	Reference
Upper Klamath Lake		+	+	Moyle 2002
Peripheral Springs				
Boulder Springs	1	Spawn	Spawn	Hayes et al. 2002
Cinder Flats	2	Spawn	Spawn	Hayes et al. 2002
Ouxy Springs	3	Spawn	Spawn	Hayes et al. 2002
Silver Bldg. Springs	4	Spawn	Spawn	Hayes et al. 2002
Sucker Springs	5	Spawn	Spawn	Hayes et al. 2002
Harriman Springs	6	Spawn*	–	59 Fed. Reg. 61744 [1994]
Barkley Springs	7	Spawn*	–	59 Fed. Reg. 61744 [1994]
Tributaries				
Wood River	8	Spawn*[b]	Spawn	Markle and Simon 1994
Lower Williamson River	9	Spawn	Spawn	Cunningham et al. 2002
Upper Williamson River	10	0[b]	0	
Sprague and Sycan	11	Spawn	Spawn	Janney et al. 2002
Lake of the Woods, OR	12	0	+*[c]	Moyle 2002
Lower Klamath Lake, CA	13	+*	+*	Scoppettone and Vinyard 1991
Clear Lake, CA[d]	14	+	+[e]	59 Fed. Reg. 61744 [1994], USFWS 2002
Willow Creek	15	Spawn	Spawn	Moyle 2002
Boles Creek	16	Spawn	Spawn	Moyle 2002
Gerber Reservoir	17	0	+[e]	59 Fed. Reg. 61744 [1994]
Sheepy Lake	18	+*	+*	Moyle 2002
Sheepy Creek	19	Spawn*	–	Moyle 2002
Tule Lake	20	(+)	(+)	USFWS 2002
Lost River	21	Spawn[f]	Spawn	59 Fed. Reg. 61744 [1994]
J.C. Boyle Reservoir	22	(+)	(+)	53 Fed. Reg. 27130 [1988]
Copco Reservoir	23	(+)	(+)[g]	Scoppettone 1988, Scoppettone and Vinyard 1991
Iron Gate Reservoir	24	(+)	(+)	Moyle 2002
Klamath River	25	(+)	(+)	59 Fed. Reg. 61744 [1994]

[a]Tributary streams and springs are listed under lakes into which they flow.
[b]R. S. Shively, U. S. Geological Survey, Klamath Falls, Oregon, personal communication, 2002.
[c]An extirpated population of *Chasmistes* in Lake of the Woods, Oregon, originally referred to as *C. stomias* (Andreasen 1975), may have been another population of shortnose suckers (Moyle 2002).
[d]Drainage for Clear Lake includes numerous small reservoirs and tributary streams that contain both species (USFWS 2002, Appendix D).
[e]Shortnose suckers in Clear Lake and Gerber Reservoir may have been confused with Klamath largescale suckers or with shortnose suckers and Klamath largescale sucker hybrids (D. F. Markle, Oregon State University, personal communication 2002), although genetic information indicates that hybridization is rare (D. Buth, University of California at Los Angeles, and T. Dowling, Arizona State University, personal communications, July, 2002).
[f]Larvae in Lost River apparently do not survive (Moyle 2002).
[g]Shortnose suckers in Copco Reservoir may have hybridized with Klamath smallscale suckers (Scoppettone and Vinyard 1991).
Abbreviations: +, currently present; +*, previously present; (+), small population, probably nonbreeding; Spawn, current or previous spawning; Spawn*, spawning inferred from fish in spawning condition; 0, not known ever to occur; –, lack of information.

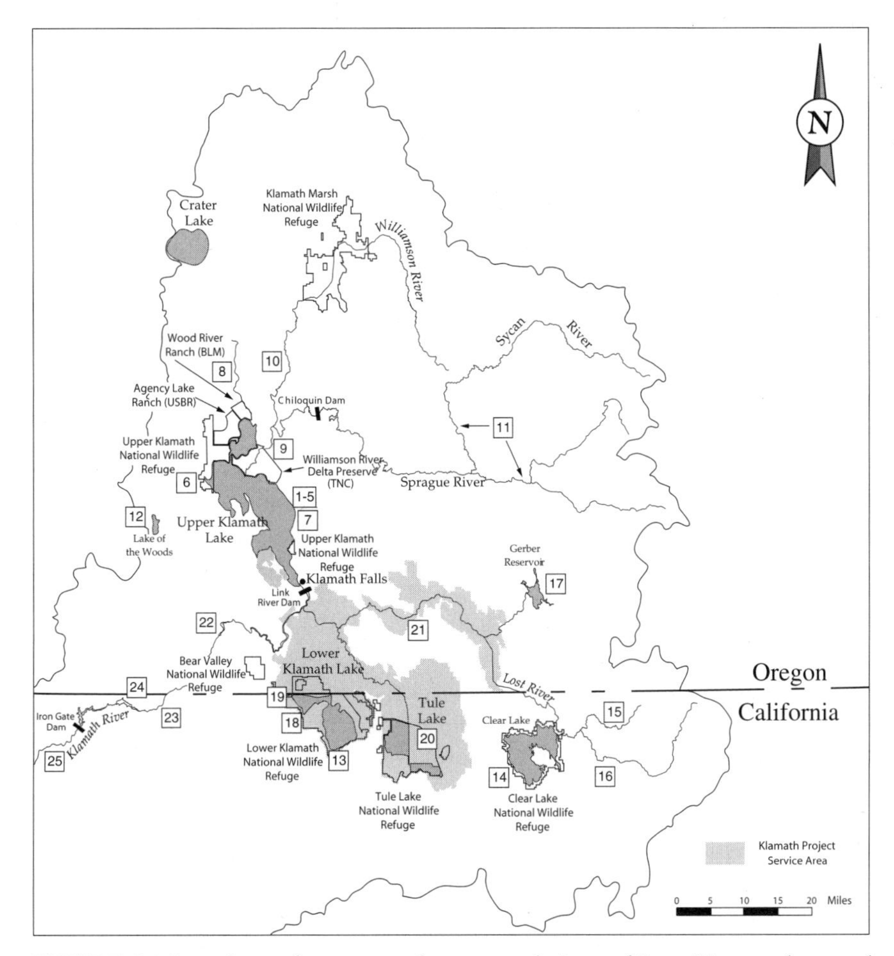

FIGURE 5-2 Locations of current and past populations of Lost River suckers and shortnose suckers. Numbers indicate current or former locations of suckers; light gray shows the area of the Klamath Project; dark gray shows standing water. See Table 5-3 for additional information.

endangered suckers, combined with their exceptional longevity, allows individual adults to contribute to multiple year classes. Successful year classes are crucial to survival of both species, as explained below.

The requirements of the two species of endangered suckers are best understood in the context of their life-history stages, as described below. Unless a species-specific difference is indicated, the description of any given life-history feature is assumed to apply to both species. The quantity and quality of information on the species have increased substantially since the fishes were listed as endangered in 1988.

Spawning

Spawning occurs in tributary streams, in springs caused by upwelling of groundwater in lakes, and around springs in rivers. The suckers may migrate as little as 2–4 mi up a stream from a lake (for example, up Willow Creek from Clear Lake), or over 20 mi (for example, up Boles Creek from Clear Lake and up the Sprague River to RM 74 from Upper Klamath Lake; R. S. Shively, U. S. Geological Survey, Klamath Falls, Oregon, personal communication, 2002). Upstream migrations commence when snowmelt leads to increases in river discharge—from early February through early April for Lost River suckers and from late February to late May for shortnose suckers (Moyle 2002). Spawning can occur at temperatures of 5.5–19°C (Moyle 2002). For example, migrations of Lost River suckers up the Williamson River in 2001 were concentrated in April and May and showed a peak in mid-April. Spawning of shortnose suckers peaked in mid-May 2001 (Cunningham et al. 2002). In any given year, some temporal separation of spawning among species may occur. Klamath largescale suckers migrate first and are followed by Lost River suckers and then shortnose suckers (Coleman et al. 1988, cited in Scoppettone and Vinyard 1991), although migrations of the three may overlap (USGS 2002).

Shortnose suckers were numerically dominant in the lower Williamson River in 2001, but Lost River suckers outnumbered shortnose suckers by more than 10 to 1 at Chiloquin Dam, about 9 mi farther upstream (Cunningham et al. 2002, Janney et al. 2002). Thus, the Lost River suckers may be more likely than shortnose suckers to migrate upriver to spawn, or perhaps the two species react differently to dams. In 2001, 30 shortnose suckers were collected at lakeshore sites, compared to 900 found in the Williamson River, whereas Lost River suckers were five times more abundant at spawning sites in the lake than in the Williamson River system (Hayes et al. 2002, Cunningham et al. 2002). This suggests that spawning by shortnose suckers in Upper Klamath Lake is relatively rare at present. Shortnose suckers that do spawn in the lake use the same spawning sites as Lost River suckers. In flowing water, the suckers spawn in riffles or runs with moderate current (less than 3.3 ft/s) over cobble or gravel bottoms at depths of 0.7–6.6 ft (Scoppettone and Vinyard 1991, Perkins and Scoppettone 1996, Markle and Cooperman 2002). Gravel appears to be preferred; patches of gravel added to a spawning area will be used if flow and depth are appropriate (Golden 1969, Scoppettone and Vinyard 1991, Moyle 2002). Spawning in the upper Sprague River appears to be concentrated around springs (L. Dunsmoor, cited in USFWS 2002). Spawning behavior is similar to that of other suckers in that one female spawns with several males and the fertilized eggs, which are 2.5–3.2 mm in diameter, drop into spaces in the gravel.

Sampling at six known spawning sites along the eastern shoreline of Upper Klamath Lake (Sucker Springs, Silver Building Springs, Ouxy Springs, Boulder Springs, Cinder Flats, and Modoc Point) indicates that Lost River suckers spawning in the lake are slightly larger than those ascending the Williamson River (lake fish were 150–200 mm longer: Hayes et al. 2002, $p < 0.05$). Nearly 80% of the fish captured at lake spawning sites occurred at three of the six sites (Sucker Springs, Silver Building Springs, and Ouxy Springs). As is common among spawning suckers, males outnumber females at spawning sites. Sex ratios at nonspawning sites in Upper Klamath Lake indicate a predominance of females; males tend to remain at spawning sites, whereas females do not (Coen et al. 2002).

Lake spawning occurs in 0.5–3.7 ft of water; 95% of successful spawnings occur in water deeper than 1.0 ft, and about 35% occurs at 1–2 ft (Klamath Tribes, in USFWS 2002). Spawning aggregations were present from mid-March to early May. Peak abundances at all sites occurred during the first 2 wk of April, and a second peak occurred at Sucker Springs, the most heavily used site, in late April. The relative spawning condition (prespawn, ripe, postspawn) of fish captured in Upper Klamath Lake from February to June 2001 suggests that some eastern regions near spawning sites, such as Modoc Point and Goose Bay, are staging areas for spawning and that some western bays are used more heavily after spawning (Coen et al. 2002). The temporal sequence of capture of the sexes during the spawning season also suggests that males move to staging and spawning areas ahead of females.

Evidence from Hayes et al. (2002) is consistent with earlier conclusions by Perkins et al. (2000b) that river spawners and lake spawners constitute subpopulations of Lost River suckers in Upper Klamath Lake, but does not prove complete segregation of populations. Of 201 Lost River suckers tagged during previous years and recaptured at springs in the lake in 2001, with some recaptures separated by as much as three yr, 198 (98.5%) were captured both times at eastern shore spawning sites. The other three fish had been tagged in the Williamson River. Also, 76% of the fish recaptured at the Chiloquin Dam fish ladder in 2001 had been tagged originally at the ladder in previous years, and 20% of the fish had been tagged at other sites on the Williamson River (Janney et al. 2002). About half the Lost River suckers caught in Upper Klamath Lake were from sites other than those where they were tagged, either for within-year or between-year recaptures; this indicates that lake-spawning fish do not restrict their breeding activities to a single lacustrine spawning site. Ten shortnose suckers captured in 2001 were recaptures from previous years; all had originally been captured at shoreline sites. Movement between lake sites was apparent, as with the Lost River sucker.

Female Lost River suckers contain 44,000–236,000 eggs, and female shortnose suckers contain 18,000–72,000 eggs. Larger females bear more eggs, as is typical of most fishes (USFWS 2002). It is unknown whether individuals of either species spawn more than once each year or whether individuals spawn every year. Recapture data on lake spawners (Perkins et al. 2000b, Hayes et al. 2002) suggest that some Lost River suckers spawn every year. Cui-ui (*Chasmistes cujus*) are known to spawn several hundred times over a period of 3–5 days (Scoppettone and Vinyard 1991); Lost River suckers and shortnose suckers might behave similarly. Coen et al. (2002) found that 75% of male but only 40% of female Lost River suckers and 69% of male but only 46% of female shortnose suckers captured in February–June 2001 were in spawning condition (see also Coen and Shively 2001). These observations suggest that a large portion of the adult population of both species is not in spawning condition during any given spawning season. Observations of tagged fish frequenting more than one lake spawning site in a year suggest multiple spawning events for individual fish. Frequency of spawning is relevant to the populations' potential for recovery.

Larvae

Embryos remain in the gravel for 2–3 wk (USFWS 2002). The subsequent larval stage lasts for about 40–50 days (Markle and Cooperman 2002). Stream-spawned larvae emerge ("swim up") from the gravel and immediately move downstream, mostly at night, in late March to early June, depending on spawning date (Moyle 2002). The abundance of larvae peaked in the Williamson River system 21 days after the peak in spawning (Coleman et al. 1988, cited in Scoppettone and Vinyard 1991). Larvae spawned in the Williamson River system pass to Upper Klamath Lake in as little as a day. More than 99% of larvae enter the lake before the caudal fin has formed and well before the yolk sac is absorbed, after which the fish must feed (Cooperman and Markle 2000). How these movement rates are related to location of spawning (lower Williamson River or Sprague River below or above Chiloquin Dam) and how different they would be if more fish spawned above the dam are unknown. Larval mortality in the Williamson River is around 93% per day (L. Dunsmoor, personal communication, in Markle and Cooperman 2002). Mortality in fishes with planktonic larvae is in general very high (Houde 1987, 1997).

Larval habitat is best described as shallow, nearshore, and vegetated in both rivers and lakes (Figure 5-3) except in Clear Lake and Gerber Reservoir, which lack vegetation (Klamath Tribe 1991, Markle and Simon 1994, Reiser et al. 2001). Larvae are most abundant in the northeastern portion of Upper Klamath Lake, including the Williamson River estuary and the lower Williamson River (Markle and Cooperman 2002). In Upper Klamath

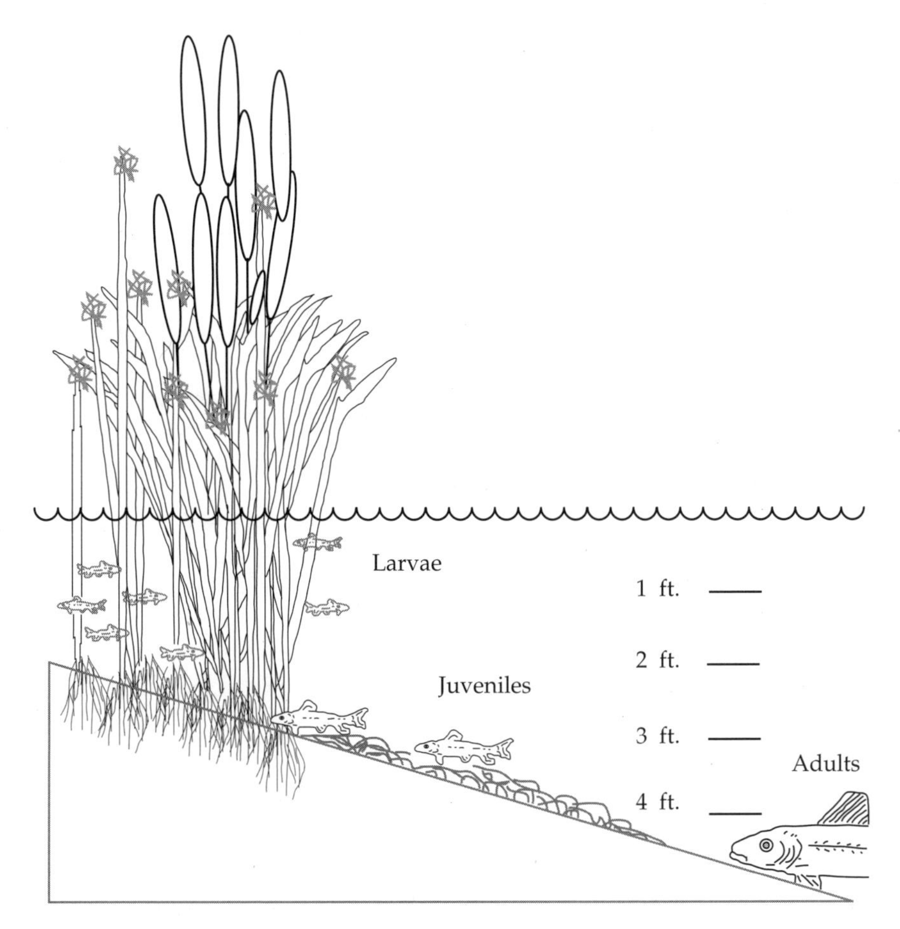

FIGURE 5-3 Generalized view of habitat of young suckers in Upper Klamath Lake. Source: USFWS 2002, p. 83.

Lake, larvae first concentrate near emergent vegetation at the mouth of the Williamson River for several weeks and then appear in other regions of the lake where emergent vegetation is found; that this process can continue for more than 2 mo is not surprising, given the protracted spawning period of the suckers (Cooperman and Markle 2000).

Studies of the larval use of habitat have focused on the importance of depth and vegetation as components of habitat. Observations by Coleman et al. (1988), Buettner and Scoppettone (1990), the Klamath Tribes (Klamath Tribe 1991; Klamath Tribe, Natural Resources Department, Chiloquin, Oregon, unpublished material, 1996), Cooperman and Markle (2000), and Reiser et al. (2001) indicate use of shallow water (less than 4.3 ft and often less than 20 in.) sometimes in areas devoid of cover but more usually near emergent vegetation, such as bullrush beds. Larvae use emergent vegetation primarily from early May through late June, although larvae may be found up to mid-July (see Reiser et al. 2001) because spawning continues into late May. Submerged aquatic vascular plants apparently are less important than emergent vegetation (Cooperman 2002), probably because macrophyte beds are seldom well developed in spring, when much of larval growth occurs. Larvae may not necessarily aggregate within dense vegetation itself but rather near it or in openings in the vegetation in areas described as "pockets of open water surrounded by emergent vegetation," or "the open water/emergent vegetation interface" (Reiser et al. 2001, p. 4–9).

Successful spawning and recruitment of suckers in Clear Lake, which is largely devoid of emergent and submerged vegetation, show that larvae can survive without such vegetation. Clear Lake is very turbid, however, and this may provide protection from visual predators. Laboratory tests show that predation on larvae by fathead minnows is highest when larvae lack cover (Dunsmoor 1993). Young and small fishes in freshwater and marine habitats worldwide often take refuge in dense vegetation when threatened by predators, although larvae of some species are entirely pelagic.

Clear Lake contains flooded annual grasses and herbs and emergent and submerged vegetation in tributaries that may be used by larvae, and it has fewer introduced predators, such as yellow perch and fathead minnows, than does Upper Klamath Lake (USFWS 2002). Thus, successful recruitment in Clear Lake does not demonstrate that vegetation is unimportant in Upper Klamath Lake. Successful spawning apparently does not occur in any of the main-stem reservoirs, which have steep shorelines, lack substantial emergent vegetation, have abundant predators, and may lack spawning areas (Desjardins and Markle 2000).

Juveniles (1–4 Inches)

Larvae are considered juveniles at a length of 1–4 in., which the suckers generally achieve by the end of July (USFWS 2002). Juveniles are termed young of the year (YOY) or age 0 through their first winter. They spend daytime near shorelines over clean, rocky bottoms composed of sand, gravel, and small boulders (Simon et al. 2000; Figure 5-3). YOY use both vegetated and unvegetated portions of shoreline, generally in water less than 4.3 ft deep (USFWS 2002). Knowledge of the extent to which vegetation is used is complicated by the difficulties of sampling juveniles in dense vegetation (Reiser et al. 2001). Abundance of YOY at first is greatest in the northeastern portion of Upper Klamath Lake; as summer progresses, young fish move southward in the lake and into deeper water and become less associated with shorelines, and they become more oriented toward the lake bottom (Gutermuth et al. 2000). Simon and Markle (2001) suggest that overwinter mortality of first-year juveniles approaches 90%. After their first year, juveniles are found throughout the lake but are most abundant in the northern one-third of the lake, as are adults, although it may be important that sampling has been concentrated on this area (Reiser et al. 2001). Juvenile Lost River suckers appear to depend less on shallow-water habitats than juvenile shortnose suckers, as shown by sampling with beach seines (Simon et al. 2000), and juvenile shortnose suckers are apparently more strongly oriented toward the lake bottom than juvenile Lost River suckers (Scoppettone et al. 1995).

Subadults (4–10 Inches) and Adults

Subadults are the least-studied age group. It is assumed that their requirements and habits are most like those of nonspawning adults but their behavior is obscure because they are too fast to catch in seines or trawls, too deep to catch in cast nets, and often too small to gillnet. Given that suckers may spend the first 3–8 yr of their lives as subadults, additional information on this stage could be important.

Lost River suckers grow rapidly for their first 5 or 6 yr to a length of 14–20 in. (Scoppettone 1988). Some males reach maturity (i.e., are capable of spawning) at 4+ yr and 15 in. and some females do so at 7+ yr and 21 in., but most fish mature at 8 or 9 yr; males often mature earlier than females. At maturity, growth slows (Scoppettone 1988, Buettner and Scoppettone 1990, Scoppettone et al. 1995, Perkins et al. 2000a). The largest and oldest fish are females. The oldest known Lost River sucker (43 yr) was obtained in Upper Klamath Lake during a fish kill in 1986 (Scoppettone 1988).

Female shortnose suckers apparently grow faster and larger than males. Both male and female shortnose suckers mature as early as 4+ yr. Males can be mature at 11 in. and females at 13 in., although maturation at 5–7 yr is more usual. The oldest known shortnose sucker (33 yr) was taken from Copco Reservoir in 1987 and was 19 in. long (Scoppettone 1988).

Adult Lost River suckers forage primarily on zooplankton and benthic (bottom-dwelling) macroinvertebrates (Coleman et al. 1988, Scoppettone and Vinyard 1991). The shortnose sucker, as could be predicted from the more terminal position of its mouth, feeds predominantly on cladoceran zooplankters (water fleas), although the guts of only a few adults have been examined (Coleman et al. 1988). The presence of detritus in the guts of shortnose suckers from Clear Lake indicates that shortnose suckers may also feed close to the bottom (Moyle 2002).

Adult suckers select water depths of 3–15 ft, as shown by daylight spring and summer observations; their strongest preference appears to be for 5–11 ft (Reiser et al. 2001, USFWS 2002). Their minimal use (1% of daytime observations) of shallower water could reflect avoidance of high light intensities and thus of aerial predators; limited use of the deepest water (about 4% of daytime observations), particularly in summer, may reflect avoidance of low concentrations of dissolved oxygen (Chapter 3).

Although adults of the Lost River suckers and shortnose suckers are captured together in many places in Upper Klamath Lake, some differences in their distribution suggest different habitat preferences. For example, in 2001, Lost River suckers were 2–3 times more abundant in trammel net samples from the western shoreline of Upper Klamath Lake, whereas shortnose suckers were 2–3 times more abundant in samples from the eastern shore (Coen et al. 2002). Possible habitat differences in these regions might be worthy of further investigation, although the differences could reflect chance encounters with aggregations of the two species.

Physiological Tolerances

Lake suckers in general are relatively tolerant of water-quality conditions that are unfavorable or even lethal for many other fishes. For example, suckers in good condition occur in Tule Lake, which periodically experiences extremes of dissolved oxygen, pH, and ammonia that are toxic to fathead minnows, a tolerant species (Dileanis et al. 1996, cited in USFWS 2002). Other lake sucker species are similarly tolerant. Endangered cui-ui evolved in the very alkaline (pH, 9.0–9.5) and saline (5 ppt) waters of Pyramid Lake, Nevada, where only five or six other native fish species persist. The only nonindigenous fish species to have successfully colonized Pyramid Lake is the Sacramento perch (G.G. Scoppettone, U. S. Geological Survey, Reno, Nevada, personal communication, 2002).

Most fishes cannot tolerate sustained pH in excess of 9 (Falter and Cech 1991). Upper Klamath Lake suckers can tolerate pH approaching 10, temperatures up to 31–33°C, concentrations of unionized ammonia up to 0.4–0.5 mg/L, and dissolved oxygen concentrations down to 1.5 mg/L. Beyond these thresholds, the suckers die in laboratory tests (typically conducted on juvenile fish); larvae are more sensitive than larger fish (Falter and Cech 1991, Martin and Saiki 1999, Saiki et al. 1999, Moyle 2002). Mortality is high in adult suckers below oxygen concentrations of about 1 mg/L (Chapter 6). Falter and Cech (1991) found that shortnose suckers had much lower tolerance of high pH (measured as pH at which swimming equilibrium was lost) than two other endemic fishes, the Klamath tui chub and the Klamath largescale sucker. Shortnose suckers lost equilibrium at a mean pH of 9.55, tui chub at 10.75, and Klamath largescale suckers at 10.73. Maximum pH in Upper Klamath Lake during summer phytoplankton blooms frequently exceeds 9.5 at the surface during daylight hours, but pH during episodes of mass mortality generally is about 7.5–8.5 (Perkins et al. 2000b), indicating that high pH does not cause mass mortality (Chapter 3). In Upper Klamath Lake in late summer, during times of physiological stress, suckers may seek higher water quality, such as that of springs and river mouths, even though such areas are otherwise avoided, probably because they are too shallow or too clear (USFWS 2002, Appendix D; Chapter 6).

Physiological tolerance tests generally are performed in a laboratory on single factors held at constant values, whereas factors in nature often vary over time and space, co-occur, and can operate synergistically. Summer conditions in Upper Klamath Lake typically involve episodes of high pH, high unionized ammonia, and low dissolved oxygen in combination with high temperatures that increase the oxygen demand of the fish. High concentrations of unionized ammonia can cause structural damage to gills, which can increase the susceptibility of fish to low concentrations of dissolved oxygen. High pH (over 9) inhibits ammonia excretion, thus creating stress (Lease 2000, cited in USFWS 2002). Susceptibility to columnaris disease, which is caused by the bacterium *Flavobacterium columnare*, increases with increasing temperature but decreases with increasing ammonia concentrations (Morris et al. 2000, Snyder-Conn et al. unpublished in USFWS 2002).

As an adjunct to laboratory studies, Martin and Saiki (1999) placed cages containing juvenile Lost River suckers in Upper Klamath Lake for 4-day periods. High mortality occurred at high pH, high concentrations of unionized ammonia, and low concentrations of dissolved oxygen; low dissolved oxygen was the strongest correlate with mortality. At sublethal temperatures and concentrations of unionized ammonia, fish were tolerant of higher pH than expected from the laboratory studies (fish tolerated pH as high as 10.8). The study suggests that laboratory tests of single factors should be viewed as being only indicative of the extremes that can be tolerated; they are not strictly predictive of responses in the field.

From the viewpoint of physiological stress on fishes generally, and especially for cold-water fishes, water-quality conditions are poor throughout much of the Klamath basin, as explained in Chapters 3 and 4. Physiological thresholds for suckers, however, are reached or exceeded less extensively than for most fishes because of the high tolerance of suckers. Harm to suckers caused by poor water quality is known for Upper Klamath Lake and may also occur in the Lost River and upper Keno Reservoir (Lake Ewauna). In other lacustrine or flowing-water environments of the basin, however, poor water quality may be much less important than other factors for suckers, although it may strongly affect some other fishes.

In Upper Klamath Lake, suckers are adversely affected by poor water quality, which is a byproduct of very high abundances of *Aphanizomenon flos-aquae*, a planktonic bluegreen (cyanobacterial) alga. Peak abundances of *Aphanizomenon* occurring in late summer or early fall cause very high pH. Under certain meteorological conditions overturn of a stratified water column and collapse of the *Aphanizomenon* population combine to cause depletion of oxygen throughout the water column and distribution of high concentrations of unionized ammonia (Chapter 3).

The adverse water-quality conditions in Upper Klamath Lake potentially have three types of effects on endangered suckers in Upper Klamath Lake: (1) mass mortality of large fish, (2) mortality, either episodic or continuous, of small fish or larvae, and (3) physiological stress on one or more age classes, which leads to physiological impairment but not necessarily death.

Poor water quality in Upper Klamath Lake is a documented cause of the episodic mass mortality of large suckers in the lake. The recent history of these episodes is given in this chapter, and the factors producing death are discussed in Chapter 3. Extensive research on the direct cause of mortality during episodes of mass mortality has led to the reasonably firm conclusion, supported by scientific evidence, that mortality is caused by inadequate amounts of dissolved oxygen. The two other potential direct causes of mortality, pH and unionized ammonia, appear not to control mass mortality. Dissolved oxygen, unlike pH and unionized ammonia, remains adverse continuously for many days during episodes of mass mortality, whereas pH and unionized ammonia do not. Thus, although additional studies of mechanisms leading up to mass mortality are warranted, the direct cause in large fish seems to be understood reasonably well.

There is insufficient evidence to show whether extreme water-quality conditions also cause mortality of juveniles and larvae. Laboratory experiments indicate such potential, but it has not been documented in the field. Field documentation, especially if mortality were steady rather than episodic, would be difficult for the smaller life stages of fish because of their quick deterioration and dispersal after death. The possibility that gradual or episodic mass mortality of small fish occurs should be studied.

Adverse water-quality conditions can affect fish indirectly, as explained above. Laboratory studies are useful, but field indicators of stress also are important in that sublethal responses to stress cannot always be produced in an interpretable way in the laboratory. Indicators of physiological stress include unusual or recurrent epizootics, poor body-condition factors, physical anomalies, and low growth rates compared with those in populations that are not exposed to adverse water-quality conditions, abnormally low fecundity or fertility of mature fish, and behavioral aberrations. Some attention has been given to the indicators—for example, physical anomalies in suckers of Upper Klamath Lake are common (USFWS 2002)—but a more comprehensive effort at evaluating indicators of stress probably is warranted.

Overall, there is no doubt that poor water-quality conditions are suppressing the endangered suckers of Upper Klamath Lake through mass mortality of large fish. Less clear is the role of potential additional suppression through mortality of smaller fish or sublethal effects of physiological stress caused by poor water-quality conditions on any or all life stages.

Population Size

Abundances of larval and juvenile suckers have been estimated from field samples over the last several years (e.g., Simon et al. 2000). Calculated population sizes of adults have been based on recapture of tagged fish during fish kills. The confidence intervals around the numbers are very large and, because many of the assumptions of mark and recapture methods are not met by these estimates, the estimates are of limited value (R. S. Shively, USGS, unpublished memo, 5 March 2002; USFWS 2002).

Newspaper reports, eyewitness accounts, and data on catch per unit effort leave little doubt that the sucker population exploited by the snag fishery in the 1960s and earlier was much larger than it was by the 1980s. Relative estimates of the size of the spawning run of suckers in the Williamson River were first based on estimated catch rates and later on standardized recapture and electrofishing methods. The estimates showed a marked decrease in abundance of fish during the middle 1980s. In 1984, the run of spawning Lost River suckers was estimated at 23,000, but it fell to 12,000 in 1985. Catch per unit effort of electrofishing fell by 57% for Lost River suckers and by 83% for shortnose suckers from 1984 to 1986 before the major fish kill of 1986 (Scoppettone 1986, Bienz and Ziller 1987, Scoppettone and Vinyard 1991). The fishery was closed in 1987. More recent estimates of abundance depend on catch per unit effort in standardized trammel-net samples and can be compared only among collections for the years 1995–2001.

No universal or absolute estimates of the size of any age class of sucker are available. Estimates are relative, limited to specific sites (e.g., spawning areas), or are otherwise qualified from the viewpoint of making an overall numerical assessment of the population. While the use of qualified or relative estimates is beneficial, efforts to make more comprehensive population size estimates in the future would be desirable (see Chapter 6). For purposes of ESA actions, the critical facts, which are known with a high degree of certainty, are that the fish are much less abundant than they originally were and that they are not showing an increase in overall abundance. Thus, the point of departure for research and remediation in the future is the need to restore abundance of the listed suckers.

Age-Class Structure

Most adult suckers in Upper Klamath Lake are large and old. The uneven age distribution has characterized the populations for several decades. Through the 1980s, the age distribution of Lost River suckers was heavily skewed to fish 19–28 yr old. In 1986, the year before fishing was banned, recruitment had apparently been poor for about 18 yr; 95% of adult Lost River suckers were 19–30 yr old (Figure 5-4; Scoppettone 1988). The data for Lost River suckers shown in Figure 5-4 are based on fish obtained during fish kills, a sampling method with unknown but multiple biases, including some evidence that older, larger fish suffer disproportionately high mortality (Chapter 6). Assuming that the fish collected during fish kills are representative of the adult population as a whole, it can be concluded that many age classes were essentially missing from the lake before 1988, when the fishery was active.

Closure of the fishery in 1987 greatly reduced mortality of spawners, after which additional mature fish began entering the spawning population (Figure 5-4B). Cessation of fishing apparently contributed to the production of a strong year class of both endangered sucker species in 1991, and to smaller but notable year classes also produced in 1990, 1992, and 1993 (Figure 5-4B; see Markle and Simon 1994, Cunningham and Shively 2001). These fish would have been expected to mature in the late 1990s, but the major fish kills that occurred in 1995, 1996, and 1997 affected not only old spawners but also probably young spawners. Spawning runs declined in the late 1990s, with little evidence of substantial recovery until 2000 (Figure 5-5). The upsurge in spawning numbers in that year and again in 2001 may represent maturation of fish from the 1991 and later year classes. It is possible that fish that lived through the fish kills of the middle 1990s were stressed by poor water quality and as a result experienced delayed maturation (e.g., Trippel 1995, Baltz et al. 1998), although Terwilliger et al. (M.R. Terwilliger et al., Oregon State University, Corvallis, OR, unpublished material, 2000) found no evidence of impaired growth associated with periods of poor water quality in juvenile suckers of Upper Klamath Lake. That spawning runs apparently increased in 1999–2001 shows that the species have substantial resilience, but this is no guarantee of recovery.

Comparisons between 2000 and 2001 data indicate a weak but significant trend toward increasing average size among all spawning shortnose suckers and female Lost River suckers in the Williamson River (Cunningham et al. 2002). A similar significant trend toward increased median size at a variety of nonspawning sites in Upper Klamath Lake was also found (Coen et al. 2002). When combined with evidence of low numbers of small river-spawning fish in recent years (Cunningham et al. 2002), the data could indicate year-class failure among fish that hatched in the middle 1990s and

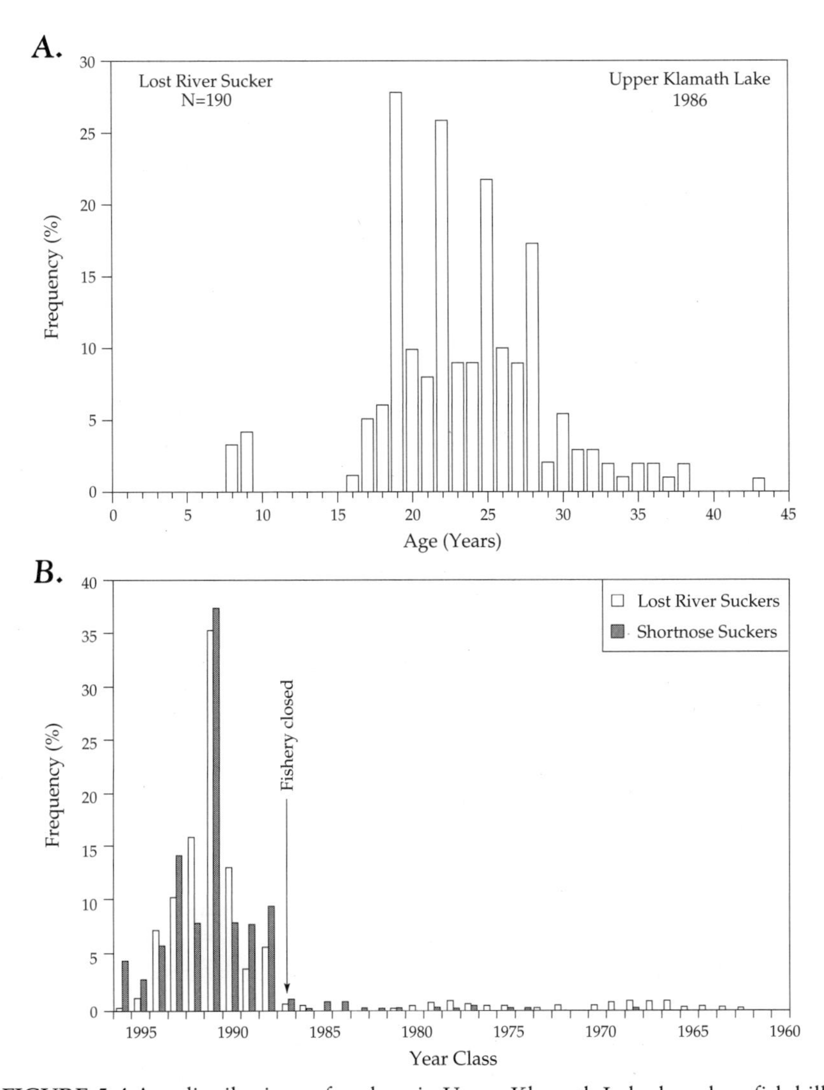

FIGURE 5-4 Age distributions of suckers in Upper Klamath Lake based on fish kills. (A) Age distribution of Lost River suckers in Upper Klamath Lake based on the 1986 fish kill. Multiple peaks indicate strong year classes estimated as 1958, 1961, 1964, 1967. Source: Scoppettone and Vinyard 1991. Pp. 359–377 in *Battle Against Extinction: Native Fish Management in the American West*, W.L. Minckley and James E. Deacon, eds. Copyright 1991 The Arizona Board of Regents. Reprinted by permission of the University of Arizona Press. (B) Age frequency distributions of Lost River suckers and shortnose suckers in Upper Klamath Lake based on fish collected from the 1997 fish kill. Effects of fishery closure in 1987 and of entry of successful 1991 year class are evident. Fish as old as 35 yr (spawned in 1962) were present. Source: Markle and Cooperman 2002, based on data from R. Shively, USGS.

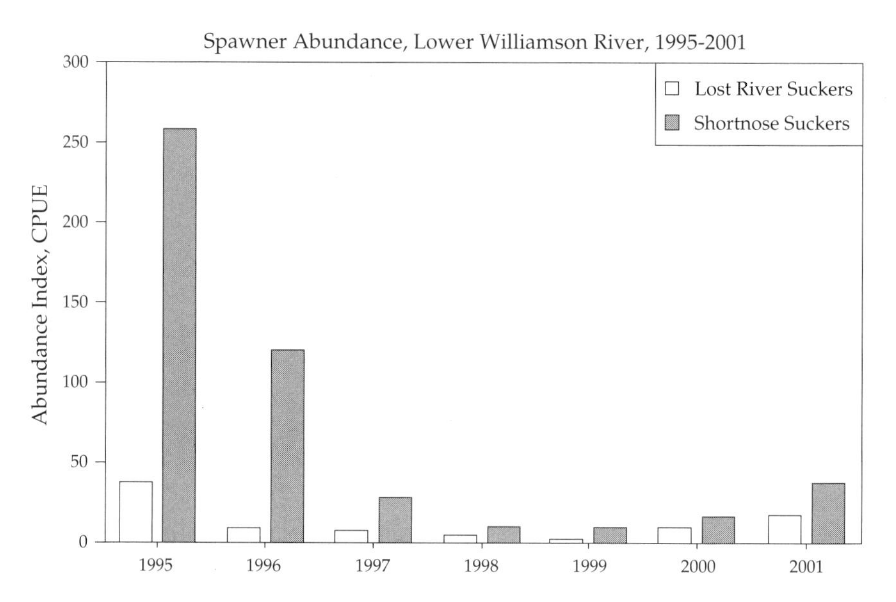

FIGURE 5-5 Spawning-run abundances of lake suckers, lower Williamson River, 1995–2001. Decline in spawners consistent with expected changes given fish kills of 1995–1997 is evident (1995 data were obtained before the fish kill that year). CPUE is a measure of catch per unit effort based on fish caught per unit of time spent fishing with trammel nets. Source: Modified from Cunningham et al. 2002, p. 30.

that would mature in the early 2000s. Concern over lost year classes might be tempered by an apparent trend in increased overall abundance among river spawners in 1999–2001 (Figure 5-5). Catches of both species from the Williamson River in spring 2002 decreased, however, by about 50% compared with 2001 (R. S. Shively, U.S. Geological Survey, Klamath Falls, Oregon, personal communication, October 8, 2002). Abundance index (catch per unit effort) for lake-spawning Lost River suckers do not indicate an increase in numbers of spawners (1999, 3.0 fish/h; 2000, 2.0 fish/h; 2001, 2.4 fish/h), and the average size of lake-spawning fish increased significantly between 2000 and 2001, suggesting lack of recent recruitment into the spawning population (Hayes et al. 2002). Catches at the shoreline areas in 2002 also decreased by about 15–20%. In fact, sampling in 2002 indicates that there has been no substantial recruitment into the adult population since 1999 (R. S. Shively, U.S. Geological Survey, Klamath Falls, Oregon, personal communication, October 8, 2002).

Observations on size of spawners since 1984 (Perkins et al. 2000b) indicates that very large Lost River suckers (over 25 in. for males, and over 28 in. for females) have been lost progressively from the population, that recent spawning aggregations are made up largely of medium-size fish (18–24 in.), and that the median age of spawners for Lost River suckers is 12 yr and for shortnose suckers is 9 yr (as judged from age-length relationships; Markle and Cooperman 2002). These findings suggest that successful year classes after 1991–1993 are largely absent, that is, that little recruitment of young spawners has occurred at the same time that the largest fish have been progressively removed by the fish kills; this raises a concern over future numbers of spawners and total reproductive output of the population.

As with Lost River suckers, knowledge of age distributions of shortnose suckers in Upper Klamath Lake comes chiefly from three fish kills in the 1990s, except that the data are even less complete and earlier data are lacking (Figure 5-4B). Indications from age distributions of fish collected after fish kills have indications similar to those for the Lost River suckers.

One other trend of note is that larger fish appear to spawn earlier in the season (Perkins et al. 2000b), but this trend may have been obscured in recent years by a relative lack of small spawners (Hayes et al. 2002). Regardless of cause, multiple strong year classes with temporal separation in spawning between year classes is potentially advantageous because it decreases the likelihood of failure of all the year's larvae if environmental factors vary for year to year during the breeding season (e.g., Trippel 1995).

Information on age distribution is a fundamental indicator of the status of a population, and it sometimes suggests reasons for failure of a species to recover. Although the 1990s, in apparent contrast with earlier years when the fishery was in place, have produced recruitment into the subadult and adult stages, the fish entering these stages have been killed in large numbers during episodes of mass mortality in Upper Klamath Lake. Thus, one reason for failure of the populations to recover is probably suppression of reproductive capacity of the population due to selective mortality of adult fish. This does not, however, rule out the possibility that part of the explanation for lack of recovery lies in suppression of the number of fish entering the subadult and adult phases. The fish collected during fish kills indicate recruitment into the subadult and adult stages in all years, and especially in some years with notably abundant year classes (such as 1991), but the amount of this recruitment may be insufficient to support overall growth of the population. Thus, one bottleneck almost certainly involves the mass mortality of large fish, and a second bottleneck could be at one or more places in the life cycle between laying of eggs and the entry of fish into the subadult and adult categories. As cited above, numerous efforts are under way to identify unusual mortality or suppression of vigor in young fish, but no conclusions are yet available on this important matter.

Perspective on Age-Class Structure and Strength, Mortality, and Reproductive Output

Most fishes experience astronomically high mortality in their early life-history stages. The millions or even billions of individuals that hatch in a population are reduced by many orders of magnitude at the time of maturation. On the average, a male and female just replace themselves over a lifetime of spawning, even though they may produce millions of fertile eggs. These facts are relevant to sucker recovery in several ways. High mortality among larvae and small juveniles is to be expected, but the rates should plummet in later years, and old fish should show low mortality. Small percentage changes in mortality of young fish can translate into large population differences later because of the high numbers of young individuals. Thus, any steps that can be taken to increase larval and juvenile survival in Klamath Lake suckers could produce great benefits.

The high mortality experienced by very old fish during the fish kills of the middle 1990s is especially alarming given the reproductive potential of these fish (e.g., Conover and Munch 2002). Large, old fish of most species produce disproportionately more eggs than smaller fish. For example, in red snapper (*Lutjanus campechanus*), which is heavily fished and depleted throughout its North American range, a single 10-yr old female (26 lb, 24 in.) can contain 9 million eggs, which is equivalent to the total egg output of 212 adult females that are 3–4 yr old, weigh 2.2 lb each, and are 17 in. long. One 26 lb old fish produces more eggs than 250 lb of younger fish. Thus, loss of larger size classes in a population can have a disproportionate effect on egg production and future recruitment (Bohnsack 1994). The value of large fish, even in small numbers, is evident in the listed suckers. The number of young produced and eventually recruited into adulthood increased greatly just after the snag fishery was closed (see Figure 5-4B), demonstrating that even low numbers of large fish can produce large numbers of recruits (Markle and Cooperman 2002).

The disproportionately high contribution of old fish is even greater than fecundity would indicate. Because the quality of eggs (size and amount of yolk) produced by old females may be greatest, larvae hatching from these eggs may be larger and more likely to survive the early periods of high mortality (e.g., Trippel 1995). Although numbers of spawning fish in the Williamson River appear to have climbed in recent years, the reproductive potential of the population is lower than it was before the fish kills because the fish are smaller (Markle and Cooperman 2002). Reproductive output of a population is determined jointly by the number of spawners and the age distribution of spawners. Two populations of equal size that contain different size distributions of fish will not be equal in reproductive value; the population with more old, large fish will have much higher reproductive potential. Any alterations that can be made in the environmental conditions that directly affect the probability or severity of fish kills should receive especially careful consideration (Chapter 3).

Species that are long lived and late to mature, such as the endangered suckers of the Klamath basin, may respond slowly both to degradation and to restoration of habitat requirements, in contrast to other species that mature more quickly. Thus, the presence of old fish is not in itself evidence of a sound population. In fact, even if old fish are numerous, their failure to propagate would render them implicitly extinct until a reversal of the situation occurs. Similarly, improvement of environmental conditions may lead to beneficial changes in the population through recruitment of young age classes, but the final evidence of progress toward recovery, which is survival of these younger classes to maturity and old age, will not be evident for a decade or more. This special perspective on the long lived, slow maturing suckers must be maintained in any evaluation of prospects for extinction and response to remediation.

Endangered Suckers in Other Klamath Basin Waters

Suckers occurred naturally in Tule Lake, Sheepy Lake, and Lower Klamath Lake, from which spawning fish ran up the Lost River (Table 5-3). All three of the lake populations apparently were extirpated when their waters were drained for agricultural purposes around 1920 (Chapter 2). During the 1930s, after farming failed in the former lake bed, the lakes were to some extent reinundated, but not to their former depths. Suckers recolonized Tule Lake but not the other two lakes. There has been no evidence of successful spawning in Tule Lake, although fish from the lake evidently spawn in the lower Lost River.

Fish of both species, but mostly shortnose suckers, have been found regularly in the reservoirs between Keno and Iron Gate Dam (e.g., J. C. Boyle, Copco, Iron Gate). Apparently, they do not spawn. Fish in these impoundments probably consist of individuals that enter the Link River from Upper Klamath Lake and survive passage at Link River Dam; they tend to be old and large (Figure 5-6). The trip out of Upper Klamath Lake is one-way, inasmuch as no fish ladders suitable for suckers are located at Link River Dam or at any of the other dams along the Klamath River (Chapter 6). The great size and age of female fish as suggested by Figure 5-6 could make such fish valuable as transplants to more favorable habitats.

Reproducing populations of endangered suckers exist in Clear Lake, in Gerber Reservoir, and in portions of the Lost River downstream (the Lost River could receive fish from Gerber Reservoir in its upper portion and from Tule Lake in its lower 7 mi, below Anderson Rose Dam). Clear Lake,

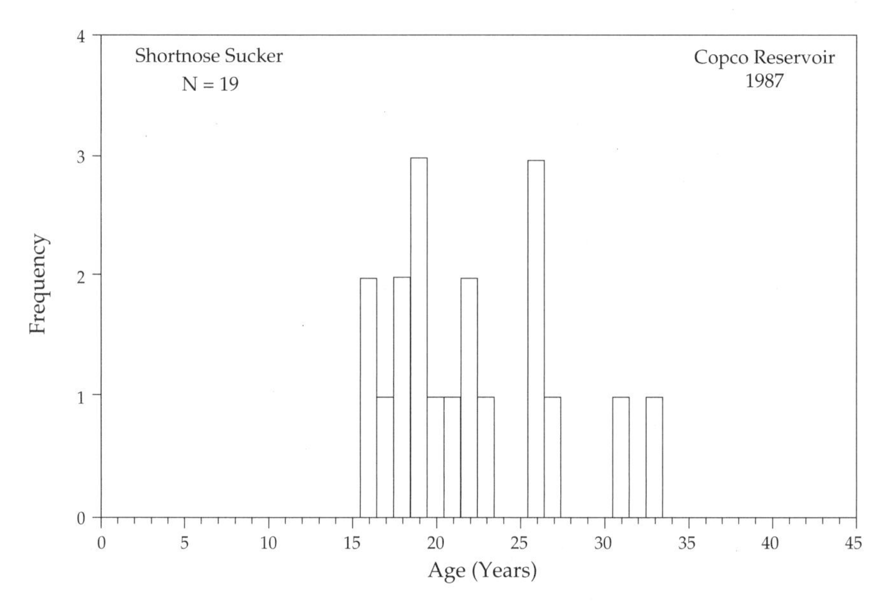

FIGURE 5-6 Age structure of a small sample of shortnose suckers taken from Copco Reservoir, 1987. Source: Scoppettone and Vinyard 1991. Pp. 359–377 in *Battle Against Extinction: Native Fish Management in the American West*, W.L. Minckley and James E. Deacon, eds. Copyright 1991 The Arizona Board of Regents. Reprinted by permission of the University of Arizona Press.

which was established in 1910, contains populations of both species (Scoppettone et al. 1995 estimated that 73,000 suckers occupied the lake), and both show recent evidence of diverse age structure and continued successful reproduction and recruitment. The reservoir is a source of irrigation water and can be drawn down during drought, which exposes the fish to multiple threats. Clear Lake was drawn down to as low as 5% of capacity during 1992, and fish collected after the drawdown and in the next spring were in poor condition, although their condition rebounded by the end of the next summer (USFWS 2002). Success of shortnose suckers and Lost River suckers in Clear Lake is encouraging in its own right and as a potential rescue population that could be used for restoring populations in other water bodies. Extreme drawdown, although prohibited by the USFWS biological opinion of 2002, is a threat if it should occur inadvertently, and the lake and its suckers presumably are vulnerable to major environmental disasters, such as a break in the dam (Moyle 2002). Unexpected changes in the spawning and rearing habitats in Willow and Boles Creeks above the reservoir also could affect sucker abundances.

Gerber Reservoir, which was created in 1925, contains shortnose suckers but not Lost River suckers. Shortnose suckers in Gerber Reservoir exhibit a wide range of size classes, indicating successful reproduction and recruitment. Gerber Reservoir is not connected to any other sucker population, so there is no possibility of genetic exchange. Condition of fish in Gerber Reservoir is known to vary from poor to good; poor condition was associated with lowest water levels in 1992 (the lake was drawn down to 1% of capacity). The population has not received a great deal of attention.

Gerber Reservoir flows into the Lost River, which flows into Tule Lake (Figure 1-2). Historical sucker runs out of Tule Lake and up the Lost River were substantial; these runs supported commercial fisheries and canneries (USFWS 2002). Today, after the construction of multiple dams, only small numbers of the two endangered species occur in the Lost River; shortnose suckers are more common than Lost River suckers. It is not known whether these populations are self-sustaining (USFWS 2002). Spawning habitat is limited, and spawning has been observed at only about three locations, although several other sites appear to provide appropriate spawning habitat. Small numbers of larvae and juveniles have been collected in the river, but these fish could originate in Gerber Reservoir. Upstream movement from Tule Lake ends at Anderson Rose Dam, 7 mi above the lake. Spawning habitat in the 7-mi reach is scarce, and rearing habitat is compromised by poor water quality from water connected with Tule Lake sumps and agricultural return flows. Water quality in the Lost River is generally poor; the river fails to meet several Oregon state-specified water-quality thresholds. Gradients in portions of the river are unfavorably steep for suckers, and seasonal dewatering is common, as are dense plant growth and algal blooms associated with poor water quality. Both summer and winter fish kills were documented for the Lost River Diversion Canal region in the late 1990s. Brown bullhead (*Ameiurus nebulosus*) and pumpkinseed (*Lepomis gibbosus*) are abundant and nine of the 16 fishes in the river are warmwater nonnatives. USFWS (2002, Appendix E, p. 31) concludes that the Lost River is highly degraded and "can perhaps be best characterized as an irrigation water conveyance, rather than a river."

Tule Lake, once larger than Upper Klamath Lake but now less than 15% of its original size, contains populations of both endangered species amounting to perhaps a few hundred fish represented by a few size classes of old fish (for example, 16–24 in.; Scoppettone et al. 1995). Suckers in Tule Lake typically have higher condition factors and lower incidence of external parasites than suckers in other parts of the basin (USFWS 2002). The Tule Lake populations historically were maintained by spawning runs up the Lost River, which for reasons listed above now are extremely limited. Conditions within Tule Lake are deteriorating because of accumulation of sediment from agricultural sources. Alterations in water-management practices, however, could arrest deterioration. Some changes might even restore spawning runs. In 1999, the U.S. Bureau of Reclamation began releasing 30 cfs during the spawning and incubation period (April–June), which led to detectable spawning activity below Anderson Rose Dam within 2 days (USFWS 2002). Such spawning could presumably lead to juvenile recruitment, but monitoring for presence of juveniles is needed. Collection of larvae reported by Shively et al. (2000a) is additional evidence of reproduction. The relatively good condition of suckers in Tule Lake makes these populations valuable for the long-term survival of both species of suckers, especially given the continuation of fish kills in Upper Klamath Lake.

Conservation Status

Lost River suckers and shortnose suckers were declared endangered by California in 1974 (Moyle 2002); Oregon placed both Lost River suckers and shortnose suckers on its protected list in 1987. USFWS first listed both sucker species as candidate (Category 2) species in 1982. They were proposed for listing as endangered in 1987 and were designated as endangered species in 1988 (53 Fed. Reg. 27130 [1988]). Despite the controversy surrounding the species in recent years, only 13 written comments were received by USFWS during the comment period before listing; 12 of the comments favored listing, one expressed no opinion, and there were no comments opposing the listing. Reasons for listing are given in Chapter 6. A federal recovery plan has been developed (Stubbs and White 1993). Critical habitat was proposed in 1994 (59 Fed. Reg. 61744 [1994]) but has not yet been formalized, nor has a recovery team been designated.

CONCLUSIONS

Human activities in the upper basin have affected not only the listed suckers, but virtually all the native species, several of which are greatly diminished in distribution and abundance. In particular, bull trout and slender sculpin have become rare in the basin in recent years. The Lost River system, which appears to have changed the most in the last 30 yr was dominated by blue chub, tui chub, and the three native sucker species, but it is now dominated by nonnative species. Upper Klamath Lake also has a high abundance of nonnative species, and most of its native species appear to be declining. A downward trend may be common, in fact, to native fishes in most aquatic habitats in the upper Klamath basin, although documentation is weak. The overall status and biology of the fishes of the basin, except for the two endangered suckers, is poorly known or at least poorly recorded. Research over the last 15 yr has produced many unpublished reports and extensive data but very few peer-reviewed papers. Thus, the utility of the available information is hard to judge. One possible remedy would be to provide funding for postdoctoral scholars to compile information and write papers by working with university and agency scientists who have collected data.

Future status of the suckers and other native fishes and the spread of nonnative species cannot be judged without periodic basin-wide survey of fishes. Monitoring is a key feature of adaptive management (see Chapter 10). Also, most information on the biology and status of the suckers and other native fishes has not been published in peer-reviewed journals or books. Also, further studies on the systematics of Klamath basin fishes are needed so that managers can avoid being surprised by the discovery of new endangered species, as are studies of the effects of nonnative species on the listed suckers and other native fishes. Introductions or spread of nonnative species already in parts of the basin are major threats to native species. The Sacramento perch in particular has the potential to spread through the canal system from the Lost River to Upper Klamath Lake, where it could become a predator of juvenile suckers and other native fishes.

Populations of the two listed sucker species in the upper Klamath basin have declined greatly in overall abundance and breadth of distribution. Stable reproducing populations of the two species occur now only in Clear Lake and Gerber Reservoir (Gerber Reservoir has only shortnose suckers). The formerly large populations of the two suckers in Upper Klamath Lake are drastically reduced, although no quantitative estimates are available for former or present population sizes. The sucker populations showed a substantial increase in recruitment, as indicated by year class strength, following the end of fishing in 1987. While the populations of Upper Klamath Lake are reproducing and all age classes are present, they are not rebounding in abundance. Episodic mass mortality of large endangered suckers is one explanation for failure of the populations of Upper Klamath Lake to rebound. Other age classes may be adversely affected in other ways, but these mechanisms are not as well documented. Prolonged low concentration of dissolved oxygen during the late summer of some years is probably the direct cause of mass mortality in Upper Klamath Lake.

The two endangered sucker species are present at other locations, but at none of these locations are substantial numbers of all age classes present. Large suckers are present in the five main-stem reservoirs of the upper Klamath basin and in the upper and lower portions of the Lost River main stem, as well as Tule Lake, but there is no recruitment. Spawning occurs in the Lost River but does not sustain a population of juveniles in Tule Lake, as once was the case. Dewatering of Tule Lake and Lower Klamath Lake and large physical and chemical changes in the Lost River almost certainly are the cause for failure of endangered suckers in the Lost River below Clear Lake and Gerber Reservoir to show recruitment or increase in abundance.

6

Causes of Decline and Strategies for Recovery of Klamath Basin Suckers

When the Lost River and shortnose suckers were listed under the Endangered Species Act (ESA), the U.S. Fish and Wildlife Service (USFWS) and others identified numerous factors that could explain their decline and their failure to recover after elimination of the sucker fishery (Chapter 5, Scoppettone and Vinyard 1991). Since the listing, many of these factors have been studied. As a result, understanding of the biology of Klamath suckers and of requirements for their recovery has improved. Information on suckers is found in over 500 articles, reports, memoranda, and critiques, although most are unpublished and so have not benefited from scientific peer review. The number of persons working on the suckers has grown from a few ichthyologists to several dozen scientists, resource managers, policy developers, consultants, and informed citizens. New information derived from the increased pace of documentation and research supports increasingly firm judgments on the current status of the species, probable causes of their decline, priorities for further study, and actions that should and can be taken to move the species toward the ultimate goal of recovery, as described in this chapter.

CRITERIA FOR JUDGING STATUS AND RECOVERY OF SUCKER POPULATIONS

Criteria for the assessment of status and recovery provide a useful point of departure for the causal analysis of decline of the endangered suckers and for evaluating proposals for their restoration. Criteria presented here are intended as a tool of convenience for present purposes; other criteria might be useful for other purposes.

Because each life-history stage of a population is linked to all other stages, unusual suppression of any life-history stage may be reflected ultimately in the suppression of the population as a whole. Thus, trends in the abundance of any stage can be chosen arbitrarily as an index of the status of a population. For the endangered suckers, the most convenient life stage to use as an index of status is the adult. As explained in Chapter 5, other stages are difficult to observe or sample, especially in large lakes, although attempts to do so are essential to the diagnosis of mechanisms that affect specific life-history stages.

If adults are used as an index of the status of the populations, three criteria, taken together, would indicate recovery: diversity in the age distribution of adults, annual entry of at least some individuals into the adult stage in most years from the younger life stages coupled with entry of large numbers of such recruits in some especially favorable years, and a population size that reflects carrying capacity for an environment that is generally well suited, although not necessarily optimal, for the suckers. The presence of multiple age classes of adults would indicate past recruitment to the adult stage and persistence of conditions suitable for the maintenance of adults. The combination of new recruitment in most years and very high recruitment in some years would indicate the general welfare of younger stages and successful spawning. The maintenance of populations at a density that approaches expected carrying capacity would indicate that growth and reproduction occur at sufficient rates to offset mortality through the life cycle as a whole.

As indicated in Chapter 5, the status of geographically defined subpopulations of the two endangered suckers varies drastically. Table 6-1 summarizes the status of various geographic subpopulations on the basis of the adults. As shown in Table 6-1, Clear Lake and Gerber Reservoir support apparently stable subpopulations and therefore provide a basis for comparison with other subpopulations. The Upper Klamath Lake subpopulations, in contrast, do not meet the criteria for recovery, nor do they indicate recovery in progress. These subpopulations took an important positive turn after elimination of fishing in 1987, through the entry of new fish into the subadult and adult populations each year and through the production of one very strong year class (1991) and several moderately strong year classes during the decade of the 1990s (Chapter 5). Indications of no recovery without further environmental change, however, include the failure of adults to show an upward turn in overall abundances and the lack of a diversified age structure among older age classes, presumably because of repeated mass mortality of large fish.

TABLE 6-1 Summary of Status of Geographic Subpopulations of Two Endangered Suckers in Upper Klamath Basin[a]

	Recovery Criteria[a]					
Geographic Subpopulations	Age Structure	New Adults	Population Density	Required Actions	Priority[b]	Specific Actions
Clear Lake	+	+	+	Protection	1	Prevent alteration of tributaries; no drawdown exceeding 1992
Gerber Reservoir[c]	+	+	+	Protection	1	Same as for Clear Lake
Upper Klamath Lake	–	+	–	Remediation	1	Numerous, see Figure 6-1
Tule Lake	–	–	–	Remediation	2	Create spawning habitat[d]
Lake of the Woods	0	0	0	Remediation	2	Remove present fish; stock suckers
Lower Klamath Lake	0	0	0	Remediation	3	Raise level; stock adults[d]
Main-stem reservoirs	–	–	–	Protection	3	Protect status quo or better

[a]According to Chapter 5 and three criteria described in text for evaluation of status and recovery.
[b]Priorities are based on the apparent ultimate value of subpopulation to recovery of population at large.
[c]Shortnose sucker only.
[d]Requires feasibility studies.
Abbreviations: +, meets criterion; –, does not meet criterion; 0, population absent.

Fishes of Tule Lake (and of the associated Lost River) show no signs whatsoever of recovery according to the criteria shown in Table 6-1. Lack of recruitment of young fish into the subadult and adult stages indicates lack of reproduction or negligible survival of young fish. Two additional locations, Lower Klamath Lake and Lake of the Woods, are listed even though they lack endangered suckers. These are locations where sucker populations conceivably could be established in the future. The main-stem reservoirs also are listed but belong to a somewhat different category because, as explained in Chapter 5 and further in this chapter, the potential for creation of suitable conditions for the entire life cycle is probably lower for these waters than for Upper Klamath Lake or the other waters where the suckers originally thrived.

REQUIREMENTS FOR PROTECTION AND RECOVERY

The ESA requires both protection and recovery of listed species (Chapter 9). Protection is accomplished by prohibitions of take and preservation of habitat. Protection alone is insufficient, however, in that the populations as a whole have shown a drastic decline over the last several decades, and there is no evidence that the populations are recovering. At the subpopulation level, as indicated in Chapter 5, the balance between protection and remediation depends on location. Because the subpopulations of Clear Lake and Gerber Reservoir are the only ones in the upper Klamath basin that meet the criteria for recovery as outlined above, their protection is of utmost importance for the long-term survival of the two endangered sucker species in the upper Klamath basin as a whole. These subpopulations appear to depend entirely on tributary spawning. Therefore, maintenance of tributary conditions suitable for spawning is an essential element of their protection. It is important that neither of the reservoirs be drawn down to extremes that would produce summer or winter mortality. Given the historical experience of the 1990s, the requirements of the 2002 biological opinion appear to be adequately protective in this respect, but it is critical for these subpopulations that no errors in judgment lead to extremes in drawdown beyond that observed in the 1990s.

The subpopulations of Upper Klamath Lake also have high priority but have different status. As explained in Chapter 5, they showed some encouraging responses to the curtailment of the snag fishery, but the numerical abundance of adults and the continuing attrition of old fish appears to be holding the population down and may even be driving it closer to extirpation. The pathway to recovery for this population is not clear. A great deal of the analysis of cause and effect in the remaining part of this chapter is devoted to the Upper Klamath Lake subpopulations because of their historical numerical importance and the lack of clarity about the means of achieving their recovery.

The Tule Lake subpopulations consist of a very small number of apparently healthy adults, but they fail to meet all three of the criteria outlined above for recovery: there is no evidence of recruitment into the adult stage, there is no diversification of age structure for adults, and abundances per unit area are low. Because the suckers are long-lived, the adults of the Tule Lake population are of high value, and also could be supplemented with salvaged individuals from other locations. The first step toward recovery of the Tule Lake subpopulations would be to establish spawning capability, which would require intensive work with tributary waters. Acquisition of water rights and steps toward the creation of (potentially artificial) physical habitat suitable for spawning and for larvae would be necessary initial steps toward recovery of these subpopulations. The Tule Lake subpopulations, although small, need not be written off as unrecoverable.

Listed fifth in Table 6-1 is Lake of the Woods. As explained in Chapter 5, this was the location of a population probably consisting of shortnose suckers, but the population was eliminated. The present fish populations of Lake of the Woods should be eliminated, and adult shortnose suckers and other native fishes should then be reintroduced. If the suckers meet the recovery criteria outlined above after a number of years, fish biologists could consider the reintroduction of game fish (fish other than suckers probably will have colonized the lake by that time in any event).

Lower Klamath Lake lacks suckers and is probably unsuitable for them (Chapters 3 and 5), but alteration of these conditions could be feasible. Steps should be taken toward acquisition of water rights suitable for maintenance of higher water levels in Lower Klamath Lake if feasibility studies support this approach. Adult suckers from salvage (as described later in this chapter) should then be transferred to Lower Klamath Lake. Water quality and habitat conditions may be unsuitable, but suitability can be determined most effectively by monitoring of trial reintroductions. To the extent that maintenance of higher water levels would interfere with agricultural use of land, its establishment would require negotiations and compensation for acquisition of private rights.

The last subpopulations mentioned in Table 6-1 are the ones in main-stem reservoirs. These reservoirs have value primarily for long-term storage of large suckers. They do not have high priority for recovery, because they are not part of the original habitat complex of the suckers and probably are inherently unsuitable for completion of life cycles by the suckers. Maintenance of adults in these locations does, however, provide some insurance against loss of other subpopulations.

Construction of fish ladders for suckers at the dams might facilitate return of fish from main-stem reservoirs to Upper Klamath Lake. A fish ladder at Link River Dam, which is scheduled for completion in January 2006, should receive high priority; movements of fish through the ladder should be monitored.

SUPPRESSION OF ENDANGERED SUCKERS IN UPPER KLAMATH LAKE: CAUSAL ANALYSIS AND REMEDIES

For several reasons, causal analysis of the suppression of endangered suckers deserves more attention for the Upper Klamath Lake subpopulations than for other subpopulations. First, despite severe suppression of endangered suckers in Upper Klamath Lake, these subpopulations still contain many fish. Second, the subpopulations in Upper Klamath Lake were large as recently as 50 yr ago, so it seems reasonable, lacking evidence to the contrary, that they could be restored by a reversal of one or more critical human-induced impairments that have occurred over the last 50 yr. Third, water management involving Upper Klamath Lake is the responsibility of the federal government through the U.S. Bureau of Reclamation (USBR), which has access to substantial resources and also has legal responsibility for reversing or moderating any adverse effects of its management of Upper Klamath Lake if causal linkages between management and harm to the suckers can be established. Fourth, even though the subpopulations of endangered suckers are suppressed in Upper Klamath Lake, all life stages are present and some recruitment appears to be occurring from one life stage to another every year; recovery seems feasible if some key factors can be identified and changed.

Actual or potential cause-and-effect relationships that explain the status of a population are hierarchical. For present purposes, *immediate* causes can be explained in terms of suppression of one or more stages of the life cycle. For example, suppression of the entire population could be explained entirely or in part by exceptionally high mortality of larvae. Suppression of more than one component of a population could prevent it from recovering. There can be more than one immediate cause of suppression of a population.

Proximate causes are environmental factors. An example is poor water quality that leads to mass mortality of adult fish. A single proximate cause may be linked to more than one immediate cause. For example, poor water quality may suppress not only adults but also other life-history stages.

Ultimate causes, in the present context, are direct or indirect results of human actions. For example, operation of unscreened canals is an ultimate cause of mortality of fish in various life stages. Human actions that have led to changes in the water quality of Upper Klamath Lake are ultimate causes of mass mortality of large fish.

Recovery of the populations of endangered suckers can be approached most efficiently through analysis of the three levels of causation that explain failure of the fish to recover. Because the possible combinations of cause and effect are numerous, remedial actions, which are expensive, must focus on chains of cause and effect that are most likely to produce recovery. Winnowing the importance of cause-and-effect relationships requires information, some of which must be quantitative to be useful. The task of the researcher or the monitoring team is to produce information, typically over a period of years, that can be used to support estimates of the suppression of the population by chains of causation involving specific life-history stages (immediate causes), specific environmental factors (proximate causes), and specific human actions (ultimate causes). Knowledge of causation can produce estimates of the beneficial effect of remediating the effects of human actions.

Intensive research on the endangered suckers has been under way for a relatively short time, especially in view of the complicating effects of natural variation caused by climate and other factors that are not under human control. Only a few causal relationships are known well enough to support remedial action with confidence, but some of these are among the most important because they explain notable mortality of one or more stages of the population. Eventually, some of the more subtle but still important types of impairment and their causes must be clarified, as indicated in the following overview and analysis of cause and effect.

The analysis of causal connectivity is summarized in Figure 6-1. The figure shows the life stages of the endangered suckers as presented in Chapter 5 and identifies potential proximate causes of suppression of each life stage. Because the life stages are interconnected developmentally, the underlying premises of the diagram are that suppression of any life stage contributes at least potentially to suppression of the overall population and that a potential remedy for the suppression of the population lies in the identification and reversal of the suppression of individual life stages. It is not a foregone conclusion, however, that reversal of a particular type of suppression on a specific life stage will move a population notably toward recovery.

Figure 6-1 shows connections between immediate, proximate, and ultimate causes as solid or dashed lines. Solid lines indicate causal connections that are well established scientifically; typically these connections involve phenomena that are easily observed or documented (such as mass mortality of adults or death due to entrainment). Dashed lines indicate causal connections that are under study and for which there is insufficient evidence to show them as unimportant, moderately important, or important.

The figure shows convergence of multiple lines on individual immediate causes in some cases. Thus, the diagram indicates the likelihood that

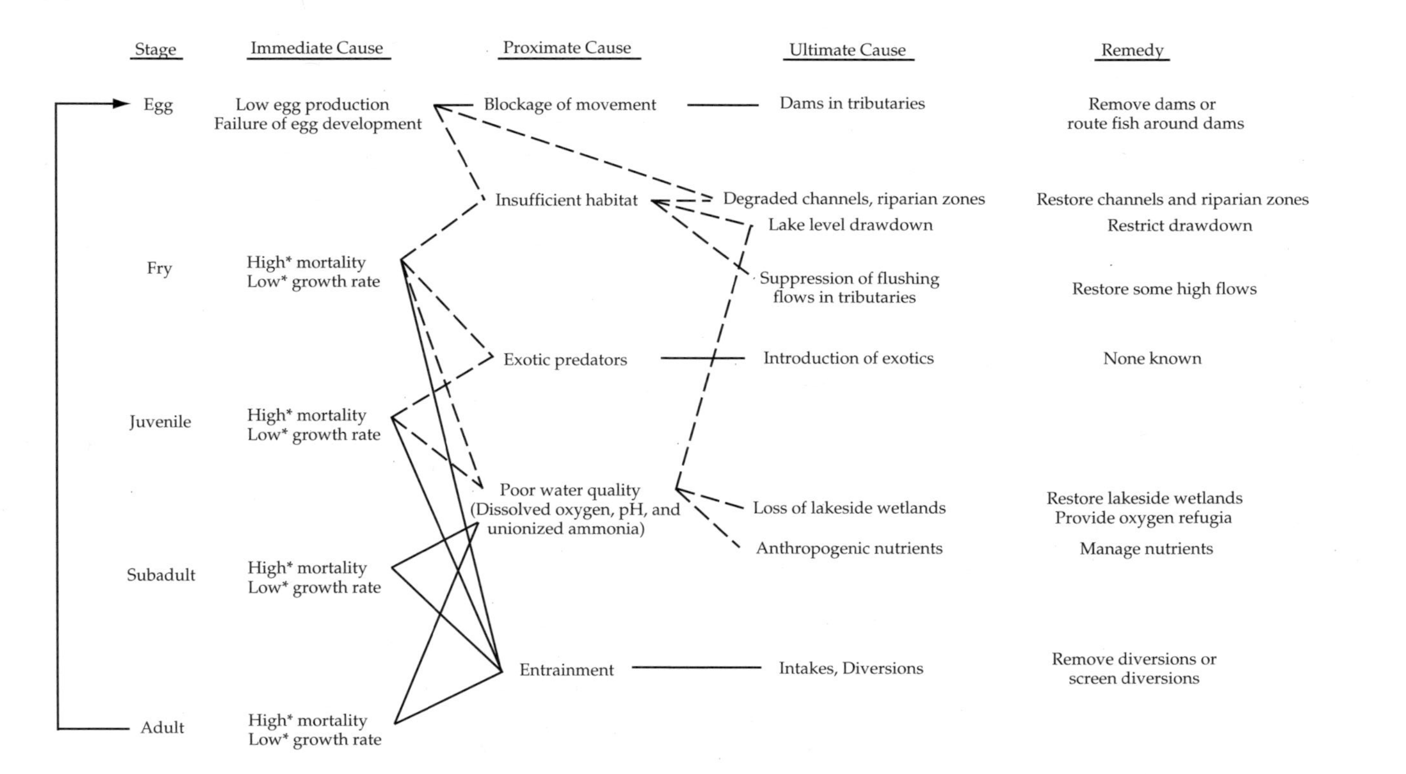

FIGURE 6-1 Diagram of possible causal connections in suppression of populations of endangered suckers in Upper Klamath Lake. Solid line = connection verified scientifically; dashed line = connection under study. "High mortality" and "low growth rate" are relative to rates in stable populations.

some immediate causes of decline are explained by multiple factors and that the factors might interact in their effects on a specific life-history stage. In addition, the diagram indicates that some environmental factors (proximate causes) have multiple connections with immediate causes; that is, they can affect more than one life stage. This is also expected from the literature on fish populations. The last column in the diagram lists remedial measures; the degree of certainty in their effectiveness is discussed below. Even though the life-history stages are interdependent and so must be considered together in the final prescriptions for recovery, it is useful to consider them individually first because each stage is affected by a distinctive suite of environmental factors. The discussion therefore follows the life-history sequence.

Production and Viability of Eggs

The production of eggs is usually discussed in terms of spawning fish, which are much more easily observed than eggs. The eggs themselves are the concern, however, and successful spawning is only one element of their final value to the population. Low viability of eggs, for example, could undermine the effectiveness of successful spawning. No researchers have attempted to make a case that the viability of eggs differs in Upper Klamath Lake or its tributaries from what would be expected in an unimpaired environment. Thus, the present discussion focuses on spawning, but it should be noted that lack of discussion of the fate of eggs after spawning is due partly to lack of information.

Dams

Small dams are found in the tributaries of Upper Klamath Lake. Where it can be shown that the dams do not allow passage of fish attempting to spawn, they should be removed or, if a dam must be retained, it should be fitted with a functional bypass.

The only moderately large dam on a tributary to Upper Klamath Lake is Chiloquin Dam, which blocks the Sprague River near its confluence with the Williamson River (Figure 1-3). Construction of Chiloquin Dam in the early 1900s (1918–1924—the exact date is unclear) may have eliminated more than 95% of the historical spawning habitat in the Sprague River (53 Fed. Reg. 61744 [1988], p. 5). This possibility is based on total river miles above the dam and does not take into account unusable portions of the river or the ascent of the dam by at least a few spawning fish via the fish ladder each year. There are more fish below than above the dam, however, and few fish enter the fish ladder (e.g., Janney et al. 2002), although the actual number is unknown. Improved access to the upper Sprague River would increase the extent of spawning habitat and expand the range of times and the conditions under which larvae enter Upper Klamath Lake.

Proposals for improving access of suckers to spawning grounds on the upper Sprague River involve two possibilities: removal of the dam and improved fish passage at the dam. Scoppettone and Vinyard (1991) recommended removal of the dam, as have others since then (e.g., Klamath Water Users Association 2001). Stern (1990) estimated the cost of removing the dam at about $500,000 and of fish passage improvements at $560,000. CH2M HILL (1996) presented detailed plans for improvement of passage and estimated the cost at $1.445 million but gave no estimate for removal of the dam. The plan of CH2M HILL includes construction of a new vertical-slot ladder on the left bank (looking upstream) that would replace the present ladder, which is ineffective. The new ladder would be based on fish passage structures through which cui-ui (*Chasmistes cujus*) move up the Truckee River and into Pyramid Lake.

CH2M HILL (1996, p. 2) dismissed removal of Chiloquin Dam because of "too many environmental concerns . . . as well as a lack of local support." The environmental concerns were not enumerated; presumably they are related to release of sediment and the difficulty of predicting how fish would respond to the new hydraulic conditions (e.g., Stern 1990). Issues related to sediments arise with virtually any dam-removal project, but often they can be resolved (Heinz Center 2002). The response of the fish is unknown, but removal of the dam is likely to result in a natural migratory response, at least by young spawners that have not already developed the habit of spawning downstream of the dam.

Lack of local support for removal of Chiloquin Dam is explained in part by water delivery via the dam to the Modoc Point Irrigation District (MPID). MPID involves about 60 farms and irrigates 3,000–5,300 acres annually, or less than 3% of the irrigable acreage in the basin. The MPID apparently has "adopted a Resolution indicating its willingness to participate in a project to restore fish passage" (Klamath Water Users Association, undated memo, about 2001) and is willing to consider moving its point of diversion away from Chiloquin Dam (E. Bartell, The Resource Conservancy, Inc., Fort Klamath, Oregon, unpublished report, 2002). Cooperation with MPID is important to the removal of Chiloquin Dam.

Removal of Chiloquin Dam has high priority and should be pursued aggressively. In the interim, spawning fish could be captured at the base of the fish ladder and released immediately above it; some of the released fish should be fitted with transmitters. Such a program would immediately give more fish access to the Sprague River and would show what upstream areas are favored by the fish. Continued monitoring below the dam also would provide information on numbers of adults returning downstream and numbers of larval fish reaching the lake. A summer sampling program could determine whether juveniles are in the river and would demonstrate the status of other native fishes in the river.

Water Level in Upper Klamath Lake

Spawning occurs at shoreline sites around Upper Klamath Lake from late February to May; maximum spawning activity occurs in March and April. More than 60% of spawning occurs in water more than 2 ft deep at locations with inflowing stream water (e.g., Reiser et al. 2001; see also Chapter 5). Inundation to a depth of at least 2 ft may be necessary for successful use of spawning substrate. At Sucker and Ouxy springs, two of the most frequently used sites (Hayes et al. 2002), lake elevations below 4,142.5 ft place 55% and 67%, respectively, of the spawning area in water shallower than 2 ft. Reiser et al. (2001, p. 7-2), in a separate analysis, concluded that lake elevations below 4,142.0 ft "severely diminish available spawning habitat"; they recommend that Upper Klamath Lake be kept at full pool elevation (4,143.3 ft) from mid-March to as late as mid-May to provide adequate water depth for spawning. Under recent operating regimes, water levels have remained above 4,143 ft for extended intervals in wet years but have fallen well below 4,143 ft in dry years (Figure 6-2).

Figure 6-2 shows the effect of water-level regulation in Upper Klamath Lake on spawning area according to the criteria proposed by Reiser et al. (2001). Under natural conditions, spring water levels would have been at or near full pool (4,143.3 ft). Under conditions prevailing in 1990–2001, full pool elevation was achieved during the spawning interval in 6 of 10 yr; in

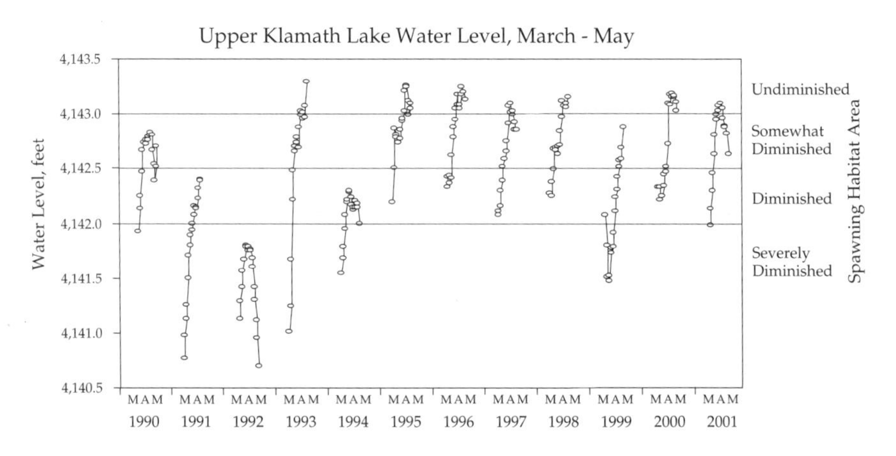

FIGURE 6-2 Water levels for 5-day intervals in Upper Klamath Lake over months of most vigorous spawning by suckers (March, April, and May—MAM), shown in context with spawning habitat designations given by Reiser et al. (2001).

the other 4 yr the water level was slightly lower to much lower, with corresponding consequences for the inundation of spawning sites.

It seems clear that drawdown of Upper Klamath Lake decreases the area of lakeside spawning habitat for the endangered suckers. Thus, a reasonable hypothesis is that lake levels below 4,143 ft, and especially those below 4,142 ft, suppress the production of larvae by reducing production of viable eggs, thus potentially affecting the population. In the absence of scientific information on the recruitment of larvae or other stages in years showing various amounts of water-level drawdown, professional judgment would be the only recourse for assigning significance of variations in spawning habitat to the relationship between production of larvae and water level in the lake. As a result of intensive study of the suckers, however, there is some direct evidence by which the hypothesis can be tested in a preliminary way.

Larval suckers have been sampled systematically since 1995 (Simon and Markle 2001). If drawdown suppresses spawning success substantially, one would expect lower relative abundance of larvae in years of extreme drawdown. The relationship between water level and abundance of larvae or juveniles would not necessarily be linear; it might involve thresholds rather than gradual changes in production of viable larvae.

Figure 6-3 shows the relationship between water level of Upper Klamath Lake in April (in the middle of the critical period) and relative abundances of larvae as shown by the standardized sampling program. Minor differences in relative abundances of larvae should not be considered significant because the sampling variance for any given year is substantial (95% confidence limits extend 50–100% around the mean in most cases).

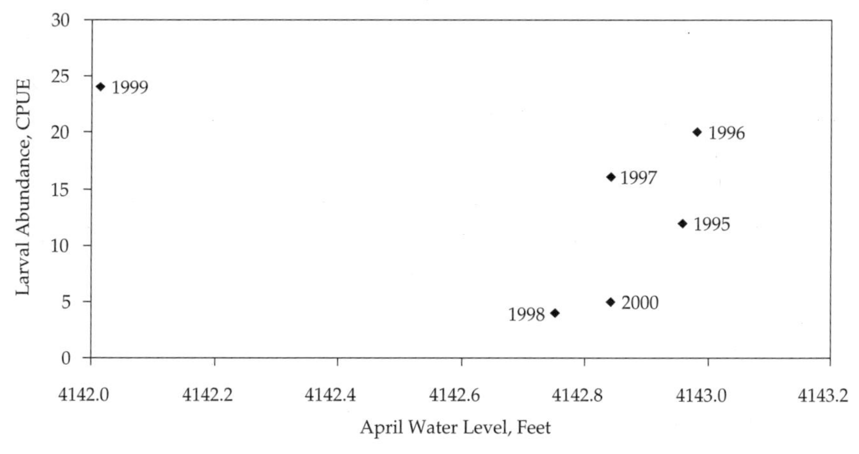

FIGURE 6-3 April water level and larval abundance (mean catch per unit effort [CPUE]) in Upper Klamath Lake. 95% confidence limits for annual means typically are 50–100% of the mean. Source: Simon and Markle 2001.

Thus, 1998 and 2000 might be considered distinctive in their scarcity of larvae, whereas 1995–1997 and 1999 belong to a second category of years involving much higher larval abundances that are virtually indistinguishable from each other because of sampling variance.

The year of lowest water levels during April was 1999, during which spawning habitat varied from somewhat diminished to severely diminished according to the criteria of Reiser et al. (2001; Figure 6-2). In all other years of the 6-yr record, the restriction of area was substantially less than in 1999. Thus, the hypothesis that diminution adversely affects production of larvae from eggs is contradicted by this test. The test is not particularly strong, because extremes of diminution and repeated years of diminution are not available in the record. Further observation might demonstrate some relationship that is not now evident. For the present there is no indication of a strong relationship between spawning success, as inferred from abundance of larvae, and water level in Upper Klamath Lake.

One other empirical test is possible. It is more remote in a life-history sense because it involves the relative abundance of adult fish. Its advantage is that it involves data that extend into different water years from those available for testing through larval abundance. As explained in Chapter 5, mass mortality of fish provides insight into the age structure of the endangered sucker populations. Specifically, the relative abundance of age classes of subadult and adult fish can be judged on the basis of their relative frequency of appearance among fish that are collected after the fish kill. As indicated in Chapter 5, any use of this information must be considered provisional because the relationship between the actual age structure of the population and the age structure reflected in the fish kill is unknown.

Given the assumption that large fish are killed in relation to their abundance in the population, relative abundance of specific year classes of fish should reflect the developmental history of each year class. If repression of larval production through restriction of spawning areas is critical in years of low water level in the lake, years affected by low level should stand out as producing a reduced population of large fish, given that large fish are ultimately a byproduct of successful spawning. The relationship between lake level and relative abundance (percentage frequency) of fish is shown in Figure 6-4. As indicated in the figure, the 2 yr of extraordinarily low water levels (1992 and 1994), which would be expected to show most strongly the negative signal involving larval production, do not indicate any repression of the year classes related to water level.

Further research may show a relationship between inundation of the spawning area and larval recruitment. Present data suggest, however, that any such relationship would be either weak or indirect. Thus, the connection does not appear to be especially important for the population. This conclusion seems counterintuitive, but there are several potential explana-

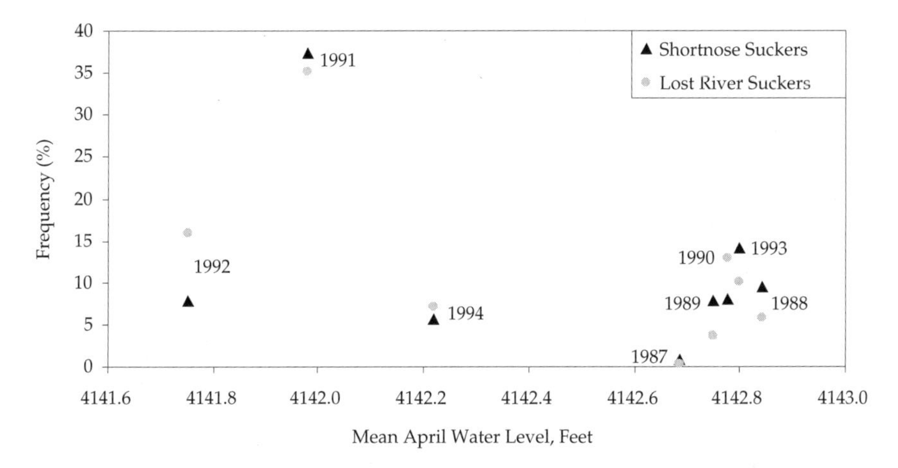

FIGURE 6-4 Relative abundance of year classes of suckers in Upper Klamath Lake, as inferred from fish recovered after mass mortality in 1997, in relation to water level during spawning interval when same year classes were produced. Source: USGS, unpublished data, 2001.

tions. First, the present population, which is much smaller than the original population, may have adequate spawning area even when spawning area is reduced, simply because it puts less total demand on the spawning area. Thus, progressive recovery of the population could produce a bottleneck related to spawning area in the future. Second, recruitment from spawning in streams may be more important than lake spawning under present circumstances. These and other possibilities cannot be distinguished at present. Overall, maintaining full pool elevation for promotion of spawning, although intuitively appealing, is difficult to defend scientifically.

Degradation of Spawning Areas

Some lacustrine spawning areas appear to be degraded, as indicated in Chapter 5. Where feasible, degraded spawning areas should be restored by introduction of additional gravel in appropriate type and size, removal of silt, or redirection of spring flows. It is unclear whether these actions will increase sucker spawning success, but they are not especially expensive and may be beneficial. Potential diminution of depth must be taken into account if restoration involves the addition of new substrate. Also, factors other than depth per se need to be studied more extensively with respect to the suitability of spawning areas. Wave action and other factors that have not yet been studied might be relevant, for example.

While lakeside spawning areas for suckers in Upper Klamath Lake have been studied extensively, tributary spawning areas have received relatively little attention. Where tributary spawning occurs, the morphometric features and substrate composition favoring the suckers should be identified, and specific efforts should be made to offset any changes in these characteristics that may have occurred through anthropogenic mechanisms. In addition, potential adverse effects of suspended load should be identified. Improvement of appropriate conditions for spawning will likely require protection of riparian zones from grazing and other disturbances, reduction in transport of suspended load related to land disturbance through agricultural and other land-use practices, and restoration of wetland near channels. Furthermore, it may be effective to protect specific spawning regions of tributaries from human presence in order to reduce the possibility of harassment and to increase public awareness of the importance of specific locations for successful spawning by suckers.

Some tributaries and lakeside spawning areas that are known to support successful spawning by suckers may not require restoration but do require vigorous protection because of their special value to the population. Even subtle changes, which might involve pumping of groundwater in the vicinity of these spawning sites, land disturbance, recreational activities, poorly managed agricultural practices, and other human activities could easily degrade or even eliminate these sources of sucker fry.

Abandonment of Spawning Areas

Some historical spawning areas have been abandoned for no apparent physical reason. Reestablishment of population components with natal affinities to the areas should be attempted. The degree of benefit cannot be estimated from present information, but the work could be accomplished without great cost. Specific locations are as follows:

1. Harriman Springs in northern Upper Klamath Lake was last used in 1974; spawning was also reported historically at Odessa Creek on the western shore (Andreason 1975, USFWS 2002). Barkley Springs on the southeast shoreline of Upper Klamath Lake was a previous spawning site but has not been used since the late 1970s (Perkins et al. 2000a), because diking, ponding, and rerouting of water associated with the construction of Hagelstein Park in the 1960s apparently blocked access of the fish to the site. Spawning substrate was added and water-control devices designed to inundate the springs were constructed in 1995, but no spawning has yet been observed.

2. Spawning suckers were reported at a spring on Bare Island (in the northern portion of Upper Klamath Lake east of Eagle Ridge) in the early

1990s, but spawning has not been observed at the site since then (Perkins et al. 2000a).

3. In the region of Agency Lake, spawning of suckers was observed in the late 1980s and early 1990s in Crooked Creek, Fort Creek, Sevenmile Creek, Fourmile Creek, and Crystal Creek. The Wood River has had the only recent spawning activity, most of it attributed to shortnose suckers. Adults were last seen in the Wood River in 1996, and larvae were last collected in 1992; no juveniles were found in 2000 (Simon and Markle 1997b, 2001; Cooperman and Markle 2003).

4. Additional, indirect evidence of abandoned spawning sites in Upper Klamath Lake itself has been obtained on the basis of lost fishing gear (Cooperman and Markle 2003). Shoreline surveys conducted during record low lake levels in 1994 revealed fishing gear on the bottom at known spawning sites, such as Ouxy and Sucker Springs. Lost gear also was found at four unnamed, flowing spring sites between Modoc Point and Sucker Springs. Failure to observe suckers spawning at these four sites during recent spawning surveys suggests that direct removal or harassment led to the elimination of the spawning aggregations.

The available evidence strongly suggests that lake and stream spawners mix only occasionally if at all and that spawning-site fidelity causes an abandoned spawning site to remain unused. Abandonment of apparently appropriate spawning sites indicates that the use of a spawning site is a social tradition, that is, that fish learn about spawning sites by following or observing other fish (e.g., Helfman and Schultz 1984). A good spawning site may remain unused by fish that show those characteristics if "teachers" are absent, as has been demonstrated for reef-spawning wrasses in the Caribbean (Warner 1988, 1990). Use of abandoned sites might be renewed spontaneously if populations of adults become substantially more abundant.

The possibility that sites are abandoned because of a break in tradition suggests a solution. Transplantation of spawning-ready fish of both sexes to historically used sites, perhaps accompanied by confinement of the fish in cages for a brief acclimation period, might initiate use of the abandoned sites. Feasibility of this approach is suggested by Warner's (1988, 1990) manipulations, which involved transplantation of fish to locales that had been experimentally depopulated, with subsequent establishment of site-specific, traditional spawning groups by transplanted individuals. Males might be attracted to caged females in spawning-ready condition; spawning readiness could even be induced, if necessary, by hormone injection. Fish could be transplanted from habitats that lack recruitment—such as Tule Lake, the Lost River, or the Klamath main-stem reservoirs—assuming that spawning-ready individuals are available. If fish from Upper Klamath Lake are used for such manipulations, they should probably be young, first-time spawners because fish with spawning experience are likely to abandon a new site for a site with which they are familiar.

Regardless of the cause of spawning-site abandonment, loss of spawning aggregations has several consequences for sucker recovery. If the aggregations at these sites represented genetically distinct groups of suckers, overall genetic diversity of the Upper Klamath Lake populations probably has been reduced. Even without genetic distinctness, the uniqueness of circumstances at each site creates potential differences in survival of larvae originating at different sites. Multiple spawning sites have a bet-hedging effect on larval survival: the more spawning sites a population uses, the more resistant the population is to exceptional loss at any one site.

Survival of Larvae and Juveniles

Mortality of larval and juvenile stages of all fishes is high, even in populations that successfully saturate their environment. High mortality in the young stages of the life history of a given fish population does not necessarily indicate that these stages are a bottleneck that leads to repression of the population. Survival of larval and juvenile stages in a repressed population could be usefully compared with those in a vigorous population; a bottleneck at the larval and juvenile stages would be indicated by substantially lower survival rates in the repressed population than in the vigorous population. However, estimation of survival rates of young life-history stages of fish is extremely difficult, and less direct indicators often are the only recourse for assessment of these stages, as is the case for sucker populations of Upper Klamath Lake.

Morphological Anomalies in Young Fish

Morphological anomalies—which may indicate parasitism, dietary deficiencies, or physiological stress during development—suggest abnormal losses of young fish during development. Where fish are not under physiological stress due to poor water-quality conditions, morphological anomalies seldom exceed 1% (Karr et al. 1986). In Upper Klamath Lake, however, the frequencies of anomalies among the larval and juvenile shortnose suckers averaged 8%, and among the Lost River suckers averaged 16% (Plunkett and Snyder-Conn 2000). The anomalies included deformities of the fins, eyes, spinal column, vertebrae, and osteocranium, as shown by Plunkett and Snyder-Conn (2000), who suspected chemical agents of human origin. These authors reviewed literature indicating high frequencies of anomalies in other fishes as well (fathead minnows and chub species) and in amphibians of the Upper Klamath Lake basin. Harmful agents have not yet been identified.

Skeletal deformities in young fish can affect their swimming performance and indirectly increase their vulnerability to predation and impair their ability to escape unfavorable habitat conditions. Plunkett and Snyder-Conn (2000, p. 2) suggest that the relatively high rate of anomalies in young suckers could result in "early elimination of anomalous 0-aged suckers from Upper Klamath Lake populations." Direct comparisons with populations in Clear Lake and Gerber Reservoir, where populations are apparently stable, would be informative.

Entrainment of Larvae and Juveniles

Entrainment at and lack of passage through Klamath River dams and other irrigation structures were added to the list of threats to the endangered suckers after the original listing (e.g., USFWS 1992a). Entrainment into irrigation and power-diversion channels is now recognized as being responsible for loss of "millions of larvae, tens of thousands of juveniles, and hundreds to thousands of adult suckers each year" (USFWS 2002, Appendix C., p. 24). Sucker larvae appear at the south end of Upper Klamath Lake beginning in late April. Millions of young fish then are swept from Upper Klamath Lake into the Link River, whence large numbers are drawn into the A Canal (USFWS 2002), from which they cannot escape.

Speculation has developed about the source of the young fish that reach the Link River. They may come from known spawning sites along the northeastern portion of Upper Klamath Lake, from such tributary streams as the Williamson River, or from unknown spawning sites farther south. Because all known spawning sites are in the northern portions of the lake, the critical question is whether currents in the lake are strong enough and of proper alignment to deliver larvae to the Link River 18 mi to the south.

Some evidence indicates that larval and juvenile fish entering the Link River originate in known riverine and lake spawning areas. Prevailing winds are from the northwest when larvae are present and establish substantial south-flowing currents, according to a numerical model developed by Philip Williams & Associates (PWA 2001). The Philip Williams model suggests that it is very feasible for larvae produced from the Williamson and Sprague system to enter the south end of the lake within a few days of swimup, the time at which larvae first leave the substrate for the water column (R. S. Shively, U. S. Geological Survey, Klamath Falls, Oregon, personal communication, 2002). Whether entrainment is caused by natural movement of fish that would historically have entered Lower Klamath Lake or is an avoidance response to poor habitat or poor water-quality conditions is unknown. Regardless, given that these larvae likely originate in known spawning aggregations and that any larvae leaving the lake to the south are permanently lost from the population, entrainment of young fish is a potentially important contributor to failure of the populations to grow.

USBR was scheduled to place fish screens at the A Canal in the summer of 2003. These screens function effectively with fish larger than 30 mm (USFWS 2002). Although retention of fish smaller than 30 mm could be achieved, the likelihood that very young, fragile fish would survive impingement (along with algae and debris) on the screens is low, and the chances of salvaging them successfully are even lower. Juvenile fish may survive impingement but, unless they move against the current, will still be lost from source populations because fish screened from the A Canal will next pass through the Link River Dam and then enter other canals, be killed by turbines, or join nonreproducing populations downstream (Figures 1-2 and 1-4). Even so, the screening does prevent loss of subadults, adults, and some juveniles through the A Canal.

USFWS (2002) recommends coordination of intake at the A Canal with timing of juvenile movements, deflection barriers that would move juveniles away from intake structures, location of intakes above the water-column strata in which young suckers usually swim, and salvage. These measures seem reasonable and should be pursued. Salvage operations may be pointless, however, if emigration from the lake is an avoidance response to poor water quality. Salvaged fish possibly could be moved to a holding facility with good water quality before return to Upper Klamath Lake or could be transplanted to other sites to establish new populations.

Adequacy of Nursery Habitat for Larvae and Juveniles

Upper Klamath Lake has lost an estimated 66% of emergent marsh vegetation and submerged vegetation (USFWS 2002). Specific changes include the apparent loss of emergent vegetation in the region between the Williamson River mouth and Goose Bay that probably once was important larval habitat; vegetation should be restored in this area as soon as possible. In general, diking, draining, and water-level management have reduced emergent and submerged vegetation along shorelines by about 40,000 acres (USFWS 2002). Remaining marginal marshes around Upper Klamath Lake are reduced, patchy, and often dewatered by middle to late summer as water level falls.

Vegetation in shallow water is a consistent aspect of larval habitat and may be important to juvenile habitat as well (Chapter 5). Abundance of this habitat feature during the larval phase, which extends from April through July, in Upper Klamath Lake is in part related to water depth. Higher water levels in Upper Klamath Lake are associated with larger amounts of emergent vegetation (Table 6-2). Ignoring emergent vegetation, total shoreline area that is at least 1 ft deep at lake water levels of 4,142–4,143 ft accounts

TABLE 6-2 Estimates of Larval Habitat Availability Calculated as Percentage of Lakeshore Inundated to a Depth of at Least 1 Ft for Lake Edge and Marsh Regions in Northeastern Upper Klamath Lake that Contain Emergent Vegetation, and Total Lake Shoreline Regardless of Vegetation

	% Larval Habitat Available		% Lake Shoreline Available	
Water Level, Lake (ft)	Dunsmoor et al. 2000	Reiser et al. 2001	Chapin 1997	Reiser et al. 2001 (All Shoreline)
4,143.0	–	–	–	85–100
4,142.8	80[a]	–	–	–
4,142.0	50[a]	100[b]	–	40–60
4,141.5	–	80[b]	–	–
4,141.2	–	80[c]	–	–
4,141.0	–	–	–	10–25
4,140.0	0[d]	0	0	–

[a]Shoreline emergent vegetation.
[b]All emergent vegetation.
[c]Marsh edge habitat only.
[d]Almost completely unavailable.

for at least 50% of the lake's perimeter, but this fraction declines rapidly with reduced water levels. Very little emergent vegetation is available to larval suckers below a lake level of 4,141 ft; emergent vegetation is essentially inaccessible below 4,140 ft (Reiser et al. 2001). Reiser et al. (2001) recommend maintaining water levels above 4,142 ft at least until July 15 to ensure access by larvae and juveniles, although the data on use of this habitat by juveniles are not clear.

Because the majority of suckers in Upper Klamath Lake now spawn in the Williamson and Sprague river system, use of habitat in the system by larvae could be important in determining production of larvae. Under current conditions (blockage of spawning migrations at Chiloquin Dam combined with a highly modified stream channel in the lower Williamson delta), a higher proportion of larvae may be produced in the lower Williamson than were produced there historically. As a result, the larvae may pass from the river to the lake more quickly and with less temporal dispersion than was the historical norm. Cooperman and Markle (2000) found that larvae left the Williamson River in as little as a single day and that 99% of larvae entering the lake had not yet developed a tail fin and so were not yet competent swimmers and feeders. The majority of larvae in the lower river sampled by Cooperman and Markle (2000) had empty guts. Thus, many larvae may be entering Upper Klamath Lake before they are ready to feed or to avoid predators (comparisons with Clear Lake and Gerber Reservoir populations would be useful but are not available). Modifications to the lower Williamson have reduced plant cover, and thus possibly reduced food production and shelter from predators. The Nature Conservancy is restoring the lower Williamson to a more natural, meandering, multiple-channel configuration that supports denser riparian and emergent vegetation. This project should be completed soon. Larvae descending from the Williamson system will find cover near the mouth of the river when vegetation and morphology have begun to recover, which may take some time.

Physical conditions that may impair spawning and support of fry in the rivers above Upper Klamath Lake have not been adequately studied. Changes in river channels have occurred as a result of removal of riparian vegetation, access of cattle to the streams, alteration of flows, and loading of the stream with fines. All of these factors should be documented and measures should be taken to reverse them on grounds that these changes are quite likely to interfere with successful spawning and larval survival.

Hypotheses about the significance of lake-level changes and capacity of Upper Klamath Lake to sustain larval suckers can be tested against information on the relative abundance of sucker larvae, as determined over the years 1995–2000. If interannual variation in lake levels is a dominant factor in the viability of larvae in the lake, years of higher lake level during the larval development period should be marked by higher larval abundance. To be of use in management, any beneficial effects of high water level should appear as higher CPUE (catch per unit effort) of larvae. This is not the case, however (Figure 6-5). In fact, the amount of larval habitat in spring varies across years much less (about 2-fold; compare Figure 6-5 with Table 6-2) than larval abundance per unit area (as indicated by CPUE—10-fold).

Additional testing is possible through use of information on relative abundance of year classes among fishes collected during episodes of mass mortality. If interannual variations in lake level correspond to relative degrees of repression of larval production, and this factor has a major effect on the populations, year classes produced in years of especially low water levels in Upper Klamath Lake should be exceptionally weak. Once again, this is not the case (Figure 6-6).

Lack of correspondence between larval abundance and indicators of year-class success based on either collection of larvae or collection of adults does not contradict the idea that inundated vegetation is critical habitat, that is, habitat that the suckers need in some unknown amount and distribution. It does call into question the idea that greater or smaller abundance of this habitat feature from one year to the next is regulating the populations. Cooperman and Markle (2003) have argued that complicating factors could mask an important relationship between water level in Upper Klamath Lake and production of larvae. From a scientific viewpoint, how-

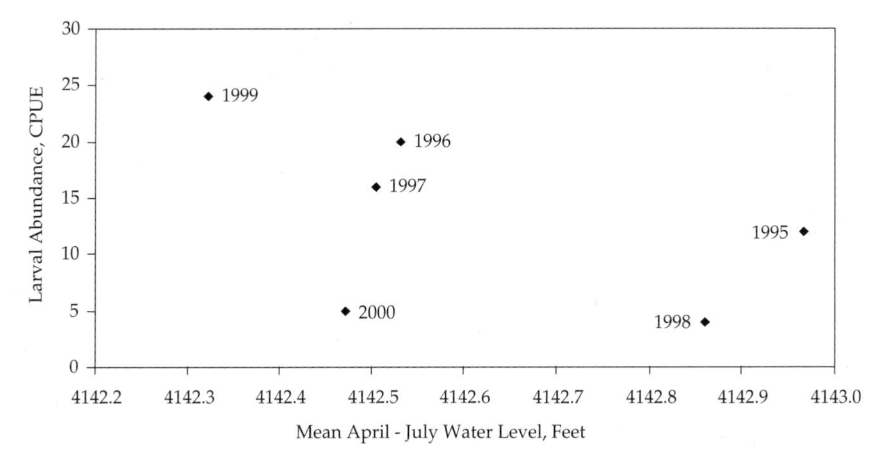

FIGURE 6-5 Relative abundance of larvae as determined by standardized sampling, shown in relation to mean water level of Upper Klamath Lake during the main interval of larval development (April–July). Source: Simon and Markle 2001.

ever, the water-level hypothesis is not supported because it fails empirical tests for the presently available data. An argument for a complex relationship involving water level would require empirical support, of which there is none. One potential line of investigation would be to examine the differences in larval production of the two sucker species. The two species appear to be responding in similar ways to environmental change, but the data

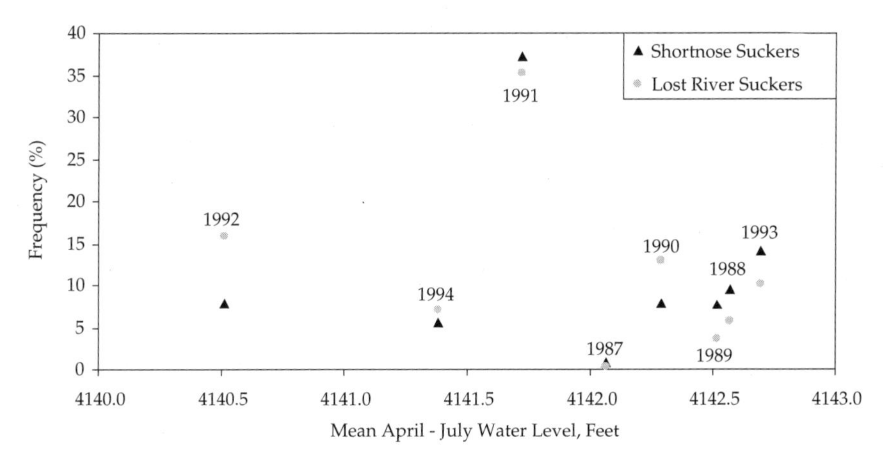

FIGURE 6-6 Relative abundances of year classes of endangered suckers collected from Upper Klamath Lake during the fish kill of 1997, shown in relation to mean water level over the interval of larval development for the same year classes. Source: USGS, unpublished data, 2001.

suggest that the responses are not exactly the same. Differences related to timing or place of spawning may be important.

From a management perspective, the difficulty with a water-level hypothesis that involves unknown complications is that observations of higher water levels at present offer no evidence that would support maintenance of higher water levels. At the same time, the lack of a relationship between observed water levels and larval abundances cannot be taken as justification for broader manipulation of water levels, which at some extreme could be notably harmful.

Monitoring of larval abundance and year-class abundances as inferred from mass mortality indicate that the explanation for interannual variability at present lies in key factors other than the amount of shallow water or emergent vegetation. This conclusion should energize the investigation of other habitat features. For example, restricted availability or poor condition of tributary spawning areas could be critical. Interannual variability of year-class abundance as affected by delivery of larvae from tributary spawning areas would be an obvious subject for further study.

The known biology of the suckers indicates that particular depths are preferred at established spawning locales and that flooded emergent vegetation is primary larval habitat. The lack of relationship between water level in Upper Klamath Lake and larval production or larval survival indicates that other factors, such as degraded water quality or poor larval habitat, override the presumed benefits of depth-related habitat availability. Recommending maintenance of particular water levels to promote sucker recovery has no clear scientific basis until the factors that override water depth are better understood and, if possible, rectified. USFWS may retain an interest in water-level manipulations as justified by the need to minimize risk. Given limitations on the legitimate use of the need to minimize risk, however (Chapter 9), it might be difficult for USFWS to justify more stringent limitations on water level as a general operating rule. One alternative is for USFWS to work with USBR in testing various water-level combinations that can be achieved through such actions as experimental use of water-bank resources or by use of the excess water that may be available in some years.

Overview of Larval and Juvenile Production

As explained above, larvae were variably abundant in trawl catches throughout the 6-yr monitoring period 1995–2000. Catches were high in 1995, 1996, 1997, and 1999, and were relatively low in 1998 and 2000. No correlations are obvious between abundance of fish in spawning runs and larval abundance (Simon and Markle 2001, USFWS 2002) or between fish kills and larval abundance. Abundances of young of the year (YOY) also have high year-to-year variation and lack any detectable relationship with abundance of spawners. The year 1999 was good for larvae and juveniles regardless of sampling locale or method (Simon and Markle 2001, USFWS 2002), and 1991 must have been favorable as well, as judged from abundance of adults (monitoring of larvae did not begin until 1995).

As in most fish populations, abundance of young suckers in Upper Klamath Lake declines progressively through each summer and fall (Simon and Markle 2001). Declines could be explained by offshore movement as the fish grow, high mortality, high emigration rates from the lake, or a combination of these. Abundance of juveniles in spring (age 1+ yr) appear to reflect a 90% overwinter mortality or emigration (Simon and Markle 2001). High incidences of physical abnormalities in these fish (Plunkett and Snyder-Conn 2000) imply that mortality or export may repress recruitment of subadults and adults, although avoidance of sampling gear by postlarval fish creates difficulties in interpretation.

Some minimal number of spawners is necessary to produce a successful year class of larvae, but the lack of correlation between numbers of spawners and abundances of larvae implies that abundant spawners are no guarantee of high larval numbers and that, given the high fecundity of suckers, a small number of spawning fish may be sufficient to produce abundant larvae if conditions for larvae are good.

Adults

Entrainment

Fish that enter water-management structures typically cannot return to the habitat from which they came or enter another suitable habitat. For Upper Klamath Lake, the A Canal has long been recognized as a source of entrainment for all life-history stages, including adults, whose loss may be especially significant because of the importance of large fish in maintaining the fecundity of the population (Chapter 5). Scheduled screening of the A Canal, which will be ineffective for small fish < 30 mm, will block entrainment of subadult and adult fish, and could thus reverse an important historical source of mortality. The benefits of this measure to the population are unknown. Entrainment of fish from Upper Klamath Lake via the Link River still occurs through intake structures of the Link River Dam, which should be screened (USFWS 2002).

Mass Mortality

Unlike most other imperiled lakesuckers, suckers of Upper Klamath Lake suffer from episodic mass mortality of reproductive-age fish. Although such mortality probably inhibits recovery, fish kills are not new to Upper Klamath Lake. Records indicate periodic kills dating at least to the late 1800s; before the 1990s, large fish kills occurred in at least 1894, 1928, 1932, 1966, 1967, 1971, and 1986 (USFWS 2002). Whether episodic mass mortality has always occurred in Upper Klamath Lake is a matter of conjecture.

The actual numbers and sizes of fish killed are difficult to estimate because of sampling difficulties, differential sampling effort, loss of small fish to birds, and loss of fish that do not float after death. Mortality may reach tens of thousands in a severe episode (Perkins et al. 2000b). The effects of fish kills on spawning populations of suckers probably have been substantial. As much as 50% of the adult populations may have died in the 1996 fish kill; sizes of spawning runs indicate that the spawning populations of both species were reduced by 80–90% from 1995 to 1998 (USFWS 2002; Chapter 5).

The largest documented case of mass mortality occurred in 1971; it involved the loss of about 14 million fish, most of which were blue and tui chubs. Water level may or may not have played a role in conditions leading to the incident, but 1971 was the year of highest recorded water level since full operation of the Klamath Project began in 1960. It is unclear whether the extent or frequency of mortality is greater now than earlier. Incidents of mass mortality in 3 consecutive recent years (1995, 1996, and 1997) are a reason for special concern, but it is impossible to determine whether such episodes now are more frequent than in the past.

It could be argued that mass deaths of suckers is a natural phenomenon caused by very high abundances of algae that have always been characteristic of Upper Klamath Lake. Or it could be argued, without particularly strong support, that mass mortality is more frequent or more severe than it used to be. It is not necessary, however, to resolve this point for ESA purposes. Because the abundances of the endangered suckers have been drastically reduced, any factor that leads to a larger population should be favored as a step toward recovery of the species, even if it involves a natural mortality mechanism. Thus, reducing mass mortality, whether natural or not, should be counted as beneficial to the welfare of the species and should be pursued.

Conditions commonly associated with fish kills include high temperature, intense blooms of bluegreen algae, high incidences of copepod (*Lernaea*) infestations (see Table 6-3), cysts, lesions, infection with *Flavobacterium columnare* (columnaris disease), high pH, high concentrations of unionized ammonia, and low concentrations of dissolved oxygen (Perkins et al. 2000b, Chapter 3). Before kills, some fish apparently move to the Link River (Gutermuth et al. 1998), and others (mainly redband trout) become concentrated in specific refuge areas, including Pelican Bay, Odessa Creek, and the Williamson River mouth. Refuges often contain springs that offer much better water quality than the lake itself (Bienz and Ziller 1987).

TABLE 6-3 Incidence (%) of Various Indicators of Stress in Suckers of Upper Klamath Lake Based on Visual Inspection

	Incidence, %				
	Lampreys Wounds	Copepods Infections	Eye Damage	Emaciation	Wounds
Lost River Suckers, Live Fish, 2001					
Lake spawning	40	22	4	0	1
River spawning	48	28	22	0	2
Lake non-spawning	51	18	8	1	2
Shortnose Suckers, Live Fish, 2001					
Lake spawning	53	30	3	0	0
River spawning	38	51	16	0	1
Lake non-spawning	48	33	8	0	4
Fish Kill					
1997		73[a]			

[a]Based on Foott 1997 and Holt 1997 in USFWS 2002; incidence of *Columnaris* disease was 92% and 80%, respectively, during the 1996 and 1997 fish kills (USFWS 2002).
Sources: Coen et al. 2002, Cunningham et al. 2002, Hayes et al. 2002.

Mortality of fish during routine sampling with trammel nets also increases during the weeks preceding a fish kill (USFWS 2002).

Although USFWS (2002) went to considerable lengths to examine the possible direct influence of high water levels in Upper Klamath Lake on sucker welfare, the data now on hand contradict the hypothesis that water level is associated with fish kills (NRC 2002, Figure 3; Chapter 3). Fish kills have occurred in years of low, average, and above-average median August lake levels. Water level may affect the accessibility of refuges that are reportedly used by large fish during periods of poor water quality and fish kills, but the data on this topic are largely anecdotal (see Buettner 1992 unpublished memo, USFWS 2002, Appendix C, and below).

High incidences of parasites, bacterial infections, and other anomalies imply that stressful conditions exist in Upper Klamath Lake for several weeks before the appearance of dead fish. Loftus (2001, cited in USFWS 2002) developed a "stress-day" index that accounts for multiple stress factors related to water quality. In 1990–1998, accumulated stress days were maximal in July and August during the fish-kill years of 1995 and 1997. The stress-day index approach is useful in that it involves regular, coordinated monitoring focused on water quality, meteorology, fish condition (parasite frequency, body condition, and so on), and attention to increased numbers of adults in the Link River or presumed refuges. When conditions and early warning signs converge, whatever remedial actions are feasible should be taken, possibly including oxygen supplementation at specific locales where suckers aggregate (Chapter 3).

In some lakes, mass mortality of fish occurs under ice ("winterkill"), usually in association with low concentrations of dissolved oxygen. Winterkill is not known to have occurred in Upper Klamath Lake or in any other lakes occupied by endangered suckers. Thus, the relevance of winterkill to Upper Klamath Lake remains hypothetical, as do management actions that would minimize its likelihood or effect.

Winter mortality (but not necessarily winterkill) has been postulated as the cause of a 90% reduction of first-year juvenile suckers in Upper Klamath Lake from late fall to early spring and population reductions in other species (Simon and Markle 2001). Comparable data are needed on winter mortality in surrounding water bodies with better water quality (such as Clear Lake) to determine whether the 90% mortality figure is extreme.

Concern over winterkill is justified, especially if water quality deteriorates further or if an exceptionally cold winter results in an unusually long period of ice cover. Improvement in water quality in the lake probably would reduce the likelihood of winterkill, but may be infeasible over the short term. Winter monitoring of oxygen should be undertaken in any event (Chapter 3).

Loss of Habitat

Adult Lost River suckers and shortnose suckers prefer open water; they use flowing waters chiefly for spawning. Total lake habitat available to suckers throughout the Klamath basin is a fraction of its original extent because of drainage and other water-management practices (Chapter 2). Even where it persists, habitat for adults may be compromised during late summer. Adult suckers appear to prefer water that is deep and turbid, and thus dark (USFWS 2002), but degraded water quality in summer apparently forces fish to use specific areas of shallow, clear water, such as the mouth of Pelican Bay in Upper Klamath Lake.

Factors Relevant to All Life-History Stages

A number of factors, some of which have already been mentioned, are potentially relevant to all life-history stages, although further research may show them to be more relevant to some stages than to others. Most prominent is poor water quality, which is linked not only to mass mortality of adults but potentially to undocumented mortality of other stages and to stress, which in turn may be a cause of anomalies, parasitism, and disease in multiple life-history stages. A second complex of factors that may apply broadly across stages, but still in unknown ways, falls under the heading of predation and competition, primarily from nonnative fishes. A final factor that cannot yet be attached to any particular life-history stage is hybridization, which may change populations genetically.

Water Quality

Suckers of Upper Klamath Lake suffer from varied deformities, parasites, lesions, cysts, and infections. The afflictions of adult suckers include eroded, deformed, and missing fins; lordosis; pughead; multiple water-mold infections; reddening of the fins and body due to hemorrhaging; cloudiness of the skin caused by low mucus production; loss of pigmentation; external parasitic infection by copepods and leeches; lamprey wounds; ulcers; gas emboli in the eyes; exophthalmia; cataracts; and a high incidence of gill, heart, and kidney abnormalities after fish kills. Plunkett and Snyder-Conn (2000) reported body-anomaly rates of 8–16% in larval and juvenile suckers. Juvenile suckers suffered infestation with copepods and trematodes of 0–7% in 1994–1996 and 9–40% in 1997–2000; shortnose suckers generally show higher rates of infestation than Lost River suckers (USFWS 2002 based on Carlson et al. 2002). Data on both species in Upper Klamath Lake and at river spawning sites also indicate relatively high frequencies of abnormalities in adults (Table 6-3). Spawning and nonspawning fish do not show substantial differences in the incidence of such indicators, except that copepod infestations appear to be higher in shortnose suckers and eye damage is higher in river-spawning fish of both species. The latter finding might reflect crowding of fish downstream of Chiloquin Dam or injuries to the fish as they attempted to negotiate the unsuitable fish ladder at the dam.

The widely used Index of Biotic Integrity (Karr et al. 1986) incorporates 1% as a threshold criterion for anomalies; sites with fish above this threshold receive the lowest metric scores for their ability to support a diverse biota. The appropriate threshold may vary geographically and by taxa, however. For the Willamette River, Hughes and Gammon (1987) identified 6% as a threshold. Hughes et al. (1998) proposed a more general threshold of 2%. Most collections from all size classes of Upper Klamath Lake suckers exceed these thresholds. It is not known why Clear Lake, with its better water quality and apparently stable population, also is characterized by "heavy parasite loads on suckers and other fish" (Snyder-Conn, personal communication cited in USFWS 2002, Appendix E, p. 38).

Even if infections and afflictions do not lead directly or even indirectly to death, they are likely to inhibit growth (e.g., M.R. Terwilliger et al., Oregon State University, Corvallis, Oregon, unpublished material, 2000) and reproduction and may compromise an individual's ability to resist other sources of stress. Without better baseline and reference values for suckers in other water bodies in and out of the Klamath basin, it is difficult to state categorically that the incidence of anomalies is extraordinary, but field researchers who work with fish seldom observe affliction rates approaching those found in Upper Klamath Lake.

Nonindigenous Species as Predators and Competitors

Eighteen of the 33 fish taxa in the upper Klamath basin are nonnative (Chapter 5). The nonnatives dominate numerically in many habitats and probably influence native species, including the endangered suckers, through predation and competition. Competition is particularly difficult to quantify in nature (Fausch 1988, 1998). Thus, it is not often possible to invoke competition as a major cause of problems in a population, and it also is difficult to moderate competition even where it can be demonstrated. In contrast, predation on native fishes by nonnative fishes is easily demonstrated; it can have devastating effects on native fishes (e.g., Fuller et al. 1999). In Upper Klamath Lake, introduced fathead minnows may prey on larval suckers, as shown in laboratory enclosures (Dunsmoor 1993, cf. Ruppert et al. 1993), although the applicability of the laboratory studies to conditions in nature is uncertain. Juvenile and adult yellow perch and juvenile largemouth bass consume larvae, as may Sacramento perch, most other centrarchid sunfishes, and the two bullhead species present in Upper Klamath Lake. Juvenile and adult largemouth bass also could feed on juvenile suckers, although adult suckers reach a body size that provides them refuge from fish predators. Comparisons of Upper Klamath Lake with other lakes in this regard could be useful. With the exception of Sacramento perch, Clear Lake apparently has been spared significant introductions of nonnative fishes, and its populations appear to be stable. A species list for Gerber Reservoir is not readily available.

The presence of numerous and diverse nonnative fishes in the Klamath system complicates recovery efforts. Nonnative species typically do well in disturbed systems (Moyle and Leidy 1992). Given that attempts to reduce abundances of nonnative fishes usually are unsuccessful, the best tactics for decreasing the success of these invaders are to discourage future introductions (especially of predators), to restore water quality if possible, and to prevent movement of nonnative fishes within the basin. Selective control of nonnative species has been pursued in some environments (Ruzycki et al. 2003), however, and should not be ruled out entirely for Upper Klamath Lake.

Hybridization and Introgression

Hybridization results in wasted spawning and loss of genetic diversity through elimination of rare alleles. Introgression (backcrossing of hybrids with parental species) can harm a rare species, as apparently has happened to the endangered June sucker, *Chasmistes liorus liorus*, which hybridizes readily with the more abundant Utah sucker, *Catostomus ardens* (Echelle 1991). The original ESA listing document for Klamath suckers (53 Fed. Reg. 27130 [1988]) cited apparently high rates of hybridization among the three Upper Klamath Lake sucker species, especially between shortnose suckers and Klamath largescale suckers, and cited hybridization as a potential contributor to loss of genetic integrity and decline of species. Apparent hybrids, as indicated by morphological intermediacy, are commonly found in the Williamson River downstream of Chiloquin Dam and in sucker populations of Clear Lake, where crosses between Lost River suckers and Klamath largescale suckers are most frequently suspected (e.g., Cunningham et al. 2002; Moyle 2002; D. Markle, Oregon State University, Corvallis, Oregon, personal communication, 2002). Recent anatomical studies of hybridization, however, imply that it is a rare occurrence. Among spawning fish captured in Upper Klamath Lake in 2001, 0.2% of fish from shoreline spawning sites, 4% from the lower Williamson River, and 6% occupying the area below Chiloquin Dam were apparent hybrids (Cunningham et al. 2002, Hayes et al. 2002, Janney et al. 2002). In contrast, one-third of fish caught at Chiloquin Dam in 2000 appeared to be anatomically intermediate. Morphological studies may overestimate hybridization; allozyme frequency and nuclear genetic data indicate that recent hybridization is rare, that nominal species are all valid, and that little genetic divergence has occurred among populations within species (D. Buth, University of California at Los Angeles, Los Angeles, California, personal communication, 2002; Dowling 2000; T. Dowling, Arizona State University, Tempe, Arizona, personal communication, 2002). Microsatellite data indicate, however, that the three species present in the Lost River (largescale, shortnose, and Lost River suckers) are significantly different from suckers in Upper Klamath Lake and the upper Williamson River (G. Tranah, Harvard School of Public Health, Boston, Massachusetts, personal communication, 2002).

Overall, morphological data indicate that hybridization has occurred, but current genetic analyses reveal that Lost River suckers and shortnose suckers are distinct and that the identity of the species has not been eroded by extensive hybridization. High priority should be attached to further genetic analysis that will give more information on hybridization and on the genetic structure of currently isolated populations.

Before the Klamath Project was completed, all sucker habitats were subject to interchange of fish (Chapter 2). Dams and irrigation canals isolated populations to an extent that could ultimately affect the genetic diversity of the species. None of the primary dams in the Klamath basin allow passage of suckers. Efforts to protect the species with regard to range fragmentation should focus on habitat protection and improvement of all subpopulations and on construction of ladders of proven effectiveness or removal of barriers to improve exchange among subpopulations.

Other Issues Relevant to Recovery

Other Natives and the Paradox of Persistent Endemics

Shortnose and Lost River suckers apparently are more susceptible to degraded habitat conditions or other factors, such as predators, than any of the 14 other native species. Blue chub and tui chub do appear in some fish kills, sometimes in large numbers, but their populations remain large in Upper Klamath Lake, as do populations of Klamath Lake sculpins and redband trout. Even the Klamath largescale suckers in the upper Klamath basin and Klamath smallscale suckers in the lower basin seem not as affected by anthropogenic change as Lost River and shortnose suckers, although the Klamath largescale sucker is listed as a species of special concern in California (Moyle 2002). Introduced species, such as yellow perch and fathead minnow, appear to be unaffected by poor water quality. Sacramento perch, which have been greatly reduced throughout their native range (Moyle 2002), apparently are doing well in the Klamath basin. Explanations for the exceptional vulnerability of shortnose and Lost River suckers could be applied to recovery efforts.

One line of evidence is related to physiological tolerances among species, but this information is limited. Falter and Cech (1991) found that shortnose suckers were less tolerant of elevated pH than were Klamath tui chub and Klamath largescale suckers (Chapter 5). Additional comparative studies of physiological responses to water-quality degradation in the Klamath basin are needed. Overall, more and better information is needed on the biology and population status of nonsucker species in the upper basin (Chapter 5). Because all native Klamath fishes are endemics, any significant declines in their populations could trigger ESA actions. Although research efforts directed specifically at native fishes other than the listed suckers would be desirable, information on them can be collected in conjunction with studies of suckers. Some of the species can be used as indicators of water quality and habitat conditions and would provide insight into the welfare of the endangered suckers, especially where differences in physiological tolerance can be demonstrated. Comparisons between endangered Klamath suckers and other catostomid species in the Klamath basin and between Klamath suckers and lake suckers elsewhere could provide additional, invaluable insight into solutions to problems in the Klamath basin.

Captive Propagation

Captive propagation is a controversial means of protecting endangered species. Successful propagation can lead to complacency about the condition of natural populations and to delay in the correction of the original causes of decline, but it also can serve as insurance against catastrophes. Although Klamath suckers have not reached the point where captive propagation is necessary, many conservation practitioners recommend against waiting until there is no alternative to captive propagation, because by then genetic resources are diminished and problems with rearing methods may be disastrous.

The Klamath Tribe has established a sucker holding and rearing facility (the Klamath Tribes Native Fish Hatchery) at Braymill near Chiloquin. The facility has been used for physiological and behavioral studies and for fertilization and larva-rearing trials (e.g., Dunsmoor 1993; L. K. Dunsmoor, Klamath Tribes, Chiloquin, Oregon, personal communication, September 3, 2002). The facility could serve as the core of a captive-propagation effort if populations continue to decline. Methods already developed there can be used, perhaps with advice based on successful propagation of cui-ui at the David Koch Cui-ui Hatchery in Sutcliffe, Nevada, if captive propagation proves necessary.

Critical Habitat

Critical habitat, as defined by the ESA (Chapter 9), was not identified for the Klamath suckers at the time of original listing, and has yet to be completed for either endangered species, although a draft proposal appeared in 1994 (59 Fed. Reg. 61744 [1994]). On the basis of established ESA criteria (for example, water quantity and quality; physical habitat appropriate for spawning, rearing, and feeding; and protection from predation and climatic stress), USFWS identified six critical-habitat units (CHUs) in the basin: Clear Lake and its watershed, Tule Lake, the Klamath River, Upper Klamath Lake and its watershed, the Williamson and Sprague Rivers, and Gerber Reservoir and its watershed. All except Gerber Reservoir are habitat units for both sucker species; Gerber Reservoir contains only shortnose suckers, but Lost River suckers presumably could live there.

The draft critical-habitat determination (59 Fed. Reg. 61744 [1994]) and its recommendations should be reviewed and revised in light of recent findings. The process of identifying critical habitat for both species needs to receive higher priority and should be more specific. In designating Upper Klamath Lake a CHU, USFWS (59 Fed. Reg. 61744 [1994]) did not identify specific areas of particular value. The CHU approach could be expanded to include the needs of specific life-history stages, for example, east coast springs for spawning, Williamson River mouth and nearby shorelines as a nursery region, Modoc Point and Goose Bay as staging areas before spawning, and west coast bays as postspawning aggregation areas (see Chapter 5). Buettner (1992) identified sites that have the greatest potential as adult refuges at low lake levels on the basis of their size, proximity to the main lake, relative water quality, and density of submerged vegetation. The issue of water-quality refuges needs more study relative to critical habitat. If the postulated patterns can be verified and the location and use of these apparent water-quality refuges can be confirmed, they might be designated as critical habitat and considered for special protection.

Although there is only weak pressure for development in the Klamath basin, the human population of the area has grown, and future growth is likely (Chapter 2). Proposals for new construction or use of the lake should take into account possible adverse effects on suckers. For example, an article in *SAIL* magazine for July 2002 identified Howard Bay, Pelican Bay, and Harriman Springs as desirable destinations for boaters. Howard Bay apparently is a preferred aggregation area for postspawning shortnose suckers (Coen et al. 2002); Pelican Bay was identified by Buettner (1992) as a refuge for suckers during the fish kills of July 1971 and August 1986 and was considered the best sucker refuge site on the west shoreline when lake levels drop; and Harriman Springs is a former spawning site. Increased boat traffic, development, groundwater pumping, or other activities may adversely affect these sites.

LESSONS FROM COMPARATIVE BIOLOGY OF SUCKERS

Of the 63 species of suckers in the world, 61 are endemic to North America. Among the few known extinctions of freshwater fishes in North America, suckers figure prominently. Previously abundant, sometimes widespread species have disappeared, including the harelip sucker (*Lagochila lacera*) and the Snake River sucker (*Chasmistes muriei*). Fully 35% of sucker species are imperiled (Warren and Burr 1994), and eight have federal endangered or threatened status (50 CFR 17.11 [1999]).

Populations of large suckers in general and lake suckers in particular have declined largely because of anthropogenic factors. Although there is an obvious need for concern about these very American fishes, comparative data indicate that they can survive long periods of interrupted recruitment and can recover from these remarkable hiatuses in reproduction if factors causing decline are reduced. For example, decline has occurred in other lake suckers: cui-ui experienced no known recruitment from 1950 to 1969; June suckers had experienced at least 15 yrs without recruitment by the middle 1980s, and that probably continued into the 1990s; some populations of razorback suckers (*Xyrauchen texanus*) experienced 20–30 yr without recruitment; and Utah suckers (*Catostomus ardens*) did not reproduce successfully between the middle 1960s and the early 1990s.

Despite extended interruptions in breeding, several species of suckers have responded successfully to recovery programs. Cui-ui successfully spawn in the Truckee River because of enhanced flows and are propagated in a hatchery managed by the Paiute Tribe, from which they are regularly transplanted into Pyramid Lake, where they are abundant (USFWS 1992b). Efforts to promote recovery of June suckers have been under way since the early 1990s and appear to have been successful; they include water-allocation agreements, refuge-population establishment, and captive breeding and release (USGS 1998). The robust redhorse, *Moxostoma robustum*, a large sucker thought to have undergone population declines in Atlantic slope drainages, is now propagated and planted and has shown successful recaptures in three southeastern rivers (Jennings et al. 1998; C. Jennings, U. S. Geological Survey, Athens, Georgia, personal communication, 2002). An extensive recovery program for razorback suckers instituted in 1988 includes captive rearing and transplantation, habitat acquisition and protection, and control of nonindigenous species; success has been mixed (Minckley et al. 1991, Mueller and Marsh 1995). This general picture of decline, public concern, multifaceted efforts at recovery, and evidence of success can suggest actions that might be successful with the Klamath basin sucker species.

All four living lake suckers (shortnose sucker, Lost River sucker, cui-ui, and June) are relatively large and long-lived (Chapter 5). High tolerance of poor water quality implies that the fishes evolved in habitats that periodically experience extremes of water quality. Long life in these suckers may reflect an evolutionary history that included harsh conditions that often resulted in reproductive failure, perhaps for many consecutive years. Exceptional longevity is a cause for optimism in that it allows the fish to recover from population declines once conditions favorable to survival are restored (Scoppettone and Vinyard 1991).

Age distributions in Upper Klamath Lake suckers, as reflected in the fish-kill data, show apparent resilience in Klamath species (e.g., Cooperman and Markle 2003). Heavy fishing pressure resulted in low numbers of old suckers until 1987, when the fishery was eliminated. Numbers of adults later increased sharply (Figure 5-4). The rapid increase demonstrates the positive effect of closing the fishery. More important, the increase shows that even after prolonged population declines brought about by overfishing, a relatively small number of large, highly fecund individuals can produce many young and help to restore a population (Cooperman and Markle 2003). Even slight improvements in conditions favorable to suckers in Upper Klamath Lake, its tributaries, and surrounding water bodies could contribute to recovery.

CONCLUSIONS

Despite elimination of fishing for the shortnose and Lost River suckers in 1987, these two listed species have failed to show an increase in overall abundance. Apparently stable populations with regular recruitment and the presence of all life-history stages at appropriate abundance are found only in Clear Lake and Gerber Reservoir. Thus, the listed suckers at these two locations require special degrees of protection, both in the lakes themselves and in tributary waters where the suckers spawn.

The two listed suckers are present in Upper Klamath Lake, where they reproduce and show the full spectrum of age classes indicating successful maturation of at least some individuals. This population has not increased in abundance, however, because of episodes of mass mortality affecting large fish and possibly other factors as well. Populations at other locations (the main-stem reservoirs, the main stem of the Lost River, and Tule Lake) are of very low abundance and consist primarily of adults; no full representation of age classes is present at these locations. Suckers have been eliminated entirely from the middle portion of the main stem of the Lost River, from Lower Klamath Lake, and from Lake of the Woods.

Small irrigation dams and the larger Chiloquin Dam across the main stem of the Sprague River impede the movement of suckers attempting to spawn in the tributaries to Upper Klamath Lake. Elimination of Chiloquin Dam could greatly expand any potential spawning area, although channel and riparian improvements to the upper Sprague might be necessary to achieve the full benefit of dam removal.

Spawning of suckers in tributaries to Upper Klamath Lake is successful in producing fry, but the spawning areas do not receive special protection and are poorly studied. Physical restoration of tributary spawning areas is a matter of high priority and will involve exclusion of livestock and other measures designed to promote conditions that favor spawning of the suckers. Physical restoration near the mouth of the Williamson River as it enters Upper Klamath Lake is also important.

Water level in Upper Klamath Lake shows no relationship to water-quality conditions that result in mass mortality of adult suckers or other potentially adverse water-quality conditions. In addition, water level shows no relationship to year-class strength or to abundance of fry or juveniles over the years during which standardized sampling is available. Thus, maintenance of water levels above recent historical levels in order to increase the abundance of suckers by maximizing the area of habitat where young suckers are found is not supported by the currently available evidence. Water levels lower than recent historical levels could have undocumented adverse effects and therefore are inadvisable. Experimental maintenance of specific water levels could be incorporated into a management plan, however, through agreements between USFWS and USBR, if USFWS sees merit in further studies of water-level control.

The two listed suckers spawn in specific lakeside areas of Upper Klamath Lake, typically in association with the presence of springs. Some spawning areas have been abandoned entirely, possibly because of the elimination, through fishing, of specific groups of fish that habitually used these areas. Some spawning areas show signs of anthropogenic degradation. Selective restoration of these areas and manipulation of stocks to encourage bonding of specific groups of suckers to the unused sites could be beneficial in spreading the reproductive risk of the sucker populations.

Suckers of all ages in Upper Klamath Lake historically have been entrained into the A Canal, which is the main supply conduit for USBR's Klamath Project. Screening of this source of mortality is scheduled for summer of 2003, but it cannot be expected to prevent mortality of very small fish. Refinement of the operation of the screens as recommended by USFWS (2002) might reduce the mortality of very young fish. The Link River Dam intake units remain unscreened, and thus remain a source of mortality for fish of all ages.

Suckers of Upper Klamath Lake and at other locations where suckers are present in the upper basins share their habitat to varying degrees with nonindigenous species, some of which are known to prey upon or compete with young suckers. Elimination of nonindigenous species over very large systems such as Upper Klamath Lake is beyond the current state of the art, but programs designed to prevent additional introductions and prevent the spread of presently nonindigenous species would be highly advisable. Because the actual effect of the nonindigenous species on the suckers is poorly known, it is difficult to judge the importance of this factor based on current information.

Hybridization among sucker species was an original concern of considerable importance to the listing of the suckers. Subsequent studies have reduced the level of this concern, but it would be advisable to have more information on the genetic identities of suckers at various locations in the upper basin.

Captive propagation is a possibility and could be conducted on a pattern that has been developed for populations of related suckers at other locations. Captive propagation is probably disadvantageous at present, however, in that it tends to undermine incentives for return of the populations to a self-sustaining basis, which may still be possible in the Klamath basin. Continued decline of the population sizes or loss of any major subpopulations would indicate a need for captive propagation.

The long life-history of suckers requires extended observation as a means of judging population trends. Benefits of restoration actions will not necessarily be evident until the fish benefiting from these actions have achieved spawning capability. Similarly, the negative effects of mortality focused on large fish may become evident only gradually, but could extinguish entire subpopulations.

7

Fishes of the Lower Klamath Basin

Native fishes of the lower Klamath basin are mainly anadromous species that use productive flowing-water habitats and a few nonmigratory stream fishes typical of cool-water environments. Because the watershed has been drastically altered by human activities, it has become progressively less favorable for anadromous fishes, including coho salmon. Given that the native anadromous fishes support important tribal, sport, and commercial fisheries and have high iconic value, there is widespread support among stakeholders, both inside and outside the basin, for restoration of these fishes to their earlier abundances. Restoration efforts would most rationally apply to all native fishes, not just those listed or proposed for listing under the federal Endangered Species Act (ESA). If broadly based restoration does not occur, additional anadromous species are likely to be listed under state and federal endangered species acts. Furthermore, because actions that are perceived to benefit one species may do harm to another, the species cannot be treated as isolated units.

The lower Klamath basin supports 19 species of native fishes (Table 7-1). Thirteen (68%) of the 19 are anadromous, and two are amphidromous (larval stages in salt water); thus, 80% of the fishes require salt water to complete their life histories. The remaining four species spend their life entirely in freshwater and show close taxonomic ties to fishes in the upper basin or adjacent basins. The species composition of native fishes supports geologic evidence that the Klamath River in its present form is of relatively recent origin. One of the resident fishes (the lower Klamath marbled sculpin), however, is distinctive enough to be recognized as a subspecies and

TABLE 7-1 Native Fishes of the Lower Klamath River and Its Tributaries

Name[a]	Life History	Status in Lower Klamath and Trinity Rivers[a]	Comments
Pacific lamprey, *Lampetra tridentata*	A	Declining	TTS, probably multiple runs
River lamprey, *L. ayersi*	A	Uncommon	Poorly known
Klamath River lamprey, *L. similis*	N	Common	Poorly known
Green sturgeon, *Acipenser medirostris*	A	State special concern, proposed for listing	TTS, important fishery
White sturgeon, *A. transmontanus*	A	Uncommon	May not spawn in river
Klamath speckled dace, *Rhinichthys osculus klamathensis*	N	Common, widespread	Most widespread fish in basin
Klamath smallscale sucker, *Catostomus rimiculus*	N	Common, widespread	Found also in Smith and Rogue rivers
Eulachon, *Thaleichthys pacificus*	A	State special concern	TTS, huge runs now gone, lowermost river only
Longfin smelt, *Spirinchus thaleichthys*	A	State special concern	Small population mainly in estuary
Prickly sculpin, *Cottus asper*	Am	Common	Larvae wash into estuary
Coastrange sculpin, *C. aleuticus*	Am	Common	Larvae wash into estuary
Lower Klamath marbled sculpin, *C. klamathensis polyporus*	N	Common?	Endemic
Threespine stickleback, *Gasterosteus aculeatus*	A/N	Common	Migratory close to ocean, anadromous; upstream forms nonmigratory
Coho salmon, *Oncorhynchus kisutch*	A	Federally threatened	Being considered for state listing, TTS
Southern Oregon-Northern California ESU			
Chinook salmon, *O. tshawytscha*			TTS
Southern Oregon-Northern California ESU	A	Commonest salmon below mouth of Trinity River	Much reduced in numbers
Upper Klamath and Trinity rivers ESU			
Fall run	A	Commonest salmon in both rivers	Much reduced, focus of hatcheries

(continued on next page)

TABLE 7-1 continued

Name[a]	Life History	Status in Lower Klamath and Trinity Rivers[a]	Comments
Late fall run	A	Possibly extinct	Presence uncertain
Spring run	A	Endangered but not recognized as ESU	Distinct life history, adults require cold water in summer
Chum salmon, *O. keta*	A	Rare, state special concern	Southernmost run of species, TTS
Pink salmon, *O. gorbuscha*	A	Extinct	Breeding in basin poorly documented, TTS
Steelhead (rainbow trout), *O. mykiss*	A, N	Common but declining; proposed for listing	Resident populations above barriers, TTS
Klamath Mountains Province ESU			
Winter run	A	Most abundant	Distinct life history
Summer run	A	Endangered but not recognized as separate ESU	Distinct life history, adults require cold water in summer
Coastal cutthroat trout, *O. clarki clarki*	A, N	State special concern	Only in lower river and tributaries, resident populations above barriers, TTS

[a]Evolutionarily significant unit.
Abbreviations: A, anadromous; Am, amphidromous; N, non-migratory; TTS, tribal trust species.

several of the anadromous species have distinct forms adapted to the special conditions of the Klamath basin.

In addition, 17 nonnative species of fishes have been recorded in the basin (Table 7-2); only two of these are anadromous. For the most part, these fishes are confined to human-created environments—such as reservoirs, ponds, and ditches—although individuals constantly escape into the streams, where they may take advantage of favorable habitats created by human activity. In addition, nonnative fishes come down continually from the upper Klamath basin.

COHO SALMON

The coho salmon (Figure 7-1) once was an abundant and widely distributed species in the Klamath River and its tributaries, although its historical numbers are poorly known because of the dominance of Chinook salmon. Snyder (1931) reported that coho were abundant in the Klamath

TABLE 7-2 Nonnative Fishes of the Lower Klamath and Trinity Rivers

Name	Life History	Status	Comments
American shad, *Alosa sapidissima*	A	Uncommon	Small annual run in lowermost reach of river
Goldfish, *Carassius auratus*	N	Uncommon	Ponds and reservoirs
Fathead minnow, *Pimephales promelas*	N	Uncommon	Invading from upper basin where extremely abundant
Golden shiner, *Notemigonus chrysoleucas*	N	Uncommon	Important bait fish in California
Brown bullhead, *Ameiurus nebulosus*	N	Locally abundant	Ponds and reservoirs, especially Shasta River; some in main stem
Wakasagi, *Hypomesus nipponensis*	N	Locally abundant	In Shastina Reservoir but a few downstream records
Kokanee, *Oncorhynchus nerka*	N	Locally abundant	Reservoirs
Brown trout, *Salmo trutta*	N, A	Common in some streams	Sea-run adults rare
Brook trout, *Salvelinus fontinalis*	N	Common	Only in headwater streams and lakes
Brook stickleback, *Culea inconstans*	N	Locally abundant, spreading	Recent introduction into Scott River
Green sunfish, *Lepomis cyanellus*	N	Common	Warm streams, ditches, and ponds
Bluegill, *L. macrochirus*	N	Common	Ponds and reservoirs
Pumpkinseed, *L. gibbosus*	N	Uncommon	Abundant in upper basin
Largemouth bass, *Micropterus salmoides*	N	Common	Ponds and reservoirs
Spotted bass, *M. punctulatus*	N	Locally common	Only in Trinity River reservoirs
Smallmouth bass, *M. dolomieui*	N	Locally common	Only in Trinity River reservoirs
Yellow perch, *Perca flavescens*	N	Locally common	Abundant in upper basin, including Iron Gate Reservoir

Abbreviations: A, anadromous; N, non-migratory.

River but also indicated that reports of the salmon catch probably lumped coho and Chinook. Historically, coho salmon occurred throughout the Klamath River and its tributaries, at least to a point as high up in the system as the California-Oregon border. It is possible that they once migrated well into the upper Klamath basin (above Klamath Falls), as did Chinook and

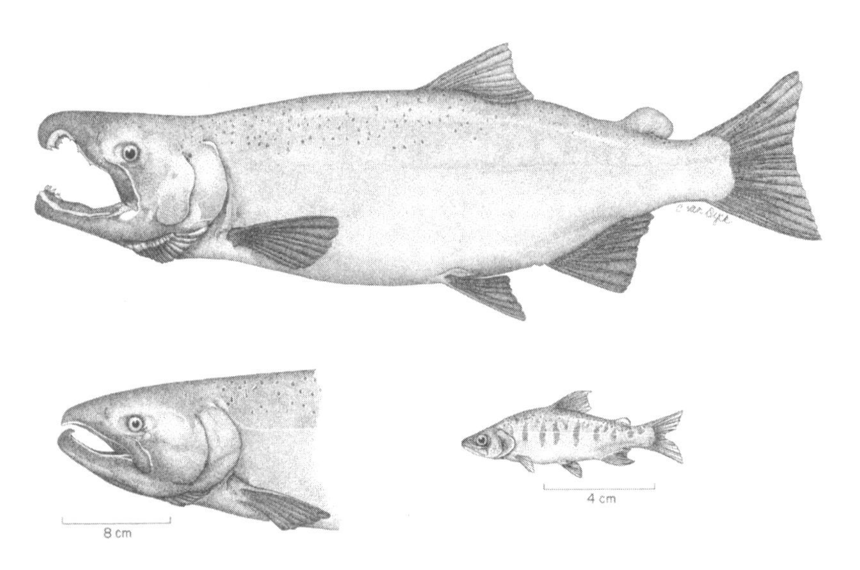

FIGURE 7-1 Coho salmon male (top), female (head), and parr. Source: Moyle 2002. Drawing by Chris M. Van Dyck. Reprinted with permission; copyright 2002, University of California Press.

steelhead, but there are no records of this, perhaps because most people are unable to distinguish them (Snyder 1931).

Today coho salmon occupy remnants of their original range wherever suitable habitat exists and wherever access is not prevented by dams and diversions (Brown et al. 1994, Moyle 2002). Because the coho salmon is clearly in a long-term severe decline throughout its range in California, all populations in the state have been listed as threatened under both state and federal endangered species acts (CDFG 2002).

Life History

Coho salmon in the Klamath basin have a 3-yr life cycle (3 yr is the time from spawning of a parent to spawning of its progeny), about the first 14–18 mo of which is spent in freshwater, after which the fish live in the ocean until they return to freshwater to spawn at the age of 3 yr. The main variation in the cycle is that a small percentage of the males return to freshwater to spawn early (in their second year, before spending a winter at sea) as "jacks." A few juveniles may also remain in freshwater for 2 yr (e.g., Bell et al. 2001), although this has not been documented for Klamath basin coho. Adults typically start to enter the river for spawning in late September. They reach peak migration strength between late October and the middle of November. A few fish enter the river through the middle of December (USFWS, unpublished material, 1998). Adult coho generally enter streams when water temperatures are under 16°C and rains have increased flows (Sandercock 1991). The presence, however, of small numbers of adult coho in the fish kill of September 2002, indicates that some coho begin migration without these stimuli. Most spawning takes place in tributaries, especially those with forested watersheds, but some main-stem spawning has been recorded (Trihey and Associates 1996). Spawning usually takes place within a few weeks of the arrival of fish in the spawning grounds. Females dig redds (nests) in coarse gravel and spawn repeatedly with large, hooknose males and with small jacks over a period of a week or more. The fertilized eggs are covered with gravel after each spawning event. Adults die after spawning.

Embryos develop and hatch in 8–12 wk, depending on temperature. Alevins (hatchlings with yolk sacs attached) remain in the gravel for another 4–10 wk (Sandercock 1991). In forested watersheds with relatively stable slopes and stream channels, mortality is lower for embryos and alevins than it is in disturbed watersheds (Sandercock 1991). Major sources of mortality include scouring of redds by episodes of exceptionally high flow and smothering of embryos by silt. When most of the yolk sac is absorbed, the alevins emerge from the gravel as fry (30–35 mm) and seek the shallow stream margins, where velocities are low and small invertebrates are abundant. Fry start emerging in late February and typically reach peak abundance in March and April, although fry-sized fish (up to about 50 mm) appear into June and early July (CDFG, unpublished data, 2000, 2001, 2002). Fry are nonterritorial and have a tendency to move around (Kahler et al. 2001); this allows them to disperse. Thus, some fry are captured in outmigrant traps at the mouths of the Shasta and Scott rivers from May to early July (CDFG, unpublished data, 2000, 2001, 2002), although most probably stay in the tributaries close to the areas in which they were spawned.

There is no sharp separation between fry and juvenile (parr) stages; juveniles are typically over about 50–60 mm and partition available habitat among themselves through aggressive behavior (Sandercock 1991). Juveniles develop in streams for a year. Typical juvenile habitat consists of pools and runs in forested streams where there is dense cover in the form of logs and other large, woody debris. They require clear, well-oxygenated water and low temperatures. Preferred temperatures are 12–14°C, although juvenile coho can under some conditions live at 18–29°C for short periods (McCullough 1999, Moyle 2002). For example, Bisson et al. (1988) planted juvenile hatchery coho in streams that had been devastated by the eruption of Mount St. Helens 3–4 yr earlier and found that they showed high rates of growth and survival in areas where maximum daily temperatures regularly exceeded 20°C and occasionally reached 29°C. Early laboratory studies in which juvenile coho were reared under constant temperatures indicated that exposure to temperatures over 25°C, even for short periods, should be lethal (Brett 1952). But laboratory studies in which temperatures were increased gradually (for example, 1°C/hr) suggest that lethal temperatures range from 24 to 30°C, depending on other conditions and the temperature to which the fish were originally acclimated (McCullough 1999). In the laboratory, juvenile coho can be reared at constant temperatures of 20–23°C if food is unlimited (McCullough 1999); but in hatcheries, they typically are reared at lower temperatures because of their reduced growth and increased mortality from disease at higher temperatures. Coho at Iron Gate Hatchery are reared at summer temperatures near 13–15°C (Bartholow 1995).

Consistent with the experiences of hatcheries, most coho develop and grow where water temperatures are at or near the preferred temperatures for much of each 24-hr cycle. For example, in tributaries to the Matolle River, California, Welsh et al. (2001) found that juveniles persisted through the summer only in tributaries where the daily maximum temperature never exceeded 18°C for more than a week. In the Klamath basin, such suitable conditions exist today mainly in portions of tributaries that are not yet excessively disturbed (Figure 1-1). NMFS (2002) has identified, in addition to the Shasta, Scott, Salmon, and Trinity rivers, six creeks between Iron Gate Dam and Seiad Valley, 13 creeks between Seiad Valley and Orleans, and 27 creeks between Orleans and the mouth of the Klamath as important coho habitat in the Klamath basin.

The explanation of seemingly contradictory information on temperature tolerance lies in the realm of bioenergetics. Juvenile coho can survive and grow at high daily maximum temperatures provided that (1) food of high quality is abundant so that foraging uses little energy and maximum energy can be diverted to the high metabolic rates that accompany high temperatures, (2) refuge areas of low temperature are available so that exposure to high temperatures is not constant, and (3) competitors or predators are largely absent so that the fish are not forced into physiologically unfavorable conditions or energetically expensive behavior (such as aggressive interactions). Thus, in the streams around Mount St. Helens cited above, food was abundant and temperatures were low much of the time. Temperatures dropped well below 15°C at night even after the hottest summer days, were below 16°C for 65–80% of the time, and rarely exceeded 25°C (Bisson et al. 1988). There were also areas of cool groundwater inflow that served as refuges on hot days, although the extent of their use by coho was not documented. And coho were the only species present. In some rivers, however, interactions of coho with juvenile Chinook and steelhead cause shifts of coho into energetically less favorable conditions (Healey 1991, Harvey and Nakamoto 1996). For example, coho juveniles occupying tributaries at the Matolle River faced not only limited food supplies but also energetically expensive interactions with juvenile steelhead (Welsh et al. 2001) and so were restricted to cool water.

Observations of juvenile coho in the main-stem Klamath River during summer suggest that juvenile coho live in the main stem despite temperatures that regularly exceed 24°C and are usually over 20°C for much of the day from late June through the middle of September (M. Rode, CDFG, personal communication, USFWS, unpublished data, 2002). Temperatures at night typically drop to 18–20°C during the warmest period. The coho occupy mainly pools at the mouths of inflowing streams where temperatures are usually 2–6°C lower than the water in the main river. The pools apparently are the only cool-water refugia in the river and occupy only a small area (B. A. McIntosh and H. W. Li, unpublished report, 1998). The coho in the pools appear to move into warmer water to forage on the abundant aquatic insects (D. Hillemeier, Yurok Tribe, personal communication). Thus, it is at least possible that coho could, from a bioenergetic perspective, occupy the main stem. Snorkel surveys of mouth pools in 2001 show, however, that juvenile coho, in contrast with Chinook and steelhead, occupied 16% of the tributary-mouth pools in June but only a single pool in August and September (T. Shaw, USFWS, unpublished material, 2002; Table 7-3).

Most of the tributary mouth pools contain juvenile Chinook salmon, steelhead, or both (Table 7-3). These fishes can compete with and prey on juvenile coho (and each other) and are somewhat more tolerant of high temperatures than coho. While many of these juveniles resulted partly from natural spawning, many of them likely came from Iron Gate Hatchery.

TABLE 7-3 Pools Containing Juvenile Coho Salmon, Chinook Salmon, and Steelhead Along Main Stem of Klamath River, 2001, as Determined in Snorkeling Surveys[a]

	No. of Mouth	No. (%) of Pools with Juvenile Fish		
Month of Survey	Pools Surveyed	Coho	Chinook	Steelhead
June	31	5 (16)	26 (84)	26 (84)
July	46	7 (15)	41 (89)	43 (93)
August	39	1 (3)	26 (67)	34 (87)
September	32	1 (3)	13 (41)	28 (88)

[a]The data are comprehensive in that they include all tributaries large enough to form a cool pool, and include some tributaries below the Trinity River (e.g., Blue Creek).
Source: T. Shaw, USFWS, unpublished material, 2002.

Many large (70–90 mm) juvenile Chinook from the hatchery move down the river from late May through July, as do large numbers of hatchery steelhead smolts in March and April. Interactions among hatchery and wild fish of all species may cause wild fish, which are smaller, to move downstream prematurely when cool-water habitat becomes limiting in summer, although this possibility has not been documented for the Klamath River. The number of pools occupied by Chinook salmon declines by August and September, as does the number of Chinook present in each pool that has fish (T. Shaw, USFWS, unpublished material, 2002); this reflects the normal outmigration of both wild and hatchery juvenile Chinook. Steelhead remain in most pools throughout the summer.

Although 2001 was a year of exceptionally low flows, Table 7-3 suggests that coho juveniles are uncommon in the main stem in early summer and become progressively less common as the season progresses. Juvenile coho are virtually absent from the main stem, including pools at tributary mouths, by late summer, even though juvenile Chinook and steelhead persist in these habitats. Although the overall rarity of coho in the Klamath basin may contribute to their absence from the mouth pools, their presence early in the summer and the reduced densities of juvenile Chinook salmon as summer progresses suggest that juvenile coho would be noticed by observers in late summer if they were present. In one respect, the near absence of coho by late summer is surprising because juvenile coho do move about and should be continually dropping into the pools from tributaries (Kahler et al. 2001). Movement of coho juveniles may be prevented by the warming or drying of the lower reaches of tributaries in late summer.

Overall, it appears that the bioenergetic demands of juvenile coho prevent them from occupying the main stem. Even with abundant food, the thermal refugia (the pools at mouths of tributaries) are inadequate: nighttime temperatures stay too high for them, and the energy costs of interactions with Chinook and steelhead, both of which are much more abundant in the pools, are probably high. Coho juveniles in the pools during June and July may die by late summer. Alternatively, they could be moving back into tributary streams, but temperatures in the lower reaches of the tributaries are similar to those of the mouth pools by late summer, and barriers to reentry (such as gravel bars) are often present. It is also possible that coho juveniles move to the estuary, perhaps traveling at night, when temperatures are lowest. Estuarine rearing of juvenile coho has been documented in other systems (Moyle 2002). A rotary-screw trap set near Orleans on the lower river for 10 yr (1991–2001) caught juvenile coho from April through July, after which the trap was taken from the river; peak numbers were observed in May and June—5 times higher than in July (T. Shaw, USFWS, unpublished data, 2002). Annual seining data from the estuary (1993–2001) indicate, however, that coho are absent from the estuary or are very rare from July through September, when temperatures often exceed 18°C (M. Wallace, CDFG, unpublished memorandum, 2002). Thus, the evidence points to the conclusion that juvenile coho are not occupying either the estuary or the main stem through the summer.

One proposal for increasing the survival of juvenile coho in the main stem in summer has been to release more water from Iron Gate Reservoir to increase the habitat for juvenile coho, as defined by analogy with habitat used by juvenile Chinook salmon, and to reduce daily temperature fluctuations in the river, thus removing the potentially lethal temperature peaks (Chapter 4). The water available from Iron Gate Reservoir, however, is quite warm in summer (18–22°C or more) and, because it is increasingly warm as it moves downstream, is unlikely to ameliorate high temperatures very much. Modeling suggests that additional flows may indeed reduce maximum temperatures some distance downstream but that they will also increase minimum temperatures (Chapter 4). From a bioenergetic perspective, increasing minimum temperatures may be especially unfavorable for coho in the main stem because nocturnal relief from high temperatures would be reduced.

The low abundance of juvenile coho in the main stem in summer, the known thermal regimes of the main stem, and the bioenergetic requirements of coho together suggest that the most crucial rearing habitat for juveniles is that of cool tributaries. Today, cool tributaries are mainly small streams that flow directly into the Klamath or into the Shasta, Scott, Salmon, and Trinity rivers. With its large, cold springs in the headwaters, the entire Shasta River was probably once favorable habitat for coho juveniles in most years, but diversions and removal of riparian vegetation have made it generally lethal thermally for salmonids in summer. If warming occurs with future climate change, it would likely exacerbate other factors that have led to warming of the tributaries (see Chapter 8).

Even a stream that has suitable summer habitat for juvenile coho may be unsuitable in winter. Studies in Oregon and elsewhere indicate that overwintering habitat is a major limiting factor where summer conditions are favorable (Nickelson et al. 1992a, b). Juveniles need refuges from winter peak flows. The refuges are side channels, small clear seasonal tributaries, logjams, and other similar areas. Simplification of channel structure through removal of woody debris or channelization eliminates much of the overwintering habitat. The condition of winter habitat for coho in the Klamath basin has not been evaluated.

Barred juveniles (parr) transform into silvery smolts and begin migrating downstream in the Klamath basin between February and the middle of June (USFWS, unpublished material, 1998) when they are about 10–12 cm long. Most smolts captured in the Orleans screw trap are taken in April and May (T. Shaw, USFWS, unpublished material, 2002) and appear in the estuary at about the same time (M. Wallace, CDFG, unpublished memorandum, 2002). Typically, coho smolts migrate downstream on the declining end of the spring hydrograph. About 60–70% of the smolts are of hatchery origin (M. Wallace, CDFG, unpublished memorandum, 2002). They are largely gone from the estuary by July. The transformation of juveniles into smolts appears to be triggered by light (perhaps moonlight) and other changing environmental conditions. Smoltification results in profound physiological and morphological changes in the fish. Smolts are compelled to move to the marine environment and will actively swim downstream to do so, especially at night. Exact timing of the downstream movement appears to be affected by flow, temperature, and other factors (Sandercock 1991). Higher flows in the river in April and May probably decrease transit time of the smolts. Low transit time could reduce predation rates and reduce energy consumption in swimming, although this has not been demonstrated in the Klamath River.

Smolts may feed and grow in the estuary for a month or so before entering the ocean (e.g., Miller and Sadro 2003). Coho entering the ocean generally have their highest mortality rates in their first few months at sea (Pearcy 1992). The first month or so after entry may be especially important due to predation, which suggests that smolts will have higher survival rates if they are large before going out to sea (C. Lawrence, UCD, personal communication, 2002). Once at sea, they spend the next 18 mo or so as immature fish that feed voraciously on shrimp and small fish, and grow rapidly.

Ocean survival depends on a number of interacting factors, including the abundance of prey, density of predators, the degree of intraspecific competition (including that from hatchery fish), and fisheries (NRC 1996). The importance of these factors in turn depends on ocean conditions (productivity, predation, and other factors), which vary widely on both spatial and temporal scales. Even relatively small changes in local and annual fluctuations in temperature, for example, can be related to changes in salmon survival rates (Downton and Miller 1998). Even more important are multidecadal (20–50 yr) fluctuations in ocean conditions, which can result in drastic changes in ocean productivity for extended periods of time (Hare et al. 1999, Chavez et al. 2003). Prolonged climatic shifts have caused significant shifts in salmonid populations to the north or south through modification of ocean temperatures (Ishida et al. 2001). Global warming thus could result in a shift in salmonid distribution to the north, and cause an overall decrease in abundance of salmonids (Ishida et al. 2001).

When the ocean is in a period of low productivity, survival rates may be low, and thus result in reduced runs coming into the streams. Commercial fishing is most likely to affect salmon populations during periods of naturally low ocean survival, but the fishery for wild coho salmon has been banned in California since 1997 and the fishery for Chinook has been greatly reduced (Boydstun et al. 2001). A fishery for coho still exists off the Oregon coast, but only hatchery fish, which are marked, can be retained.

Historically, the abundance of coho spawners reflected a balance between ocean survival and freshwater survival (Figure 7-2). A year of especially poor conditions for survival in freshwater (e.g., created by drought) could be compensated for if conditions in the ocean (e.g., high regional productivity: Hobday and Boehlert 2001) enhanced survival there. Persistently poor conditions in freshwater, such as exist throughout the Klamath basin today, make the recovery of populations difficult, however, even when ocean conditions are favorable and fisheries have been shut down or reduced. When ocean conditions are poor, the positive effects of restoring of salmonid habitat in streams may be masked (Lawson 1993, NRC 1996). Thus, only long-term monitoring can reveal effects of restoration.

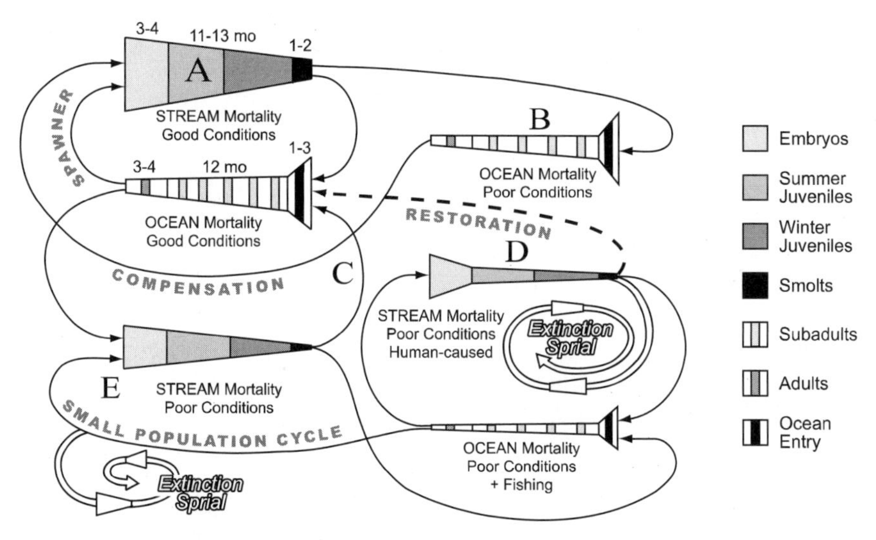

FIGURE 7-2 Population cycles of coho salmon in California. If conditions are favorable in spawning and rearing streams (A) and conditions are also favorable for high survival rates in the ocean, large populations of salmon will result. Even if conditions for survival are relatively poor in the ocean (B), large populations of coho may be maintained (although not as well as in cycle A) as long as production of coho in freshwater is high. Likewise, poor conditions in freshwater from natural causes (C) can be partially compensated for if ocean survival rates are high. If coho streams are degraded by human activity (D) and ocean conditions are poor, combined mortality may result in downward spiral of population size. If conditions in both fresh and salt water result in low survival (E), extinction may occur. Source: Based on information in Moyle 2002.

Hatcheries

Coho salmon have been an important part of the Klamath basin fish fauna since prehistoric times (CDFG 2002), and many attempts have been made to augment their populations in the Klamath basin. The first attempt occurred in 1895, when 460,000 fish from Redwood Creek—part of the same evolutionarily significant unit (ESU) as Klamath River coho—were stocked in the Trinity River. It is not known whether these fish, which were taken from a small stream, survived and contributed to later populations. Hatchery production of coho salmon in the Klamath basin began in the 1910–1911 season and continued for another 5 yr. From 1919 to 1942, six additional plants of hatchery-reared fish, all apparently of local origin, were conducted (CDFG 2002). The principal hatcheries today are the Iron Gate Hatchery (operating since 1966) on the Klamath and the Trinity River Hatchery (operating since 1963) on the Trinity River. Faced with a declining egg supply, operators of the two hatcheries at various times brought in fertilized eggs from the Eel and Noyo rivers in California and the Cascade and Alsea rivers in Oregon (CDFG 2002). Thus, present hatchery stocks probably are of mixed origin. Although a few hatchery fish have been planted in tributaries, hatchery fish are for the most part released as smolts into the main stem on the assumption that they will head directly to the sea.

Genetic studies of the contribution of hatchery coho to wild populations in the Klamath basin are not available. Brown et al. (1994) inferred that most wild coho stocks in the basin were partially mixed with hatchery stocks because the two hatcheries are at the far upstream end of coho distribution and produce large numbers of fish. In recent years, the Trinity River Hatchery has released an average of 525,000 coho per year and the Iron Gate Hatchery about 71,000 per year (CDFG 2002), although historically the Iron Gate Hatchery has released about 500,000 coho per year (CDFG, unpublished data, 2002). The coho typically are reared to the smolt stage and marked with a maxillary clip before release, which occurs between March 15 and May 1. They reach the estuary in concert with wild smolts, which peak in late May and early June, but typically are longer than the wild fish—about 170–185 mm vs 135–145 mm (M. Wallace, CDFG, unpublished data, 2002). Although the effect of large numbers of hatchery coho on wild coho is not known for the Klamath, hatchery fish may dominate wild fish when the two are together (Rhodes and Quinn 1998). In any event, hatchery fish are apparently more numerous than their wild counterparts. In 2000 and 2001, 61% and 73%, respectively, of the smolts captured in the estuary were of hatchery origin (M. Wallace, CDFG, unpublished data, 2002).

The percentage of hatchery fish in the spawning population has not been estimated directly, but Brown et al. (1994) estimated that 90% of the adult coho in the system returned directly to the hatcheries or spawned in the rivers in their immediate vicinity. Other hatchery coho no doubt stray into other streams, but the percentage is not known (CDFG 2002). In a survey of spawning coho in the Shasta River in 2001, individuals from the Iron Gate and Trinity River hatcheries were identified; seven of 23 carcasses examined were hatchery fish (CDFG, unpublished data, 2001). Regardless of origin, natural-spawning coho in the basin's tributaries have managed to maintain timing of runs and other life-history features that fit the basin's hydrologic cycle well.

Status

Coho salmon populations in California in general and in the Klamath basin specifically have declined dramatically in the last 50 yr (Brown et al. 1994, Weitkamp et al. 1995, CDFG 2002). The Southern Oregon-Northern California Coast (SONCC) ESU, of which Klamath stocks are part, was listed as threatened by the National Marine Fisheries Service (NMFS) as a consequence. The California Department of Fish and Game (CDFG 2002) recommended listing the ESU as threatened under the California state endangered species act, and the recommendation was adopted by the Fish and Game Commission as official state policy. Analysis by CDFG (2002) suggests that SONCC populations have stabilized at a low level since the late 1980s but could easily decline again if stream conditions change. Surveys in 2001 indicated that 17 (68%) of 25 historical coho streams in the Klamath basin contained small numbers of juvenile coho (CDFG 2002). In the Trinity River, wild coho stocks have experienced reduction of about 96% (USFWS/HVT 1999). The role of coho spawners of hatchery origin in maintaining these populations is not known, but marked fish of hatchery origin have been found among the spawners.

CHINOOK SALMON

Chinook salmon were and continue to be the most abundant anadromous fish in the Klamath basin, and their management potentially influences the abundance of coho in the basin and vice versa. They support important commercial, sport, and tribal fisheries. Annual runs have ranged from about 30,000 to 240,000 fish in the last 25 yr (CDFG, unpublished data, 2002), although runs were much larger historically (Snyder 1931). Chinook salmon spawn and grow primarily in the main stem of the Klamath River, in the larger tributaries (such as the Salmon, Scott, Shasta, and Trinity rivers), Bogus, Indian, Elk, and Blue creeks, and also in some smaller tributaries. Large numbers once spawned in the Williamson, Sprague, and

Wood rivers above Upper Klamath Lake, but these runs were eliminated by the construction of Copco Dam in 1917 (Snyder 1931).

Two ESUs are recognized for Klamath basin Chinook: the Southern Oregon and Coastal (SOCC) ESU and the Upper Klamath and Trinity rivers ESU (Myers et al. 1998). The SOCC ESU consists only of fall-run Chinook that spawn in the main-stem Klamath roughly from the mouth of the Trinity River to the estuary and is tied to other runs in coastal streams from Cape Blanco, Oregon, to San Francisco Bay. The Upper Klamath and Trinity rivers ESU encompasses the rest of the Chinook in the basin, including Trinity River fish. It consists of three runs (fall, late fall, and spring). Runs are named for the season of entry and migration up the river, which is not necessarily the same as the spawning time. Thus, spring-run Chinook migrate upriver during the spring, but spawn in the fall. The spring run differs in its life history from other runs and diverges slightly from them genetically as well; it may merit status as a separate ESU (Myers et al. 1998). Because studies of Chinook salmon and fisheries in the Klamath basin do not separate fish from the two ESUs (e.g., Hopelain 2001, Prager and Mohr 2001), Chinook salmon are treated here as either fall-run or spring-run. The late fall-run Chinook in the basin is either extinct or poorly documented (Moyle 2002). The vast majority of the fish today are fall-run fish of both wild and hatchery origin.

Fall-Run Chinook Salmon

Life History of Fall-Run Chinook Salmon

Fall-run Chinook in the Klamath have the classic ocean type of life-history pattern: juveniles spend less than a year in freshwater (Healey 1991). This pattern allows the salmon to take advantage of streams in which conditions may become unfavorable by late summer (Moyle 2002). Adult Chinook salmon that have the ocean type of life-history pattern also typically spawn in the main channels of large rivers and their major tributaries. Historically, the fall run in the Klamath was known as a summer run because fish started entering the estuary and lower river in July, peaked in August, and were largely finished by late September (Snyder 1931). Today, the run peaks in early September and continues through late October (Trihey and Associates 1996; USFWS, unpublished material, 1998). The 2- to 4-wk shift in run timing suggests that the main-stem Klamath and Trinity rivers have become less favorable to adult salmon in summer, presumably because of high temperature (Bartholow 1995), or perhaps because of excessive harvest of early run fish. Even with the shift in timing, temperature during the time of the spawning run probably is stressful to the migrating salmon and may result in increased mortality of spawning adults. Literature reviewed by Bartholow (1995) suggests that temperatures under 14°C are optimal for adult migration and that chronic exposure of migrating adults to 17–20°C can be lethal, although they can endure temperatures as high as 24°C for short periods. McCullough (1999, p. 75), commenting on adult migration primarily with data from the Columbia River, concludes that spring Chinook migrate at 3.3–13.3°C, summer Chinook migrate at 13.9–20.0°C, and fall Chinook migrate at 10.6–19.4°C.

Fall-run Chinook reach upstream spawning grounds 2–4 wk after they enter the river; they then spawn and die (USGS 1998). In 2001, adult Chinook were first recorded entering the Shasta River on September 11; the run peaked on October 1, and 95% of the run had entered the system by October 27 (CDFG, unpublished data, 2001). In 1993–1996, spawning in the reach between Seiad Creek and within 40 mi of Iron Gate Dam on the main stem began in the second week of October, peaked in the last week of October, and was completed by the middle of November (USGS 1998). This spawning period coincides with declining temperatures, which by early November are within the optimal range for incubation of developing embryos (4–12°C); 2–16°C is the range for 50% mortality (Healey 1991, Myrick and Cech 2001).

Time to emergence from the gravel varies with the temperature regime to which the embryos are exposed. In the main-stem Klamath River, alevins can emerge from early February through early April, but peak times vary from year to year (USGS 1998). In the Shasta River, newly emerged fry have been captured as early as the middle of January (USGS 1998). After they emerge, fry disperse downstream, and many then take up residence in shallow water on the stream edges, often in flooded vegetation, where they may remain for various periods. As they grow larger, they move into faster water. Some fry, however, keep moving after emergence and reach the estuary for rearing (Healey 1991). This pattern seems to be common in the Klamath River, although the small juveniles in the estuary leave, apparently for the ocean, after only a few weeks (Wallace 2000). The time that juveniles spend in the estuary may depend on upstream conditions (Wallace and Collins 1997). When river conditions are relatively poor (for example, warm), the juveniles move into the estuary when smaller and stay there longer. In other systems, juveniles may live in the estuary through the smolt stage and this can be important for allowing juvenile Chinook of the ocean life-history pattern to grow to larger sizes before entering the ocean (Healey 1991). Juveniles are found in the Klamath estuary from March through September (the sampling season), over which time new fish constantly enter and older fish leave (Wallace 2000; unpublished data 2002).

Other juvenile fall-run Chinook rear in the river or large tributaries for 3–9 mo, but downstream movement is fairly continuous. During June and July, movement of wild fish may be stimulated by the release of millions of juvenile salmon from Iron Gate Hatchery; the hatchery fish probably compete with wild fish for space. An outmigrant trap set at Big Bar, near Orleans, for 10 yr (1991–2001) captured juveniles from late February through late August, although the trap was usually set only from early April through July (T. Shaw, USFWS, unpublished material, 2002). Time of peak catch varied from year to year but usually was between late May and the middle of July. Outmigrant traps on the Scott and Shasta rivers catch Chinook fry, parr, and smolts from early February through July in most years. Peak numbers occur in March or early April for the Shasta River and from the middle of April to the middle of May in the Scott River. A survey of main-stem pools at the mouths of creeks in 2001 indicates that juveniles can be found in the main stem from January through September, but abundances are considerably reduced by August and September (T. Shaw, USFWS, unpublished material, 2002). Thus, there appears to be a steady movement of fish down the main stem throughout summer; the fry stay for various periods in the main stem at temperatures of 19–24°C. That pattern is consistent with the thermal tolerances of juvenile Chinook salmon, which can feed and grow at continuous temperatures up to 24°C when food is abundant and other conditions are not stressful (Myrick and Cech 2001). Under constant laboratory conditions, optimal temperatures for growth are around 13–16°C. Continuous exposure to 25°C or higher is invariably lethal (McCullough 1999). Juveniles can, however, tolerate higher temperatures (28–29°C) for short periods. Depending on their thermal history, fish in wild populations may experience high mortality at temperatures as low as about 22–23°C (McCullough 1999). In the lower Klamath River, the presence in late summer of refuges that are 1–4°C cooler than the main stem and lower temperatures at night may increase the ability of the fry to grow. The abundance of invertebrate food also makes the environment bioenergetically favorable, although intense intraspecific competition may occur around the refuge pools.

What limits the survival of Chinook fry in the main stem is not known. Food is apparently abundant, and summer temperatures, although potentially stressful, are rarely lethal. It is possible that shallow-water rearing habitat is limiting for fry, especially if there is competition for space with other salmonids, including hatchery-reared Chinook and steelhead (e.g., Kelsey et al. 2002). Fry (under 50 mm) require shallow edge habitat for feeding and protection from predators. Thus, increasing flows to increase edge habitat may be desirable for as long as small fish are present. Some fall-run Chinook apparently remain in the river long enough to become smolts before they migrate to the sea; the rest do not (migration to the estuary is known to occur without smoltification in some cases). Timing of migration may be critical. Baker et al. (1995) indicated that prolonged exposure of outmigrating smolts to temperatures of 22–24°C in the Sacramento River resulted in high mortality. Juvenile Chinook salmon that transform into smolts at temperatures over 18°C may have low ability to survive in seawater (Myrick and Cech 2001).

Once the Chinook are at sea, they grow rapidly on a diet of shrimp and small fish (Healey 1991). They can move widely through the ocean but typically are most abundant in coastal waters, where growth and survival are strongly influenced by ocean conditions. They return to the Klamath mainly as 3-yr-old fish, but jacks (2-yr-old males) and 4-yr-old fish also are common.

Hatcheries

Hatcheries for Chinook salmon have been operating continuously since 1917. Both the Iron Gate Hatchery and the Trinity River Hatchery produce large numbers of spring-run (13%) and fall-run (87%) juvenile Chinook of native stock (Myers et al. 1998). The hatcheries release 7–12 million juveniles into the river each year (about 70% from the Iron Gate Hatchery, all fall-run). The fish generally have been released over 2–3 days in late May or early June and take 1–2 mo (mean, 31 days) to reach the estuary (M. Wallace, CDFG, unpublished data, 2002), although some fish probably remain in pools for most of summer. Smaller fish take longer than larger fish to reach the estuary, but because they are feeding and growing on the way downstream, all juveniles are about the same size when they reach it. About 40% of the juvenile fish in the estuary in 2000 were of hatchery origin (CDFG, unpublished data, 2000); this is presumably a fairly typical figure. Adult Chinook returning to the hatcheries are roughly one-third of the total run—30% in 1999, 44% in 2000, and 28% 2001 (CDFG, unpublished data, 2001). There has been an increase in the percentage of hatchery fish in the run in recent years—up from 18% in 1978–1982, and 26% in 1991–1995 (Myers et al. 1998). Their contribution to natural spawning is not known, but estimates for the Trinity River suggest that it is roughly the same as the percentage of hatchery returns (Myers et al. 1998).

Status

The fall-run Chinook salmon in the Klamath basin overall probably has declined in abundance, but it is still the most abundant salmonid in the basin. In the first major study of Klamath salmon, Snyder (1931) stated that "the actual contribution of the river to the entire salmon catch of the state is not known, nor can it be known. . . . The fishery of the Klamath is particularly important, however, because of the possibility of maintaining it, while that of the Sacramento probably is doomed to even greater depletion than now appears." Snyder did not provide estimates of run sizes, but the river harvest alone in 1916–1927 was 35,000–70,000 fish (as estimated from Snyder's data showing an average weight of 14 lb/fish and a harvest of 500,000–1,000,000 lb each year). If, as Snyder's data suggest, the river harvest was roughly 25% of the ocean harvest in this period, annual total catches were probably 120,000–250,000 fish. This in turn suggests that the number of potential spawners in the river was considerably higher than the number spawning in the river today. Since 1978, annual escapement has varied from 30,000 to 230,000 adults. In both 2000 and 2001, runs were over 200,000 fish. If it is assumed that fish returning to the hatcheries are, on the average, 30% of the population and that 30% of the natural spawners are also hatchery fish, then roughly half the run consists of salmon of natural origin (including progeny of hatchery fish that spawned in the wild).

Additional evidence of decline is the exclusion of salmon from the river and its tributaries above Iron Gate Dam in Oregon, where fairly large numbers spawned, and the documented decline of the runs in the Shasta River. The Shasta River once was one of the most productive salmon streams in California because of its combination of continuous flows of cold water from springs, low gradients, and naturally productive waters. The run was probably already in decline by the 1930s, when as many as 80,000 spawners were observed. By 1948, the all-time low of 37 fish was reached. Since then, run sizes have been variable but have mostly been well below 10,000. Wales (1951) noted that the decline had multiple causes, most related to fisheries and land use in the basin, but laid much of the blame on Klamath River lampreys: the lampreys preyed extensively on the salmon in the main stem when low flows delayed their entry into the Shasta River.

In some respect, it is remarkable that fall-run Chinook salmon in the Klamath River are doing as well as they seem to be. Both adults migrating upstream and juveniles moving downstream face water temperatures that are bioenergetically unsuitable or even lethal. As explained later in this chapter, the vulnerability of the run to stressful conditions was dramatically demonstrated by the mortality of thousands of adult Chinook in the lower river in late September 2002.

Spring-Run Chinook

Life History

Like coho, spring-run Chinook have a stream type of life history, which means that juveniles remain in streams for a year or more before moving to the sea (Healey 1991). In addition, the adults typically enter freshwater before their gonads are fully developed and hold in deep pools for 2–4 mo before spawning. In California, this strategy allows salmon to spawn and develop in upstream reaches of tributaries that often are inaccessible to fall-run Chinook because of low flows and high temperatures in the lower reaches during fall (Moyle 2002). Major disadvantages of such a life-history pattern in the present system are that low flows and high temperatures during the adult and smolt migration periods can prevent the fish from reaching their destinations or greatly increase mortality during migration (Moyle et al. 1995, Trihey and Associates 1996).

Spring-run Chinook enter the Klamath system from April to July, although the fish that appear later apparently are mainly of hatchery origin (Barnhart 1994). The Chinook aggregate in deep pools, where they hold through September. Temperatures below 16°C generally are regarded as necessary for spring-run Chinook because susceptibility to disease and other sources of mortality and loss of viability of eggs increase as temperature increases (McCullough 1999). In the Salmon River, temperatures of pools holding spring-run Chinook often exceed 20°C (West 1991, Moyle et al. 1995). Spawning peaks in October. Fry emerge from the redds from March to early June; the fish reside through the summer in the cool headwaters (West 1991). Because most of the streams in which they reside also are likely to be used by juvenile coho salmon, interactions between the two species are likely (see O'Neal 2002 for information specific to the Klamath). Some juveniles may move down to the estuary as temperatures decline in October, although most do not move out until the following spring (Trihey and Associates 1996); they spend summer in the same reaches as the holding adults. More precise details of the life history of spring-run Chinook in the Klamath basin are unavailable.

Status

Spring-run Chinook may once have been nearly as abundant as fall-run Chinook in the Klamath basin. Perhaps 100,000 fish spread into tributaries throughout the basin, including the Sprague and Williamson rivers in Oregon (Moyle 2002). The Shasta, Scott, and Salmon rivers all supported large runs. Spring-run Chinook suffered precipitous decline in the 19th century caused by hydraulic mining, dams, diversions, and fishing (Snyder 1931). The large run in the Shasta River disappeared coincidentally with the construction of Dwinnell Dam in 1926 (Moyle et al. 1995). In the middle to late 20th century, the decline of the depleted populations continued as a result of further dam construction (for example, of Trinity and Iron Gate Dams) and, in 1964, heavy sedimentation of habitat that resulted from catastrophic landslides due to heavy rains on soils denuded by logging (Campbell and Moyle 1991). By the 1980s, spring-run Chinook had been largely eliminated from much of their former habitats because the cold, clear water and deep pools that they require were either absent or inaccessible. In the Klamath River drainage above the Trinity, only the population in the Salmon River and Wooley Creek remains; it has annual runs of 150–1,500 fish (Campbell and Moyle 1991, Barnhart 1994). Numbers of fish in the area continue to decline (Moyle 2002). Because the Trinity River run of several thousand fish per year is apparently sustained largely by the Trinity River Hatchery, the Salmon River population may be the last wild (naturally spawning) population in the basin. The Trinity River Hatchery releases over 1 million juvenile spring-run Chinook every year, usually in the first week of June. Apparently, all spawners in the main-stem Trinity River below Lewiston Dam are of hatchery origin.

NMFS debated designation of the Klamath spring-run Chinook as a distinct ESU, but decided that it was too closely related to fall-run Chinook to justify separation (Myers et al. 1998). Nevertheless, the presence of genetic differences and of great differences in life history suggest that it should be managed as a distinct ESU (as was done for the Sacramento River spring-run Chinook) or as a distinct population segment. Protection and restoration of streams used by spring-run Chinook salmon would provide additional protection for coho salmon because the two salmon have similar temperature and habitat requirements.

STEELHEAD

Steelhead (anadromous rainbow trout) are widely distributed and common in the Klamath basin. They consistently co-occur with coho salmon in streams, and the juveniles of the two species can have strong interactions (e.g., Harvey and Nakamoto 1996). All populations are considered by NMFS to be part of the Klamath Mountains Province ESU. Besides having genetic traits in common, the populations share a life-history stage called the half-pounder, which is an immature fish that migrates to the sea in spring but returns to spend the next winter in freshwater (Busby et al. 1994, Moyle 2002). Two basic life-history strategies are recognized in the basin: summer steelhead (stream-maturing) and winter steelhead (ocean-maturing). Barnhart (1994) and Hopelain (1998) divide the winter steelhead further into early (fall-run) and late (winter-run), but the two forms have similar life histories and will be treated together here as winter steelhead.

Winter Steelhead

Life History

Winter steelhead are the most widely distributed anadromous salmonids in North America. Key factors in their success in a wide variety of habitats include an adaptable life history, higher physiological tolerances than those of other salmonids, and ability to spawn more than once (Moyle

2002). The flexibility in life-history pattern is reflected in the fact that most populations have juveniles that spend 1, 2, or 3 yr in freshwater and adults that spend 2–4 yr in the ocean and return one to four times to spawn. This variability virtually ensures that runs can continue through periods of adverse conditions unless the stream habitat becomes chronically unfavorable to survival of steelhead.

Winter steelhead enter the Klamath River from late August to February (Barnhart 1994). They disperse throughout the lower basin and spawn mainly in tributaries but also show some main-stem spawning. Snyder (1933) noted that fish entering the Shasta River in 1932 came in bursts of 2–3 days over a 7-wk period. Spawning, which can take place any time from January through April, apparently peaks in February and March. Mature fish first return to spawn after a year, at 40–65 cm; the smallest fish are those that spent a winter in freshwater as half-pounders (Hopelain 1998). Up to 30% of the mature fish spawn a second time, after another year at sea; up to 20% spawn a third time; and a very few a fourth time (Hopelain 1998).

Fry emerge from the gravel in spring and most (80–90%) spend 2 yr in freshwater before going to sea. The rest spend either 1 or 3 yr in freshwater (Kesner and Barnhart 1972, Hopelain 1998). The juveniles occupy virtually all habitats in the basin in which conditions are physiologically suitable. They can tolerate minimal depths and flows and so can be found in the smallest accessible tributaries and in the main river channels. Although spawning occurs mainly in tributaries, the juveniles distribute themselves widely, and many move into the main stem. For example, large numbers of parr have been observed moving out of the Scott and Shasta rivers in early July (W.R. Chesney, CDFG, unpublished reports, 2000, 2002). Habitat preferences change with size: bigger fish are more inclined to use pools or deep runs and riffles, and the larger juveniles prefer water at least about 50–100 cm deep with water-column velocities of 10–30 cm/s and deep cover (Moyle 2002). Juveniles feed primarily on invertebrates, especially drifting aquatic and terrestrial insects, but fish (including small salmon) can be an important part of the diet of larger individuals. Aggressive 2-yr-old steelhead (14–17 cm) often dominate pools.

A key to the success of steelhead in freshwater is their thermal tolerance, which is higher than that of most other salmonids. Preferred temperatures in the field are usually 15–18°C, but juveniles regularly persist in water where daytime temperatures reach 26–27°C (Moyle 2002). Long-term exposure to temperatures continuously above 24°C, however, is usually lethal. Steelhead cope with high temperatures by finding thermal refuges (springs, stratified pools, and so on) or by living in areas where nocturnal temperatures drop below the threshold of stress. Persistence in thermally stressful areas requires abundant food, which steelhead will shift their behavior to find. Thus, Smith and Li (1983) found that juvenile steelhead persisted in a small California stream in which daytime temperatures sometimes reached 27°C for short periods by moving into riffles where food was most abundant; these fish, however, were at their bioenergetic limits for survival. Overall, the ability of steelhead to thrive under the summer temperatures experienced in the lower Klamath and the different habitat requirements of juvenile steelhead of different sizes indicate that they will benefit from the expansion of habitat created by increased flows in the main-stem Klamath and tributaries, as long as water quality, especially temperature, remains suitable for them.

Steelhead juveniles become smolts and move into the estuary from early April to the middle of May (Kesner and Barnhart 1972). Small numbers continue to trickle into the estuary all summer (M. Wallace, CDFG, unpublished data, 2002). A majority of the early fish that return each year to the river in September are immature (half-pounders, 25–35 cm). These fish usually stay in the lower main stem of the Klamath through March before returning to the sea. This life-history trait allows the steelhead to consume eggs of the large numbers of Chinook salmon that enter the river at the same time (USGS 1998). Half-pounders that return to spawn in the following winter are much smaller (40–50 cm), however, than the first-time spawners that skipped the half-pounder stage (55–65 cm) (Hopelain 1998).

Hatcheries

The Iron Gate Hatchery produces about 200,000 and the Trinity River Hatchery about 800,000 winter steelhead smolts per year (Busby et al. 1994). The fish are released into the rivers in the last 2 wk of March, and most reach the estuary about a month later (M. Wallace, CDFG, personal communication, 2002), coincident with the emigration of wild smolts. Diets of outmigrating smolts are similar to those of wild smolts, although the consumption of a greater variety of taxa and fewer organisms by the hatchery fish than by wild fish suggests that they have lower feeding efficiency than wild fish (Boles 1990). Otherwise, the interactions between hatchery and wild fish in the Klamath are not known, although hatchery steelhead released into a stream will dominate the wild steelhead (McMichael et al. 1999), potentially increasing the mortality in wild fish from predation, injury, or reduced feeding. Hatchery steelhead also can have adverse effects on juveniles of other salmonids, especially Chinook and coho salmon, through aggressive behavior and predation (Kelsey et al. 2002).

In the 1970s and early 1980s, adults of hatchery origin made up about 8% of the run of Klamath River steelhead and 20–34% of the run in the Trinity River (Busby et al. 1994). As numbers of wild steelhead decline, the percentage of hatchery fish in the population presumably will increase.

There is some indication that the runs most heavily influenced by hatchery steelhead in the Trinity River have a lower frequency of half-pounders in the population than do wild populations (Hopelain 1998).

Status

Historical numbers of winter steelhead in the Klamath River are not known, but total run sizes in the 1960s were estimated at about 170,000 for the Klamath and 50,000 for the Trinity (Busby et al. 1994). Historical numbers for the Klamath River above the Trinity undoubtedly were much higher because by 1917 all access to the upper basin was eliminated and habitat in the tributaries was greatly degraded or blocked. In the 1970s, Klamath River runs were estimated to average around 129,000; by the 1980s, they had dropped to around 100,000 (Busby et al. 1994). Similar trends were noted for the Trinity River. Numbers presumably are still declining, although all estimates of abundance, past and present, are very shaky. NMFS considered winter steelhead in the Klamath to be in low abundance and to be at some risk of extinction (Busby et al. 1994) but has not listed them under the ESA.

Summer Steelhead

Life History

Summer (spring-run) steelhead have the same relationship to winter steelhead that spring-run Chinook salmon have to fall-run Chinook salmon in the Klamath River. They are closely related but have different life histories. Summer steelhead enter the Klamath River as immature fish from May to July and migrate upstream to the cool waters of the larger tributaries (Barnhart 1994, Moyle 2002). They hold in deep pools roughly until December, when they spawn. Temperature requirements of adult summer steelhead are not well documented, but maximum daytime temperatures of less than 16°C seem to be optimal, and temperatures above 20°C increase stress substantially (Moyle et al. 1995) through susceptibility to starvation (they do not feed much while holding) and disease. High temperatures also decrease viability of eggs inside the females. Juveniles probably occupy mainly the same upper stream reaches in which they were spawned, that is, above the areas in which most winter steelhead spawn and rear but where coho are likely to be present. Other aspects of their life history are similar to those of winter steelhead, including a predominance of 2-yr-old smolts and the presence of half-pounders (Hopelain 1998). There is some evidence, however, that summer steelhead have higher repeat spawning rates and grow larger in the ocean (Hopelain 1998). As is the case with spring-run

Chinook salmon, major disadvantages of the summer steelhead's life-history pattern in the present system are that reduced flows and increased temperatures during the adult and smolt migration periods prevent the fish from reaching their destinations or greatly increase their mortality during migration (Moyle et al. 1995, Trihey and Associates 1996).

Status

Summer steelhead once were widely distributed in the Klamath and Trinity basins and were present in most headwaters of the larger tributaries (Barnhart 1994). In the 1990s, estimated numbers were 1,000–1,500 adults divided among eight populations; the largest numbers were in Dillon and Clear creeks (Barnhart 1994, Moyle et al. 1995, Moyle 2002). Numbers presumably are still declining because of loss of habitat, poaching in summer, and reduced access to upstream areas during migration periods as a result of diversions. Summer steelhead and winter steelhead probably are different ESUs. NMFS considers the stocks depressed and in danger of extinction (Busby et al. 1994). Summer steelhead are not produced by Klamath basin hatcheries.

OTHER FISHES

Pink Salmon

Small runs of pink salmon probably once existed in the Klamath River and elsewhere on the coast. The pink salmon now appears to be extirpated as a breeding species in California, although individuals stray occasionally into coastal streams (Moyle et al. 1995, Moyle 2002).

Chum Salmon

Periodic observations of adult chum salmon and the regular collection of small numbers of young suggest that this species continues to maintain a small population in both the Klamath and Trinity rivers (Moyle 2002). It was more abundant in the past and occasionally was harvested, but it has never been present in large numbers. The run in the Klamath basin is the southernmost of the species. The life history of this species in the Klamath basin, including timing of spawning runs and outmigration of juveniles, is probably similar to that of fall-run Chinook salmon.

Coastal Cutthroat Trout

Because of their similarity to the more abundant steelhead, coastal cutthroat trout have been largely overlooked in the Klamath basin. They occur mainly in the smaller tributaries to the main stem within about 22 mi of the estuary. They also have been observed further upstream in tributaries to the Trinity River (Moyle et al. 1995). Their life history in the Klamath River is poorly documented but is apparently similar to that of winter steelhead. Adults enter the river for spawning in September and October, and juveniles grow in the streams for 1–3 yr before going to sea. Cutthroat trout can spawn two to four times. Competition for space by spawners and juveniles with the dominant steelhead is reduced by the ability of cutthroat to use habitats higher in the watersheds than are typically used by steelhead (Moyle 2002). Voight and Gale (1998) suggest that in small tributaries in the lower 22 mi of the Klamath River, cutthroat may actually be more abundant in headwater streams than they were historically because they have become resident above migration barriers created by human activities, such as log jams and debris flows. The life history of one such population on the nearby Smith River is documented by Railsback and Harvey (2001).

The general absence of cutthroat trout from streams higher in the Klamath basin presumably results from their general intolerance of water that exceeds 18°C (Moyle 2002) and from competition with the more tolerant steelhead and perhaps other salmonids. Juveniles move downstream when they reach 12–20 cm during April through June, coincidentally with the outmigration of juvenile Chinook salmon, a major prey (Hayden and Gale 1999, Moyle 2002). Adults apparently do not move far once they reach salt water and some may return to overwinter in freshwater; others may move up in summer. Movements into freshwater by nonbreeding fish may be triggered by abundance of juvenile salmon, which are prey; the timing of such movements into the lower Klamath appears to vary greatly from year to year (Gale et al. 1998). Large numbers of adult cutthroat are observed every summer in lower Blue Creek, where they seek refuge from poor conditions in the main-stem Klamath (Gale et al. 1998).

Eulachon

The eulachon or candlefish is a smelt (Osmeridae) that reaches the southern extent of its range in the Mad River, Redwood Creek, and the Klamath River (Moyle 2002). Historically, large numbers entered the river to spawn in March and April, but they rarely moved more than 8 mi inland. Spawning occurs in gravel riffles, and the embryos take about a month to develop before hatching and being washed into the estuary as larvae. The eulachon in the Klamath River once was an important food of the American Indians in the region (Trihey and Associates 1996). Since the 1970s, their numbers have been too low in most years to support a fishery. The causes of the decline are not known but probably are tied to changing ocean conditions and poor habitat and water quality in their historical spawning areas (Moyle 2002).

Green Sturgeon

Probably 70–80% of all green sturgeon are produced in the lower Klamath River and Trinity River, where several hundred are taken every year in the tribal fishery, which is the principal source of life-history information on this species (Moyle 2002). Green sturgeon enter the Klamath River to spawn from March to July; most spawning occurs from the middle of April to the middle of June at temperatures below 14°C. Spawning takes place in the lower main stems of the Klamath and Trinity rivers in deep pools with strong bottom currents. Juveniles occupy the river until they are 1–3 yr old, when they move into the estuary and then to the ocean. Optimal temperatures for juvenile growth in the river appear to be 15–19°C. Temperatures above 25°C are lethal (Mayfield 2002). After leaving the river, green sturgeon spend 3–13 yr at sea before returning to spawn and often move long distances along the coast. They reach maturity at 130–150 cm and are repeat spawners. Large adults (250–270 cm) typically are females that are 40–70 yr old (Moyle 2002). There is some evidence that green sturgeon populations are in decline, but reduction of the marine commercial fishery for them may have alleviated the decline somewhat (Moyle 2002). In 2003, NMFS rejected a petition to have them listed as a threatened species.

Pacific Lamprey

Lampreys once were so abundant in the coastal rivers of California that they inspired the name Eel River for the third largest river in the state. They supported important tribal fisheries. Today, their numbers are low and declining (Close et al. 2002, Moyle 2002). Their biology is poorly documented, but they probably have multiple runs in the Klamath basin. Most adults (30–76 cm) enter the river from January through March to spawn from March to June, although movement has also been observed in most other months (Moyle 2002). How far upstream lampreys moved historically is not known, but it is certain, as shown by the genetics of resident lampreys, that they entered the upper basin above Klamath Falls at least occasionally. Most spawning appears to take place in the main stem or larger tributaries. Like salmon, lampreys construct redds for spawning in gravel riffles, although the tiny larvae emerge from the gravel in just 2–3 wk. They are washed downstream once they emerge, and they settle in sand and mud at the river's edge. The larvae (ammocoetes) live in burrows in these quiet areas for probably 5–7 yr and feed on algae and other organic matter. During the larval stage, they move about frequently, so they are commonly captured in salmon outmigrant traps. Factors limiting the survival of ammocoetes are not known, but it is likely that rapid or frequent drops in flow deprive them of habitat and force them to move into open water, where they are vulnerable to predation. They do not appear to be limited by temperatures in the basin, but anything that makes their shallow-water habitat less favorable (such as pollution and trampling by cattle) is likely to increase mortality.

The blind, worm-like ammocoetes undergo a dramatic transformation into eyed, silvery adults when they reach 14–16 cm, after which they migrate to the sea (Moyle 2002). Downstream migration usually is coincidental with high flows in the spring, but movement has also been observed during summer and fall (Trihey and Associates 1996). In the ocean and estuary, they prey on salmonids and other fish for 1–2 yr before returning to spawn. The Pacific lamprey is a tribal trust species with a high priority for recovery to fishable populations (Trihey and Associates 1996). Its cultural importance to American Indians is largely unappreciated (Close et al. 2002).

Native Nonanadromous Species

Speckled dace, Klamath smallscale sucker, lower Klamath marbled sculpin, threespine stickleback (some of which are anadromous), and Klamath River lamprey are quite common in the lower river and its tributaries of low gradient. With the possible exception of the sculpin, these species probably all have fairly high thermal tolerances (Moyle 2002). In the reaches within 30 mi or so of the ocean, marbled sculpin apparently are replaced by the two amphidromous species, prickly sculpin and coastrange sculpin. With the exception of the lamprey, which feeds on fish, all the resident fishes feed mainly on aquatic invertebrates. The relationship between the native nonanadromous and anadromous species has not been worked out in the Klamath, but the dace, stickleback, sculpins, and suckers are probably subsidized by nutrients brought into the streams by the anadromous fish and may suffer heavy predation, especially in the larval stages, by juvenile salmon and steelhead.

Nonnative Species

The lower Klamath basin is still dominated by native fishes, but other species have a strong presence in highly altered habitats, such as reservoirs and ponds. The Shasta River, once a cold-water river, now supports large populations of brown bullheads and other warm-water, nonnative species because summer flows consist largely of warm irrigation-return water. There also is a continuous influx of nonnative fishes from the upper Klamath basin, where they are extremely abundant. Because there is a positive relationship between degree of habitat disturbance and abundance of nonnative fishes

(Moyle and Light 1996), improving habitat for native fishes should reduce the likelihood that nonnative species will become more abundant.

MASS MORTALITY OF FISH IN THE LOWER KLAMATH RIVER IN 2002

During the last half of September 2002, mass mortality of fish (fish kill or fish die-off) occurred in a reach of the Lower Klamath River extending about 30 mi up from the confluence of the river with its estuary (Figure 1-1). In responding to the general need for a timely assessment of the conditions leading to this mortality, CDFG released in January 2003 a report that describes the extent of the mortality and its distribution among species, hydrologic and meteorological conditions that accompanied the mortality, some aspects of water quality, and the results of physical examination of both living and dead fish. A second CDFG report will deal with long-term consequences of the mortality. Also during 2003, USGS released a report dealing with the mortality of September 2002 (Lynch and Risley 2003). The USGS report documents environmental conditions that coincided with the mortality, but does not attempt to reach conclusions as to its cause.

The sponsors of the NRC study on endangered and threatened fishes asked the NRC committee to study information on the fish kill of 2002 and include the analysis in its final report. While it is reasonable that this issue be covered in the committee's report, it is also important to note that the fish kill primarily affected Chinook salmon, for reasons that are explained below, and not the threatened coho salmon that is the focus of attention for the NRC committee in its work on the lower Klamath basin. Furthermore, the NRC committee was only able to consider the two reports cited above and unpublished records on weather and temperature; other reports to be issued in the future might provide additional information that would influence conclusions about the cause of the fish kill. The fish kill of 2002 in the Klamath lower main stem is unprecedented in magnitude. It raises questions as to whether human manipulation of the Klamath River or the adjoining estuary was directly or indirectly responsible and, if so, what might be done to prevent its recurrence. A full and final explanation of mortality probably is not possible, however, given that the fish kill was not anticipated and therefore the conditions leading to it were not well documented.

Extent of Mortality

CDFG, quoting USFWS, has estimated the total mortality of fish in the last half of September 2002 at about 33,000. This estimate, which is subject to revision, is likely to be conservative. The projected run size of fall-run Chinook salmon, which was the most abundant of the fish that died, was estimated at 132,000. Thus, regardless of any adjustments that might be made in the final estimate of mortality, a substantial portion of the Chinook salmon run was lost before spawning.

Both CDFG and USFWS estimated the species composition of the fish kill, which extended beyond salmonids to other taxa, including the Klamath River smallscale sucker, but percentage estimates from CDFG are limited to the salmonids. A sample of 631 dead fish collected under the supervision of CDFG showed 95.2% Chinook salmon, 4.3% steelhead trout, and 0.5% coho salmon. These estimates differ only slightly from the USFWS estimates. Further details may appear in reports yet to be issued. Among both Chinook and steelhead, nonhatchery fish appeared to have died in greater numbers than fish of hatchery origin. A similar determination for coho salmon is complicated by the fact that only small numbers of coho were found. If the coho had been in peak migration at the time when mortality occurred, more dead coho probably would have been found. The coho migration occurs later than the Chinook migration, which probably explains why few coho were affected.

Direct Causes of Mortality

CDFG has given infection as the direct cause of death of the fish. Both living and dead fish were infected with *Ichthyopthirius multifilis,* a protozoan, and *Flavobacter columnare,* a bacterium. As indicated by CDFG, these two pathogens are widespread and, when they become lethal to fish, typically are associated with high degrees of stress. Crowding may be considered an additive agent to stress in that it facilitates efficient transmission of pathogens from one fish to another. A combination of crowding and stress thus would be especially favorable for the development of these pathogens in sufficient strength to cause mortality of fish. Potential agents of stress, which may have acted in combination rather than alone, include high temperature, inadequate concentrations of dissolved oxygen (undocumented), and high concentrations of unionized ammonia (undocumented).

Indirect Causes of Mortality

Low flow in the Klamath River main stem is the most obvious possible cause of stress leading to the lethal infections of fish in the lower Klamath River during 2002. Low flow can cause crowding of the fish in their holding areas as they await favorable conditions for upstream migration and can be associated with high water temperature and with lower than normal concentrations of dissolved oxygen. CDFG therefore reviewed information on flow in the main stem, as did USGS (Lynch and Risley 2003).

The flow of the Klamath River at Klamath, which is just a few miles above the estuary, is shown in Figure 7-3 for dry yrs used by CDFG in its overview of low flows in the river. The flows at Iron Gate Dam, about 185 mi upstream, are given for comparison. For an extended span of years not restricted to drought, September flow at Iron Gate Dam is about one-third of the flow at Klamath. For example, mean September discharge at Klamath was 2,973 cfs for 1988 through 2001 (excluding 1996, 1997) and the same statistic for the Klamath River at Iron Gate Dam is 1,130 cfs, as determined from USGS gage records.

The USGS elected not to use data for the Klamath gage because the accuracy of the gage at low flow is subject to errors greater than 15%. Figure 7-3 shows the sum of the gages at Orleans (main stem above the Trinity) and at Hoopa (on the lower Trinity), both of which produce discharge readings within 10% of the true value, for comparison with the flows in the main stem at Klamath. The two sets of values are separated by some additional discharge (undocumented) that accumulates below the

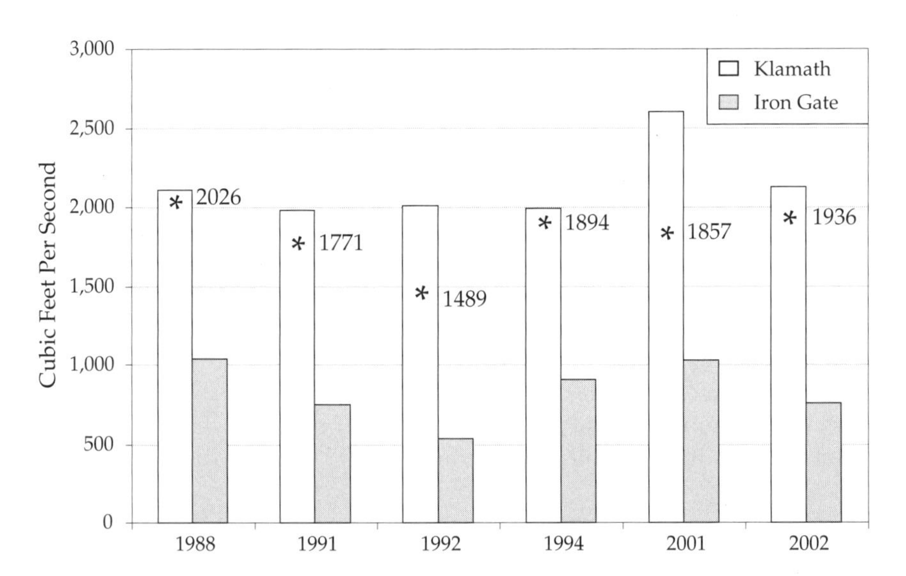

FIGURE 7-3 Mean flows of the Klamath main stem at Klamath (near the site of the 2002 fish kill) and at Iron Gate Dam (about 185 mi upstream) in September for 6 low-flow years considered by CDFG in its analysis of the fish kill. The asterisk shows the sum of flows for the Klamath at Orleans and the Trinity at Hoopa, as a check on the Klamath gage (this sum omits small tributaries below the Trinity). Sources: Data from CDFG 2003 and USGS gages.

Trinity. The Klamath gage data and the sum of the two gages above it show essentially the same picture qualitatively, as does the analysis by USGS based on the Orleans gage alone. Also, USGS restricted its analysis of flows to 1–24 September, which coincides better with observed mortality than 1–30 September, but the mean gage readings for these differing intervals are essentially identical (< 1% difference at Klamath). All data indicate that flows comparable with those of 2002 have occurred a number of times over the last 15 yr without causing mass mortality of salmonids. This does not rule out the possibility that low flow was a factor, but it does suggest that the occurrence of flows similar to those of 2002 has not in the past been sufficient by itself to cause mass mortality.

The USGS analysis adds a new dimension to future concerns related to flow in that it shows a substantial increase in distance to the water table over 2001 and 2002, both of which were dry years. Because shallow alluvial water reaches the tributaries and main-stem Klamath as groundwater, which supports flow in dry weather, drawdown of the water table by pumping should be taken into account in any future evaluation of low flows, particularly if pumping is a growing response to water scarcity during drought. Flow could be related to crowding on a conditional basis through run size or timing of run. CDFG considered this possibility by using estimates of run timing and run size of Chinook salmon, which accounted for most of the fish biomass in the river during the last half of September. The analysis showed that the run of Chinook was only slightly larger than average and that it was bracketed by run sizes both smaller and larger for other comparably dry years. Thus, run size does not show evidence of being a conditional influence related to flow.

The August–October run of Chinook appears to have peaked earlier in 2002 than in other years of record, and this suggests a conditional relationship with low flow in causing mortality. CDFG was reluctant, however, to attribute great significance to this possible relationship, given the small amount of information on which it is based. The data available to CDFG indicated that air temperatures were not unusually high during September 2002 compared with other years of low flow when no fish kills occurred. Information on water temperature is sketchier, but also indicates that average maximum water temperatures fell within the range of water temperatures in previous years of low water when there were no fish kills. The USGS made comparisons of the Klamath River with the Rogue River, which is located nearby and has more comprehensive temperature records. Both water and air temperatures on the Rogue River were approximately 2°F higher in 2002 than the mean for the period of record. While the difference is small, the threshold for harm to salmonids lies close to September temperatures, even in years of average flow. The USGS analysis, like the CDFG analysis, did not suggest that temperatures in 2002 were extreme by comparison with other years of low flow when no fish kills occurred. Thus, if temperature is a factor governing mortality, it would involve coincidence of high temperatures with some other factor, the nature of which is not clear from the presently available information.

Tests of water quality did not indicate the presence of toxicants, although the water was not sampled until seven days after the onset of the first observation of dead fish (CDFG 2003). It is always possible that toxicants not tested were involved, but this seems unlikely, given that the fish kill occurred over an extended period and that there is no circumstantial evidence of the role of toxicants other than possibly ammonia generated by the fish themselves.

CDFG also considered fish passage. According to CDFG, high flows in 1997 and 1998 may have caused aggradation and expansion of channel bars that inhibited fish passage during extremes of low flow. These changes did not result in fish kills during the low-water year of 2001, but flows in 2001 were not as low as those in 2002. Thus, a current hypothesis of CDFG is that a change in channel geometry has created new conditions that are detrimental to fish at low flows even though such flows previously did not lead to high mortality. The hypothesis is speculative in that changes in channel conditions have not been established by measurement, but it should remain under consideration until further relevant evidence is collected.

Summary of Explanations

The possibility that passage is inherently more difficult at low flows now than it was before 1997–1998 was the only explanation of unique conditions leading to the fish kill that CDFG could not rule out in preparing its January 2003 report. Because of the limited data about conditions before and during the kill, other hypotheses probably will emerge as other reports are prepared. One hypothesis that has not been evaluated by CDFG involves the effect of temperature extremes during the fish kill. As explained earlier in this chapter, mean water temperature is less important for salmonids than extremes of water temperature. Thus, for example, the failure of temperatures to decline sufficiently at night when mean temperature is high could place unusual stress on salmonids but could be overlooked in a consideration of mean and maximum temperatures alone. Such conditions could occur, for example, when back radiation is so low (perhaps as a result of cloudiness or high humidity) that a typical amount of cooling would not occur at night.

A sequence of events involving daily minimum temperature rather than fish passage might be a cause of mortality. A large number of salmon moved up the river coincident with a series of days in which water temperatures were high enough to inhibit migration. McCullough (1999) states that, based on studies in the Columbia River, Chinook salmon cease migrating when maximum water temperatures exceed 21°C. Lynch and Risley (2003) indicate that during the time of the kill, maximum water temperatures in the river at Orleans, 30 mi upstream of the kill, averaged 20.3°C, and that the average minimum was 19.7°C. Thus it seems likely that temperatures in the Klamath River at the site of the kill reached or approached the inhibitory temperatures. As they commonly do, the salmon held in pools when the temperatures were high, waiting for conditions to improve before continuing upstream. The temperature and flow data given by Lynch and Risley (2003) indicate, however, that conditions did not improve and that nocturnal temperatures were not much lower than daytime temperatures. Because salmon are more vulnerable to infectious diseases at higher temperatures (McCullough 1999), crowding encouraged the disease outbreak that resulted in the kill.

The fish-passage hypothesis of CDFG or the minimum temperature hypothesis given above may or may not justify additional release of flow from Iron Gate Dam. It is unclear whether low flows actually blocked upstream migration or, as suggested by the literature, that most of the fish stopped moving because of high temperature (CDFG cites evidence that at least a portion of the run was capable of moving upstream during these low-flow conditions). The emergency release of 500 cfs of additional water from Iron Gate Dam by USBR, which arrived long after the fish kill had ended, lacked any specific justification. For relief of physical blockage, if it occurs, only a large amount of water (e.g., 1,500 cfs) would be of use. Additional water from the Trinity could be especially valuable in that it would be cooler, if released in quantity.

If passage is the key issue, the recurrence of low flows similar to those of 2002 will probably be accompanied by mass mortality of fish. If other explanations, including minimum temperature, are the key explanation of mortality, recurrence is less likely, although higher temperatures over the long term caused by climate change could increase the likelihood that such kills would occur. Aggressive pursuit of some recommendations related to coho salmon (see information on augmentation of cold-water tributary flows in Chapter 8) could, if successful, reduce the risk of mass mortality of Chinook salmon. In any case, it is clear that increased monitoring of water quality and channel conditions in relation to flows in the lower main stem is needed in support of measures that may be necessary to prevent loss of Chinook salmon.

CONCLUSIONS

The lower Klamath basin is a geologically dynamic region that historically had large runs of anadromous fishes with diverse life histories. The fishes were widely distributed in the basin; some even entered the rivers that fed Upper Klamath Lake. The Salmon, Scott, Shasta, and Trinity rivers—all of which are major tributaries of the Klamath River—were major salmon and steelhead producers. The Shasta River in particular, with its cold flows and high productivity, was once especially productive for anadromous fishes. In the Klamath basin as a whole, Chinook salmon were and are the most abundant salmonid, followed by steelhead. Coho salmon rank third, but are well below Chinook and steelhead in abundance.

Virtually all populations of anadromous fishes have declined considerably from their historical abundances, although documentation for some species, such as Pacific lamprey and green sturgeon, is poor. Three of the most distinctive forms—coho salmon, spring-run Chinook, and summer steelhead—are on the verge of extinction as naturally maintained populations in the basin. It is significant that these three are the most dependent on summer water temperatures below 18°C and that they historically spawned and developed in tributary streams, many of which now are too warm for them. The anadromous fishes have been in decline since the 19th century when dams, mining, and logging severely altered many important streams and shut off access to the upper basin. The declines continued through the 20th century with the development of intensive agriculture with its dams, diversions, and excessively warm water both inside and outside the basin. Continued logging in headwater areas and commercial fishing also have contributed to the decline.

The main-stem Klamath River has become a challenging environment for anadromous fishes because of decreased flows and increased summer water temperatures. Although it is inhospitable to juvenile coho, it is still important for the rearing of juvenile Chinook salmon and steelhead, but increases in temperatures in July–September of 1–3°C may make it unsuitable even for them in the future. Increased flows down the river in summer are likely to benefit anadromous fishes only if temperatures can be kept within bioenergetically favorable ranges. This is particularly true for the lowermost reach of the main stem, below the Trinity River, which may be either cooler or warmer in late summer than the main stem, depending on the amount of water being released from Lewiston Dam.

Millions of juvenile fish, including Chinook salmon, steelhead, and coho are released into the Klamath and Trinity rivers each year by the Iron Gate and Trinity River hatcheries, which were built to mitigate salmonid losses created by large dams. These hatcheries have helped to maintain fisheries for coho and Chinook salmon, but their effect on wild populations of salmonids in the basin is not well understood. It is likely, however, that interactions between the hatcheries and wild juveniles in the river are having an adverse effect on the survival of wild juveniles through competition for space and food and aggressive interactions (e.g., McMichael et al. 1999,

Kelsey et al. 2002), to the extent that the contributions of hatchery fish to fisheries are at least partially offset by the decreased contribution of wild fish (Levin et al. 2001). A high percentage of naturally spawning adult coho and Chinook salmon are of hatchery origin.

Native nonanadromous fishes are widespread and common in the drainage, but their relationships to anadromous fishes are not known. Nonnative fishes are uncommon in the lower basin except where drastically altered habitats favor them. If habitat degradation continues, the Klamath River and its main tributaries will probably favor nonanadromous native and nonnative fishes increasingly at the expense of anadromous fishes. The hierarchical nature of watersheds assures that many environmental changes, some of which are quite small individually, collectively affect fish populations not only in their immediate vicinity but also both upstream and downstream because of the extensive movement of fishes (Fausch et al. 2002).

The problems with coho salmon are a reflection of larger problems with poor habitat and water quality for anadromous fishes generally in the basin. Restoration efforts that benefit coho salmon should benefit most, but not necessarily all, declining species. Prevention of further listings under the ESA requires a systematic, basin-wide approach to restoration and management. Some major gaps in knowledge are as follows:

1. Information on the biology of coho and other salmonids in the basin is largely unsynthesized; synthesis and interpretation of data on historical trends and present conditions would be especially valuable.
2. Studies on anadromous fishes other than fall-run Chinook, winter steelhead, and coho are very limited or lacking, particularly for summer steelhead, spring-run Chinook, and Pacific lamprey. It cannot be assumed that management strategies favoring species of primary interest also favor other species.
3. The biology of nonanadromous native fishes and macroinvertebrates in the basin is largely unknown, including basic descriptions of life histories and environmental requirements and their relationships to coho salmon and other anadromous fishes.
4. The potential effects of global climate change on the Klamath basin and its fishes, especially coho, are poorly understood, including the relationship between changing ocean conditions and the abundance of coho and other anadromous fishes. Climate warming would almost certainly be disadvantageous to coho.
5. The thermal consequences of stream and watershed restoration actions, including increasing summer flows down the main-stem Klamath River, are not well documented, especially in relationship to coho salmon.
6. The effects of hatchery operations on wild populations of coho and

other salmonids in the basin are not understood, including the effects of hatchery steelhead and Chinook on juvenile coho salmon.

7. Strategies for improving tributaries for spawning and rearing of coho and other anadromous fishes are not yet well defined.

8. The lower 30–40 km of the main-stem Klamath seems to be increasingly unfavorable to anadromous fishes, for reasons that are not known. The effect on the lower river of changing flows from the Trinity River needs to be evaluated, as do the potential benefits of comanaging flow releases from the dams on the Trinity and Upper Klamath rivers.

9. Reliable abundance estimates and habitat affinities of juvenile coho and other salmonids are largely lacking.

8

Facilitating Recovery of Coho Salmon and Other Anadromous Fishes of the Klamath River

Restoration of anadromous fishes to higher abundances in the Klamath basin will require multiple interactive initiatives and will take many years to reach full effectiveness. This chapter emphasizes actions needed for recovery of coho salmon; the same actions likely will benefit other species as well. Remedial actions to be evaluated here include restoration of tributary habitat, restoration of main-stem flows and habitats in the Klamath River, removal of dams, changes in land use and water management, changes in operation of hatcheries, and creation of an institutional framework for fisheries management. Research and monitoring programs are the means by which remedial actions should be evaluated and adjusted.

RESTORATION OF TRIBUTARIES

Coho salmon, spring-run Chinook salmon, and summer steelhead depend heavily on tributaries to complete their life cycles and sustain their populations (Chapter 7). Thus, restoring large, self-sustaining runs of anadromous fishes in the basin requires restoration of the tributaries to conditions that favor spawning and rearing of anadromous fishes. For most of the tributaries, restoring low summer temperatures probably is the most important action (Table 8-1). Removing barriers, improving physical habitat, and increasing minimum flows also are important and are strongly linked to the objective of lowering summer temperatures.

Because the four main tributaries differ from each other, a uniform approach to management and restoration in their watersheds is unlikely to

TABLE 8-1 Factors Likely to Limit Production of Coho and Other Salmonids in the Shasta, Scott, Salmon, and Trinity Rivers and Their Tributaries

Limiting Factors	Shasta, Main	Shasta, Tributaries	Scott, Main	Scott, Tributaries	Salmon, Main	Salmon, Tributaries	Trinity, Main	Trinity, Tributaries
Migration Barriers								
Dams, weirs, diversion structures	x	x	o	x	–	–	x	o
Low-flow blockage	x	x	x	x	o	o	x	o
Thermal barriers	x	x	x	–	o	–	–	–
Hydrologic Changes								
Low summer and fall flows	x	x	x	x	–	–	–	o
Reduced peak winter flows	o	–	–	–	–	–	x	–
Reduced spring flows due to diversions	o	o	o	o	–	–	x	o
Reduced base-flow support from ground water	x	x	x	x	–	–	–	–
Water Quality								
High temperature	x	x	x	–	o	o	o	o
Low dissolved oxygen (DO)	x	x	o	–	–	–	–	–
pH, alkalinity, dissolved solids	–	–	–	–	–	–	–	–
Suspended solids	–	–	o	o	o	o	o	o
Geomorphology								
Loss of spawning gravel	x	o	x	x	o	–	x	x
Fine sediment deposition	x	x	x	x	o	–	x	x
Channel aggradation and instability	x	o	x	x	–	–	x	x
Reduced in-stream cover	x	o	x	o	–	–	x	x
Loss of riparian cover	x	x	x	o	–	–	o	x
Land Use Constraints								
Timber management practices	–	o	x	x	o	o	x	x
Grazing and pasture in riparian areas	x	x	x	o	–	–	o	o
Grazing in upslope areas	–	o	–	o	–	–	o	o
Management of fuels	–	o	–	x	x	x	o	o
Land conversion for agriculture	x	x	x	o	–	–	–	o
Unscreened diversions	x	x	x	x	–	–	–	o
Tailwater return flows	x	–	x	–	–	–	–	–
Water development	x	x	x	o	–	–	x	o
Urbanization	o	o	o	–	–	–	–	–

Abbreviations: o, common and of moderate concern or significance; x, widespread or important; –, probably not a limiting factor.

succeed. The following discussion outlines key issues that confront restoration of salmonids in each watershed. This review is not exhaustive; it focuses on the most important factors that appear to limit coho salmon and other anadromous species in the basin.

Shasta River

The Shasta River once was one of the most productive salmon streams in California (Snyder 1931, Wales 1951). It supported large runs of Chinook salmon, coho salmon, and steelhead. Over 80,000 Chinook salmon spawned in the river in the 1930s, by which time the population probably was already in serious decline as a result of habitat changes caused by placer mining and agriculture starting in the 1850s. Snyder (1931, p. 73) referred to it as early as 1931 as a "stream once famous for its trout and salmon." The historical runs of coho salmon and steelhead are not known but were probably large, given the apparent quality of the habitat. An assessment of the river in the 1960s suggested that runs of coho averaged around 1,000 fish per year and runs of steelhead averaged around 6,000 fish per year (CDFG 1965). The productivity of the Shasta River is related to its unusual hydrology and geologic setting (Chapter 4). Unlike the Scott and Salmon rivers, the Shasta River is dominated by groundwater discharge, principally through numerous cold-water springs. The headwaters of the Shasta watershed lie primarily on the northern and western flanks of Mt. Shasta. Rainfall and snowmelt recharge an extensive groundwater system that feeds the Shasta River. Historically, the river flowed at a minimum of about 200 cfs all year. The water was cool in summer and, in comparison with its companion watersheds, warm in winter. The exceptional thermal stability of the Shasta made it one of the most important tributaries for support of salmonids in the Klamath watershed.

Today, agricultural development of the Shasta valley (principally alfalfa and irrigated pasture) and the construction of Dwinnell Dam (which impounds the Shastina Reservoir) have fundamentally changed the hydrology and productivity of the Shasta River. The largest diversion of water is to the Shastina Reservoir, constructed in 1926, which loses a substantial part of its storage each year through seepage and blocks access to about 22% of the historical salmonid habitat. Surface diversions and loss of spring flow to the channel because of groundwater withdrawals have reduced summer flows to about 10% of their historical rates. The low volume of flow, high contribution of warm agricultural return flows, and loss of riparian shading lead to summer water temperatures that consistently exceed acute and chronic thresholds for salmonids. Because of high water temperatures, the Shasta River in summer supports mainly nonsalmonid fishes, such as the brown bullhead and speckled dace. Juvenile fall-run

Chinook salmon have emigrated by summer, and juvenile steelhead and coho persist mainly in the upper reaches of a few tributaries.

Given its former productivity, the Shasta River has exceptional potential as a restoration site for coho salmon as well as steelhead and Chinook salmon. Although multiple factors limit the abundance of salmonids in the Shasta (Chapter 4), the key to their recovery is to restore enough cold-water flow to keep the daily mean temperatures of the river below 20°C throughout summer. This would allow juvenile salmonids, including coho, to reoccupy the main stem of the Shasta, where they could take advantage of the river's naturally high productivity. Flows must also be restored in several key tributaries (such as Parks Creek and Big Springs) to improve their connectivity with the main river and to provide access to spawning sites.

The restoration of cold-water flows to the Shasta River presents many difficulties. The science behind restoration of the system, however, is relatively simple. Given the magnitude of the groundwater recharge area that is connected to the Shasta River, there appears to be ample potential to restore cool flows (Chapter 4). Additions of cool water to the relatively small volume of current summer flows are likely to have a substantial beneficial effect on temperature and habitat. Modest changes in the timing and magnitude of surface diversions and groundwater pumping, particularly in the upper reaches of the Shasta River and the tributaries between Dwinnell Dam and Big Springs, would have a large beneficial effect on the volume and temperature of water in the river during summer. Because the thermal mass of present flows is small, the benefits of cooling the water may be limited to the upper reaches of the river. If new water-management programs are linked to programs that seek restoration of riparian zones and channels, however, it is very likely that a substantial portion of the Shasta River can be restored to highly productive rearing habitat for coho and other salmonids. It is also appropriate to consider removal of the aging Dwinnell Dam. It loses more water to seepage than it provides for irrigation (Chapter 4), and its removal would restore flows, increase gravel recruitment, and allow access of salmonids to 22% of their historical habitat.

Numerous stakeholder groups and several federal and state agencies are now addressing habitat issues for salmonids in the Shasta watershed. Although not as well funded as the Scott River programs, the Shasta River restoration efforts are making progress, particularly in riparian fencing and management of tailwater return flow. To restore habitat effectively, these groups must develop methods for augmentation of the Shasta River with cool water during summer. Habitat restoration efforts that fail to deal with this issue are unlikely to succeed. A federally organized program promoting technical review of private habitat restoration efforts could make such efforts more successful.

Scott River

The Scott River originates in forested headwaters of the Marble, Scott, and Trinity mountains, meanders through the broad, agriculturally rich Scott valley, and then passes through the steep Scott River Canyon before joining the Klamath River (Chapter 4). The surrounding mountains are largely national forest, including the Marble Mountains Wilderness Area. The Scott River valley is private agricultural land, and the canyon reach below it is a state Wild and Scenic River (CDFG 1979b). The Scott River exhibits strongly seasonal flows derived from numerous tributaries that drain the western and southern edges of the watershed. The tributaries were and are critical for spawning and rearing of coho and steelhead, and the meandering river on the valley floor was important for spawning of fall-run Chinook and Pacific lamprey. It is likely that in all but the most severe drought years the main stem originally provided important and productive habitat for juvenile salmonids, including coho, throughout the summer, especially in the sloughs and pools of the numerous beaver dams that once were characteristic of the streams on the valley floor (CDFG 1979b).

The Scott River is still an important spawning area for salmonids, as indicated by the annual outmigrant trapping by the California Department of Fish and Game (e.g., Chesney 2002). Numbers of fish are severely diminished, however, and habitat is poor for one or more stages of the life history of all anadromous salmonids (CDFG 1979b). The decline in habitat for salmonids in the watershed has multiple, linked causes (summary in Chapter 4). In the forested western and southern margins of the watershed, intense logging and associated road building on highly erosive soils has produced high sediment yields. Tributaries draining that portion of the watershed have been degraded by deposition of fine sediments. In the lower portion of the tributaries, extensive diversions for irrigation remove water from streams during summer. In the valley, grazing and farming have reduced riparian cover on tributaries and on the main stem. In addition, historical placer mining in the main stem and some tributaries has severely degraded spawning habitat, and has formed migration barriers during low-flow years. The most important effect on salmonid habitat is associated with high water demand for alfalfa and irrigated pasture. Surface diversions and groundwater pumping lead to extensive low-flow and no-flow conditions during summer on the main stem and the lower tributaries. Increased reliance on irrigation wells since the 1970s and changes in cropping patterns appear to be the cause of declining flows between late summer and early fall. Low flows reduce or degrade rearing habitat and limit migration during fall. Low-flow conditions on the Scott also are accompanied by poor water quality (Chapter 4). The low volume of water in the river, coupled with the accrual of tailwater return flows, leads to high summer temperatures. Typical maximum weekly average temperatures are well above acute or chronic thresholds for salmon from summer into early fall.

Despite widespread decline in suitability of habitat, the Scott River retains high potential for becoming once again a major producer of anadromous fishes, especially coho salmon. The lower reaches of the tributaries on the west side of the basin, and the south and east forks, are still used extensively by coho and steelhead despite considerable degradation of the habitat. In addition to continuing efforts to reduce sedimentation and restore riparian vegetation cover in the streams, the key to restoring coho and other salmonids is to improve access of fish to the upper basin tributaries and to enhance cold-water flows. Improving access will require additional screening of diversions and removal of blockages but also will require more aggressive management of adjudicated surface diversions and groundwater to maintain sufficient flows for fish passage. Restoration of habitat for salmonids on the main stem of the Scott River also remains a considerable challenge. Low flows and associated high temperatures have the greatest effect on fall-run Chinook and lamprey but may also affect coho, particularly during dry falls. High water temperatures and loss of riparian vegetation probably have eliminated holding and rearing habitat for coho in the main stem. Restoring summer and fall conditions suitable for coho in the main stem will require careful and creative management of existing surface-water and groundwater resources in the Scott River valley. Water leasing and conjunctive use of groundwater and surface water may be the only means of reducing diversions and groundwater pumping during critical low-flow periods.

Multiple stakeholder groups and the local Resource Conservation District in the Scott valley have conducted a number of well-funded efforts to restore habitat in the Scott watershed. Cooperation between these groups and the state and federal agencies that support them appears to be the most effective way of restoring habitat in the basin. To date, however, the groups have not attempted to resolve the most important but intractable issue: increasing the amounts of cold water entering the tributaries and the main stem.

Salmon River

The Salmon River has a steep gradient, is largely forested, and lacks broad alluvial valleys. About 98% of the watershed is in federal ownership, and more than 48% is designated as wilderness. The main stem, forks, and Wooley Creek are designated Wild and Scenic Rivers (CDFG 1979a). Wooley Creek is in nearly pristine condition, which is unique in the Klamath watershed. Most strikingly, the Salmon River is free of dams and is not subject to depletion of flow by diversions.

The Salmon River watershed contains about 140 mi of channel suitable for spawning and rearing of fall-run Chinook salmon and 100 mi of steelhead and coho habitat (CDFG 1979a). Other fishes in the community include spring-run Chinook salmon and summer steelhead, which, like coho salmon, require deep pools and cold water throughout the summer. The principal habitat for spring-run Chinook salmon and summer steelhead in the Salmon River drainage today is Wooley Creek, although small numbers are also found in the forks of the Salmon River as well (Moyle et al. 1995, Moyle 2002).

Despite natural flow conditions and absence of agriculture, salmonid populations in general are low in the Salmon River, and coho salmon in particular are scarce (Olson and Dix 1993, Brown et al. 1994, Elder et al. 2002). Records are poor, but salmonids most likely were considerably more abundant in the past (CDFG 1979a). Olson and Dix (1993) estimated that only about 25% of the available spawning habitat was used by Chinook salmon and steelhead. The causes of decline and the status of current populations are not clear.

A variety of natural and anthropogenic factors may suppress salmonid populations in the Salmon River. Unlike the Shasta and the Scott rivers—which have alluvial valleys that formed favorable habitat for holding, spawning and rearing of salmon—the Salmon River has a bedrock channel of high gradient that limits the total amount of suitable habitat as defined by depth and velocity. The high rates of uplift in the watershed, coupled with unstable rock types, produce naturally high erosion rates that are associated principally with mass movements (CDFG 1979a). High erosion rates, which are accompanied by high sediment yields, have been accelerated by human activity in the last century (Elder et al. 2002).

In addition to naturally high sediment yields, the Salmon River watershed exhibits strong seasonal variations in flow, including large winter floods and low base flow during the last half of the summer. Low-flow conditions in the summer, particularly during drought, and the scarcity of cold springs may have naturally produced sufficiently high summer temperatures (maximums, 20–26°C) in some tributaries and in the main stem to limit production of salmon within the basin. Thus, the Salmon River watershed, although nearly pristine, may have geologic and hydrologic characteristics that are suboptimal for salmon. Under these conditions, human activities that increase sedimentation or raise stream temperature in the basin could have an especially large effect on salmon and steelhead.

The first major anthropogenic disturbance to the Salmon River was placer mining and other forms of gold mining, which peaked in the basin between 1850 and 1900 but continue today on a small scale (CDFG 1979a, Chapter 4). Placer mining disturbs the channel and disrupts sediment transport processes that sustain spawning gravels and maintain pools. A more important disturbance in recent years has been a combination of logging and fires. Logging and its associated road-building have greatly increased erosion on the steep and fragile slopes of the watershed and have reduced shading of small tributaries, thus increasing water temperatures. Stream crossings also significantly impair tributary streams in this basin by forming barriers to migration and local sources of erosion. Large fires may have exacerbated the effects of logging in the basin. Almost 30% of the basin has burned in the last 25 yr, and most fires have occurred in the logged portions of the basin (Salmon River Restoration Council 2002). These catastrophic fires, coupled with extensive logging that follows fires ("salvage logging"), have greatly increased the number of logging roads and increased the frequency of landslides (CDFG 1979a, Elder et al. 2002). Elder et al. (2002) estimated that from 1944 to 1988 about 216 mi of stream in the basin were scoured by debris flows caused by landslides. In addition, poaching of the vulnerable adult summer steelhead and spring-run Chinook may be important in reducing their populations (West et al. 1990, Moyle et al. 1995).

Factors outside the basin—including ocean or estuary conditions, harvest, and conditions on the Klamath main stem—may have reduced adult populations of salmonids in the Salmon River. Overall, however, it is likely that land-use activities in the Salmon River watershed have had the largest adverse effects on production of salmon and steelhead in the Salmon River basin.

Because the Salmon River watershed is owned principally by the federal government, there has been comparatively little controversy surrounding management and restoration efforts within the basin. A small but growing stakeholder group is cooperating with state and federal agencies and tribal interests in the Salmon River basin. High priority has been placed on monitoring of salmon and steelhead runs, improvements in riparian habitat, management of fuels, and assessment and rehabilitation of logging roads (Elder et al. 2002). Given proper funding and agency participation, these efforts may be sufficient to improve conditions for coho and other salmon and steelhead in the watershed.

Trinity River

Because the Trinity is the largest tributary of the Klamath River and enters only 43 mi upstream of the estuary, management and investigative efforts by the agencies have regarded it as if it were a separate river system. The creation in 1963 of the Lewiston and Trinity dams combined with the transbasin diversion of a significant proportion of the annual flow further enforces this impression of separation. Even so, the Trinity River flows influence water temperature and quality in the lower Klamath River and its estuary.

The Klamath River below Iron Gate Dam and the Trinity River have the same fish fauna, including the runs of salmon, which belong to the same ESUs (Moyle 2002). Chinook salmon, for example, have two ESUs: the Upper Klamath and Trinity ESU and the Southern Oregon and California ESU, the latter of which includes salmon in the lower Klamath and Trinity rivers. Both genetic evidence and marked hatchery fish demonstrate that salmon and steelhead from the two systems continuously mix. In addition, both systems have large hatcheries that produce coho salmon, Chinook salmon, and steelhead. Immigrating spawning adults and emigrating smolts from the Trinity River rely on lower Klamath River water temperature and quality to support their success in terms of egg quality, osmoregulatory ability, and survival. Thus, efforts to conserve coho salmon and other declining fishes must take both systems into account.

Data on the numbers of salmon and steelhead returning each year to the Trinity River and its tributaries are fragmentary and incomplete. There is general agreement, however, that populations of the most sensitive salmonids (coho, spring-run Chinook, and summer steelhead) have declined considerably (perhaps 90% or more) to a few hundred individuals of wild origin (Moyle et al. 1995, Moyle 2002, CDFG 2002). Populations of winter steelhead and fall-run Chinook also are much lower than they historically were, but there are few estimates before 1977. Between 1977 and 1999, fall-run Chinook salmon escapement was estimated to range from about 7,000 to 125,000 fish; fewer than 25,000 spawners were present in 12 of 23 years (CDFG 1999). From 1992 through 1996, only about 1,900 adult steelhead were recorded in the river above the confluence with the North Fork River each year; this is only 5% of goals set in 1983, which were based on estimates of historical abundances (USFWS 1999). The Trinity River Hatchery releases large numbers of juvenile coho, steelhead, and fall-run Chinook each year, but its role in maintaining the present runs is not well understood. Although the hatchery has been in operation since 1964, it has failed to prevent the continued decline of salmon and steelhead populations. In years when the numbers of returning Chinook salmon are low, percentages of hatchery Chinook in the run can be as much as 40–50%.

Causes of the decline in coho and other anadromous fishes are similar to those elsewhere in the Klamath basin (USFWS 1999). Some of the most important probable causes of decline specific to the Trinity River include construction of dams and associated regulation, enhancement of erosion associated with logging and grazing practices, placer mining, and hatchery operations. Construction of Lewiston and Trinity dams in the main stem in 1963 blocked access to over 109 mi of salmonid spawning habitat (cold water, good gravels), including 59 mi of spawning habitat for Chinook salmon. The dams and associated water diversion also reduced flows downstream, blocked recruitment of gravel to areas downstream of the dam, and reduced rates of channel-forming geomorphic processes. Extensive poorly managed logging and road building on steep slopes with highly unstable soils, followed by large fires, have resulted in a high frequency of landslides and erosion that cause high sediment loads in the river and its tributaries. Massive erosion triggered by the floods of 1964 in particular resulted in large-scale destruction of spawning and rearing habitat. In addition, extensive placer mining for gold in the 19th century, and to some extent into the 20th century, resulted in loss of spawning and rearing habitat that still persists in many places. Finally, the Trinity River Hatchery has a major effect on wild populations of coho salmon, Chinook salmon, and steelhead, given that marked hatchery fish are frequently observed spawning in the wild. It is possible that hatchery production is suppressing populations of wild fish (e.g., Kostow et al. 2003), especially of coho salmon, but this has not been studied in the Trinity basin.

The South Fork Trinity River is one of the largest tributaries within the Klamath basin. Although poorly documented, historical salmon and steelhead runs within the South Fork were very large, and included coho. Poor logging and grazing practices on unstable soils in the South Fork Trinity coupled with highly destructive floods in 1964, destroyed most spawning and rearing habitat within the South Fork. Although habitat conditions appear to be improving, this tributary adds little to the overall salmon and steelhead productivity of the basin.

Recognition that runs of anadromous fish in the Trinity River are declining and in need of recovery has led to many restoration projects throughout the basin. Friends of the Trinity River, for example, estimate that nearly $100 million was spent on restoration projects in the basin from 1983 through 2000 (FOTR 2003). The 1999 EIS/EIR on dam operations indicated that reduced flows below Lewiston Dam, especially in spring, had significantly altered salmonid habitat in the Trinity River. As a result, the Secretary of the Interior in December 2000 issued a Record of Decision (ROD) recognizing that long-term sustainability of the Trinity River's fishery resources requires rehabilitation of the river. The ROD called for specific annual flows designed to vary with water-year type and patterned to mimic natural variability in annual flows. The ROD also specified physical channel rehabilitation, sediment management, and watershed restoration efforts throughout the basin (USFWS 1999, 2000). Additionally, the ROD called for an Adaptive Environmental Assessment and Management (AEAM) program, guided by the Trinity Management Council, to use sound scientific principles in guiding the course for recovery in the Trinity River basin. Because of lawsuits by Central Valley water users challenging the EIS/EIR, however, the new flow regime has not yet been fully implemented.

Poor land-use practices and water diversions have reduced the capacity of the Trinity River to support coho salmon and other anadromous fishes.

There are no quick fixes for problems that are so severe and pervasive. Some of the measures that could be taken to improve the situation for salmonids both in the Trinity and the lower Klamath River already have been identified in the ROD, and in sediment TMDLs for the main stem and the South Fork (EPA 1998, 2001). The proposed flow schedule for the main-stem Trinity attempts to manage releases in a flexible manner that benefits aspects of the life histories of multiple species while responding to interannual variability in runoff conditions. Coho may benefit less than other species from main-stem flow alterations, however, due to their affinity with smaller tributaries.

Only large-scale restoration projects can reverse the adverse effects of logging, grazing, mining, and fires in the Trinity basin. Effective actions include removal of roads; elimination of logging, grazing, and off-road vehicle use from sensitive areas; planting and protection of trees to reduce erosion and restore riparian zones; and use of any other means to reduce erosion rates. Channel restoration and rehabilitation projects need to focus on restoring key geomorphic attributes of alluvial channels. These actions are called for by the ROD and are to be guided by the Trinity Management Council. Given that 80% of the lands within the Trinity basin are federally managed, large gains could be realized. It is unclear, however, whether these efforts will be restricted only to the areas immediately downstream of Lewiston Dam or, more appropriately, will be applied throughout the entire watershed, including the South Fork. A watershed approach is likely in the long run to be more successful than localized restoration. For coho salmon, physical restoration and protection of cold-water sources in tributaries that were historically important for spawning and rearing are of key importance.

Estimates of numbers of spawners of coho and other salmonids are needed as an index of the effectiveness of restoration efforts. The concept of numerical restoration goals, as set in 1983 and adopted by the 1999 EIR, is valid, but should be reviewed using information from such sources as the Indian fishery and extent of original habitat. The restoration goals must apply to fish spawning in tributaries as well as in the main stem. Goals should include minimum numbers (e.g., following years of poor ocean conditions) as well as numbers for years of average conditions.

The many small restoration projects in the basin should be continued, but should be viewed as experiments in adaptive management that ultimately will demonstrate the most effective treatments for Trinity River problems. Coordination of existing projects with those outlined in the ROD should be expanded.

It is vital that management of the Trinity River, including releases from Lewiston Dam, be viewed in the context of the entire Klamath watershed. The two systems are inextricably linked and are dependent upon each other for long-term success. Efforts presently are under way to use enhanced flow releases from the Trinity to reduce the likelihood of fish kills in the lower Klamath. This represents an important step forward in cooperative management for the sake of the entire basin, rather than a single component.

Small Main-Stem Tributaries

About 50 permanent streams, many of which are quite small, flow into the main-stem Klamath between Iron Gate Dam and the mouth of the Klamath. The streams formerly supported substantial runs of steelhead, coho, and other anadromous fishes (Kier Associates 1998). The watersheds of most of the tributaries have been extensively logged, and many roads have been constructed in them. Irrigation diversions in the largest of the tributaries have reduced their summer flows. The status and trends of fish populations in individual tributaries for the most part are not well known, although Blue Creek and other nearby streams are being monitored by the Yurok Tribe (e.g., Gale et al. 1998, Hayden and Gale 1999). Most of these tributaries probably support far fewer adult and juvenile anadromous fish than they once did, because of changes to habitat caused by logging, mining, agriculture, and road construction, and as a result of water diversions. Restoration of habitat, low temperatures, and flows in these small streams would be of major benefit to tributary-spawning species—especially coho salmon, steelhead, and cutthroat trout—and potentially could improve rearing conditions for juvenile salmonids in the Klamath main stem by cooling the pools at the mouths of small tributaries. The emphasis on these restoration efforts should be on those tributaries that have existing or potentially significant sources of cold water.

THE MAIN-STEM KLAMATH RIVER

Modeling of Habitat Availability in Relation to Flow

The National Marine Fisheries Service (NMFS) has sponsored habitat availability monitoring in the Klamath main stem in support of the preparation of its biological opinions (NMFS 2001, 2002). The modeling work was reported by Hardy and Addley (2001) in a document commonly referred to as the Hardy Phase II draft report. The NRC Committee was encouraged to consider the final version of this report, but was cautioned against excessive reliance on the draft report on grounds that the final report would contain more thorough model calibration and possibly other changes that might alter the results.

The NRC committee read and discussed the draft Hardy Phase II report. The committee saw the modeling approach as flawed by heavy reliance on analogies between habitat requirements for Chinook salmon and habitat requirements for coho salmon. Habitat requirements for Chinook salmon are better known, but the behavior and environmental requirements of Chinook salmon differ substantially from those of coho salmon (Chapter 7). To the extent that this approach is carried forward into the final report, the NRC committee's skepticism about the validity of the analogy would also be carried forward. In addition, the NRC committee, as explained elsewhere in this chapter, concludes that rearing of coho in the Klamath main stem is much less important than rearing of coho in tributaries, which are the preferred rearing habitat of coho. Thus, the importance that can be attached to regulation of flows in the main stem is probably less, in the viewpoint of the committee, for coho than it would be for Chinook, for example. Because the Hardy Phase II draft report does not deal with tributaries, the analysis in the draft Hardy Phase II diverged from the committee's analysis of the critical requirements for coho.

The committee recognizes that main-stem flow may directly affect the coho population at the time of downstream migration of smolts. While it is unclear whether additional water would favor the success of this migration, it is also clear, even in the absence of modeling, that NMFS can argue, given the absence of data to the contrary, that there is some probability of benefit for the smolts to be derived from minimum flows at the time of smolt migration, as expressed in the NMFS biological opinion of 2002. Adaptive management principles could be applied to this issue.

Management of Flow at Iron Gate Dam

In its biological opinions of 2001 and 2002, NMFS (2001, 2002) called for increases in minimum flows from Upper Klamath Lake via Iron Gate Dam for the benefit of coho salmon. NMFS reasoned that increased flows would increase rearing habitat for juvenile coho salmon, thus increasing their growth and survival in the river. For bioenergetic and ecological reasons (Chapter 7), it is unlikely that increased summer flows would benefit juvenile coho salmon. Additional water would likely be too warm for them (Chapter 4), and their principal habitat affinities during rearing are with the tributaries rather than the main stem. Additional flows would probably benefit Chinook salmon, steelhead, Pacific lamprey, and other more thermally tolerant fishes in the river by providing them with additional rearing habitat.

There is limited flexibility for managing the temperature of releases from Iron Gate Dam. Some cool water flows into Iron Gate Reservoir from springs and tributaries, but it is of little value for cooling the river in summer because of the large volume of the reservoir relative to these accretions. Because the deep waters of Iron Gate Reservoir store cool (hypolimnetic) water throughout the summer, however, it would seem that the construction of a deep withdrawal, coupled with selective aeration of the hypolimnion during the summer, could make available a pool of water for cooling the Klamath main stem below Iron Gate Dam. Unfortunately, the cool summer water has a volume of only about 15,000–18,000 acre ft (M. Deas, Watercourse Engineering, Inc., personal communication, July 16, 2003), which is sufficient to cool the reservoir release for only seven to ten days. Use of the water for cooling would not provide sustained benefits for the fish, and also would remove the source of cool water for the Iron Gate Hatchery, which relies on the deep water of Iron Gate Reservoir for hatchery operations. Furthermore, information from thermal modeling shows that introduction of cool water would provide benefits only for a relatively short distance downstream of the dam, given that summer thermal loading of the main-stem Klamath is high and that accretion of flow from tributaries consists primarily of warm water in summer.

Higher summer flows from Iron Gate Dam appear to increase minimum temperatures by reducing the effect of nocturnal cooling (Chapter 4). Higher flows also may raise the temperatures of the few cold-water refuges available in the main stem, the pools into which cool tributaries flow. Juvenile salmonids seek these pools during the day but disperse at night as the water cools (M. Deas, Watercourse Engineering, Inc., personal communication, November 25, 2002; unpublished data, USFWS). Even small disturbances to these pools (for example, by anglers) cause the fish to move into unfavorably warm water (M. Deas, Watercourse Engineering, Inc., personal communication, November 25, 2002), potentially harming or killing them. A natural-flow paradigm now commonly referenced in fisheries management is based on the premise that ecosystem functions and processes and the aquatic communities of rivers are affected by deviations from natural flows, including specific seasonal patterns and specific interannual ranges of variability by season (Poff et al. 1997). In the Klamath River, for example, the native fishes evolved with an annual sequence of winter pulse flows (principally from tributaries), high spring flows (from tributaries and the upper basin), and low flows in late summer and fall (principally from the upper basin). Base flows varied with climatic conditions. Some years provided strong winter and spring flooding that connected the channel with the floodplain, redistributed sediment, cleaned gravel, and re-formed the habitat features of the channel; other years had lower flows with much smaller effects. The timing of the flows and the ambient warming of the main-stem Klamath occurred in synchrony with tributary conditions; salmon smolts emigrating from a tributary did not leave a cool, springflow condition to enter a main stem experiencing a warm, summer base-flow condition. Thus, managing stream flows in ways that reflect timing and duration of the unregulated hydrograph is a holistic approach that recognizes climatological reality but can still be consistent with extensive human use of water resources. Such an approach would not demand high base flows in years of drought but could capitalize on years of high flow to maintain and restore habitat. It is also worth noting that historically the upper Klamath basin supplied only a portion of the flows of the lower Klamath River. Thus, increasing flows from the Scott and Shasta rivers would not only have thermal benefits to the main stem but mimic natural sources of flow more closely. Temperature in the lower basin will likely be increasingly important as global climate change occurs (Parson et al. 2001).

THE LOWERMOST KLAMATH AND OCEAN CONDITIONS

The lowermost Klamath is important to coho as an entry and exit point for the main stem. In addition, any substantial change in the hydrograph at the mouth of the Klamath could be expected to influence conditions in the estuary. While it may be attractive to use Trinity flows to influence conditions in the lower Klamath River, it must not occur at the expense of Trinity River restoration goals. Within the ROD for the Trinity River EIS/EIR, watershed restoration and monitoring that benefits fishery resources below the confluence of the Trinity and the Klamath rivers may be considered for action by the Trinity Management Council.

As explained in Chapter 4, total annual flow in the lower Klamath and its estuary has been altered only to a small degree by water development in the upper basin, even though water development has had drastic effects on hydrographs in a number of headwater areas. Thus, changes in total flow are not sufficiently large to suggest significant biological effects on the estuary strictly related to amount of flow. Furthermore, fall flows, even in years of average or above average moisture, tend to be higher than they were historically at the mouth of the Klamath (USFWS/HVT 1999, Hardy and Addley 2001), which would indicate that fall migrations probably have not been impaired by flow depletion per se. Warming of the water and poor water quality have greater potential significance, particularly near the mouth of the Klamath (see the section in Chapter 7 on fish mortality in 2002).

Estuary and ocean conditions undoubtedly induce variation from year to year in the strength of coho migrations. In part these variations are natural (i.e., they may be related to synoptic changes such as those associated with Pacific decadal oscillation or with shorter-term climate variability affecting ocean conditions). In addition, as mentioned in Chapter 4, the estuary and river mouth have undergone chemical changes because of anthropogenic influences upstream. The extent to which these factors are affecting coho populations is unknown at present, however. While favorable ocean conditions may magnify the strengths of certain year classes, any such favorable effects should not be used as a reason for reducing emphasis on improvement of watershed conditions for coho, given that especially good ocean conditions inevitably alternate with poor ocean conditions (NRC 1996).

REMOVAL OF DAMS

Dams often have major adverse effects on native fishes, especially anadromous fishes (Moyle 2002). There is growing national and international recognition that removal of some dams may provide substantial benefits to fish and downstream ecosystems by increasing flows, improving the flow regime, and providing access to upstream habitat (Heinz Center 2002, Hart and Poff 2002). Dams that have been removed so far in the United States primarily have been small and have had low or even negative economic value, although some larger dams have been proposed for removal on grounds that the benefits of removal outweigh the value of the dams and the cost of removal.

All dams (including both large public or corporate dams and small private dams) and diversions in the lower Klamath basin need to be systematically evaluated for their effects on anadromous fishes; those with strong adverse effects should be investigated further for modification or removal. Specifically, Iron Gate Dam should be evaluated for removal in conjunction with recapture of flows in Jenny Creek that are now diverted out of the Klamath basin to the Rogue River. Iron Gate Dam was built in 1962 to reregulate flows from Copco Dam. Copco Dam was built in 1917 to generate power, mostly at times of peak demand. Water released from the dam on demand caused major daily fluctuations in downstream flows that were harmful to the fish and other ecosystem components (Snyder 1931). Iron Gate Dam was intended to allow more uniformity in the release of water. The reservoir behind the dam flooded about 6 mi of the Klamath River. The flooded main-stem reach and its tributaries apparently were excellent spawning habitat for Chinook, coho, and steelhead (Snyder 1931), probably because of cool water in the tributaries. To mitigate this loss, the Iron Gate Hatchery, which uses water from the reservoir, was built to provide a source of young salmon. The hatchery releases several million juvenile Chinook, coho, and steelhead into the river each year (only about 70,000 per year are coho salmon; see Chapter 7). Iron Gate Reservoir supports a recreational fishery mainly for nonnative yellow perch and stocked rainbow trout.

There has been no systematic evaluation of the benefits and costs associated with the removal of Iron Gate Dam, but removal of the dam would recapture about 6 mi of lost habitat in the main stem of the dam and substantial tributary habitat; the 6-mi reach could also have lower summer water temperatures than most of the main stem. Removal of Iron Gate Dam would require operation of Copco Dam in a more uniform manner, which would result in loss of power revenues from Copco Dam. An alternative water supply also would be needed for the Iron Gate Hatchery. Opportunities for removal of Iron Gate Dam could be considered in the near future under the Federal Energy Regulatory Commission (FERC) relicensing process. The current license for operation expires in 2006; a draft application is due in 2003 (FERC Relicensing Number 2082).

CHANGES IN OPERATION OF HATCHERIES

The reason for building the hatcheries on the Trinity River and at Iron Gate Dam was to ensure that fisheries could be sustained at levels at least as high as they were before the construction of the dams. Despite the operation of the hatcheries, commercial fisheries for Klamath basin fishes have largely been shut down, and sport fisheries have declined; the principal remaining fishery is the tribal subsistence fishery for salmon and sturgeon. Overall, anadromous fish in the basin now reach only a small fraction of their historical abundance. Abundance has declined despite the release of millions of juvenile Chinook, coho, and steelhead into the rivers each year by the hatcheries (Chapter 7). There is growing evidence from numerous river basins that large-scale releases of hatchery fish have an adverse effect on remaining populations of wild fish and do not contribute as much to fisheries as generally supposed (e.g., Hilborn and Winton 1993, Knudsen et al. 2000, Levin et al. 2001, Moyle 2002). Adverse effects can occur even when hatchery coho are stocked in streams ostensibly to help rebuild wild populations (Nickelson et al. 1986).

The effect of the hatchery fish on populations of wild salmonids in the Klamath basin is not well understood, but it probably is negative. For example, the release of millions of juvenile Chinook salmon every June floods the river with fish that are larger than the wild fish. The hatchery fish are likely to displace or stress wild Chinook and coho salmon (Rhodes and Quinn 1998). If food and space are not limiting factors in the river (that is, if the environment is not saturated with fish), hatchery fish would not make much difference in the growth and survival of wild fish. But this is probably not the case, especially as the water warms and fish seek the cool pools at the mouths of tributary streams. Furthermore, not all hatchery fish emigrate as assumed when stocked. Some of the stocked fish may remain in the river, potentially until the following spring, through the process of residualization. Residualization occurs when the smoltification process stops and a juvenile fish reverts to the parr stage (Viola and Schuck 1995). The smoltification process can stop when fish are exposed to temperatures beyond the physiological tolerance for smoltification. In some instances, large fractions of fish remain and compete with wild fish for limited habitat (Viola and

Schuck 1995). Residualization has not been studied in the Klamath basin, but its potential for harm to wild fish indicates that it should be studied.

The Klamath and Trinity basins provide an unusual opportunity for large-scale tests of hypotheses relating the effect of hatchery operations to the welfare of wild salmon and steelhead populations. The two basins can be regarded as a paired system in many respects. Because both have production hatcheries for coho, Chinook, and steelhead at the top of the accessible reaches for the species, comparative manipulations of hatchery practices are possible through an adaptive-management framework. For example, the Iron Gate Hatchery could be shut down for 6–8 yr (two Chinook and coho life cycles) while the Trinity River Hatchery remains operational (with the requirement that all fish be marked when released). Such a large-scale experiment would be informative if accompanied by intensive monitoring of juvenile and adult populations. An ecological risk analysis of the costs and benefits of hatchery programs should be conducted (Pearsons and Hopley 1999), especially in relation to coho salmon. If hatchery production results in a net loss of wild coho salmon, hatchery operation should be modified or even terminated.

LAND-MANAGEMENT PRACTICES

Throughout the distribution of coho salmon in the Klamath basin, the effects of land-use practices on the welfare of coho must be closely examined and, where damage to salmon habitat has occurred, restoration must be undertaken. Undesirable practices from the viewpoint of the welfare of coho include augmentation of suspended load through any agricultural practices that enhance erosion, forestry that does not incorporate best management practices, and mining that does not involve strict controls on sediment mobilization or that occurs directly in a stream channel. Coho would almost certainly benefit from regulation of grazing to an extent that involves exclusion of cattle from riparian zones and stream channels. The practice of flash grazing (exposure of riparian zones only for short intervals), while showing the appropriate intent, should be reviewed for actual effectiveness in terms of environmental objectives. Complete exclusion of livestock may be necessary in many instances, at least until woody vegetation is well established, and streambank conditions may never be consistent with the presence of large numbers of cattle, even on a short-term basis. Plans to restore stream channels, while laudable in intent, should be reviewed by federal and state agencies for effectiveness; government should assist landowners in finding the technically most desirable ways of achieving their restoration objectives. Review of channel and riparian conditions and their linkages to land-use practices should be included in a recovery plan for coho salmon (see Chapter 9).

CREATION OF A FRAMEWORK FOR FISH MANAGEMENT

Management of fish in the lower Klamath basin must deal with both harvest and habitat. For most of the history of the basin, regulation of harvest was the primary management tool, and it was complex in that it involved tradeoffs between ocean and river fisheries and among commercial, sport, and tribal fisheries (Pierce 1991). Despite harvest management, salmon and steelhead populations declined. Today, commercial fisheries are banned, and the sport and tribal subsistence fisheries are restricted. Reduced fishing pressure on wild fish populations, especially of coho salmon, is clearly part of the solution to restoration of the populations, but management of harvest does little good if spawning and rearing habitat is inadequate. The Klamath basin requires habitat restoration.

Numerous state and federal laws provide a basis of aquatic-habitat management and drive the policy of government agencies (Gillilan and Brown 1997). Examples of such legislation relevant to the welfare of fish in the lower Klamath basin are as follows:

- The Fish and Wildlife Coordination Act of 1934, which requires federal agencies to consult with state and federal wildlife agencies before any water development or modification project is undertaken;
- The National Environmental Policy Act of 1970, which requires all federal agencies or holders of federal permits to file reports on the potential environmental effects of their actions;
- The Wild and Scenic Rivers Act of 1968, which identifies rivers with special public values and prohibits construction of new dams on designated rivers;
- The Clean Water Act of 1972, which promotes having the natural waterways of the United States be "drinkable, swimmable, and fishable." Under this act, many streams in the Klamath basin have been declared impaired in water quality;
- The Endangered Species Act of 1973 (ESA), which requires the designation of "critical habitat" for endangered and threatened species (see Chapter 9);
- The National Forest Management Act of 1976 (NFMA), which requires national forests to be managed to provide viable, widely distributed populations of all native vertebrates, including fish;
- The Sustainable Fisheries Act of 1996 (SFA), which requires fisheries agencies to identify "essential fish habitat" (EFH) for managed species;
- The Trinity River Stream Rectification Act (1980), which is intended to control erosion and deposition problems that arise from the Grass Valley Creek watershed;
- The Trinity River Basin Fish and Wildlife Management Act (1984), which directed the Secretary of the Interior to develop a management pro-

gram to restore fish and wildlife populations in the Trinity basin to levels approximating those that existed immediately before the TRD construction;

- The Central Valley Project Improvement Act (1992), section 3406(b), which called for interim flows until the completion of the 12-yr Trinity River Flow Evaluation Study (USFWS/HVT 1999). The provision Congressionally requires the Secretary to implement recommendations resulting from the study.

Collectively, these laws provide a strong mandate to protect and improve fish habitat in the Klamath basin. Occasionally, they have resulted in major shifts in land use or policy to favor fish. For example, the NFMA resulted in the creation of a process that greatly altered management of public forest lands in the Pacific Northwest (Thomas et al. 1993, FEMAT 1993). A number of Klamath River tributaries have been designated "key watersheds" through this process, indicating their importance to anadromous fishes, and steps needed to enhance their ability to support fish have been outlined. For the most part, however, the laws do not require actions; rather, they provide for consultation and documentation of problems and can stimulate action. Their effect usually is to raise public awareness of problems and thus lead to protection or improvement of habitat through legal and social channels or through changes in agency policies.

An example of the potential of federal legislation to influence remedial action without actually requiring it is the EFH provision of the SFA. Like the ESA for listed species, the EFH provision directs fisheries management agencies to look beyond harvest management to habitat management. The provision recognizes that fisheries can be sustained only if habitat is available to support all life-history stages of the harvested species (Fletcher and O'Shea 2000). It does not mandate habitat management, but it does require the identification, by regional fisheries management councils, of the habitat for each species and of the factors adversely affecting the habitat. The results of the identification process are presented to other federal agencies, which are advised to consider them when they undertake activities that might affect the habitat. Implementation of the EFH is a large task, given that hundreds of species are harvested, but virtually no funding has been provided for it (Fletcher and O'Shea 2000). Even so, the EFH provision has been useful in calling public attention to the importance of habitat to the maintenance of fisheries.

The EFH designations made by the Pacific Fisheries Management Council are generic (PCFFA 2002). In this respect they resemble the critical-habitat designation made by NMFS for Southern Oregon/Northern California Coast coho salmon, which includes all existing and historical habitat (Fed. Reg. 64 (86): 24061–24062 [1999]). For the Klamath basin, there is only a general indication that EFH encompasses all anadromous salmonid habitat, present and historical, without regard to species, with a generic description of the habitat requirements of each life-history stage of each species. Despite the lack of enforcement provisions in the EFH requirement of the SFA, it would be worthwhile to designate species-specific EFH in the Klamath basin as a means of assisting decision-making in the many federal, state, and local agencies engaged in land and water management. Ideally, the EFH should be used in setting priorities for conservation and restoration of habitat.

POSSIBLE FUTURE EFFECTS OF CLIMATE CHANGE

Records relevant to the hydrologic cycle in the Klamath watershed are based on about 100 yr of rainfall and runoff records. Probabilistic analyses of the records are used in planning future water-resource management and in designing strategies for restoration of species at risk. Such use of the historical record is based on the assumption that the hydrologic cycle of the past is a general predictor of the hydrologic cycle of the future.

The rapid and substantial rise in atmospheric mixing ratios of carbon dioxide and other greenhouse gas in the industrial era could contribute to a measurable increase in global mean temperatures (IPCC 2001, NRC 2001). Global circulation models (GCMs) indicate that global mean temperatures will rise over the next century and that regional climates will be affected in variable ways (IPCC 2001, Strzepek and Yates 2003).

Regional climate change would probably affect the hydrologic cycle of the Klamath watershed (Snyder et al. 2002, Kim 2001, NAST 2001), but there appears to be no substantial effort on the part of government or private entities to plan for climate change. Planning, if it were to take place, faces two important hurdles. Climate change apparently is assumed to be a distant possibility, to be dealt with after more immediate issues are resolved. It is worthwhile to note, however, that regional climate change could occur over a period considerably shorter than the history of the Klamath Project. A second hurdle is that the current GCMs operate on a spatial scale that cannot resolve regional topographic features, which influence climate in most parts of the West (NAST 2001, Redmond 2003). Multiple efforts are under way to downscale the models so that they project regional climate change more accurately, but current GCMs are not suitable for planning on a watershed scale. Even so, several regional models have sufficient spatial and temporal resolution to allow realistic forecasts of the kinds of changes that are likely in a watershed (e.g., Snyder et al. 2002, Kim 2001, Lettenmaier et al. 1999, Lettenmaier and Hamlet 2003); these models are potentially useful to resource managers even though they might not accurately quantify the magnitude and timing of regional change.

A detailed model of the Klamath basin region at 25 mi resolution has been developed by Snyder et al. (2002). Use of the model demonstrates three important kinds of changes in the hydrology of the Klamath watershed that could occur over the next century: (1) warming, especially at high elevation in spring (April, May); (2) higher total precipitation, especially in spring; and (3) an increase in the ratio of rainfall to snowfall and large decreases in spring snowpack. The changes modeled by Snyder et al. (2002) and others have strong implications for management of water resources and all aquatic species, but especially salmonids (NAST 2001, O'Neal 2002). For salmonids, the most important potential changes include altered timing of snowmelt, lower base flows, and additional warming of water in summer.

Large reductions in snowpack coupled with higher precipitation would increase winter runoff and decrease spring runoff. Land use and water management already have shifted peak runoff (Figure 4-2), and climate change could increase the shift. Decline in spring runoff would have important implications for spring migration of coho salmon and other salmonids. Base flows during summer and fall would most likely decline in response to climate change because of increased evapotranspiration associated with higher temperatures and the concentration of annual runoff in winter. Base flows, especially in tributaries, already are too low and would decline further.

Increases in water temperature, particularly during summer low-flow periods would probably harm coho salmon and anadromous fishes in general (Chapter 7). Climate change could make temperature an even more urgent issue than it is now for the future of salmonids in the Klamath basin.

The effects of climate change in the Klamath basin would probably vary spatially within the basin. For example, the Wood River and the Shasta River both have headwater and groundwater recharge areas that lie at sufficiently high elevation to be more resilient than most stream reaches in the event of temperature increases and associated changes in precipitation. Conservation of cool-water sources in these and similar tributaries is likely to be even more critical in the future than it is now.

Uncertainty in the magnitude and timing of climate change in the Klamath basin and the uncertainty about its timing have discouraged resource managers from developing comprehensive, specific strategies to cope with it. It is important that climate change be addressed in the framework of adaptive management (Chapter 10) through programs that anticipate changes that would accompany warming.

CONCLUSIONS

Conditions in tributary waters are of paramount importance for rearing of coho salmon, as is also the case for spring-run Chinook salmon and summer steelhead, which is in contrast to other stocks of anadromous salmonids, including fall-run Chinook. Tributary waters include both the four main tributaries and numerous small tributaries that enter these main tributaries or enter the Klamath main stem directly. Small tributaries offer exceptional potential for restoration of coho salmon.

Remedial actions intended to promote the welfare of coho salmon are not uniform in type and priority across all tributaries. The Shasta River, which probably has the single largest potential for restoration of coho salmon and anadromous fish in general, shows depression of salmonid stocks caused by extensive diversions and blockage of flows at small dams as well as Dwinnell Dam; diversion of spring flows for agriculture leading to warming of these waters during the critical summer months; loss of riparian vegetation; reduction of base flow through diversions and excessive pumping of groundwater; and possible episodes of low oxygen concentrations. The Shasta River also shows loss of substrate characteristics consistent with successful spawning and has significant channel degradation associated with land-management practices. Practices leading to the degraded state of the Shasta include timber management, grazing, agriculture, and water management.

The Scott River also has high potential for restoration of coho salmon. Groundwater flows from springs are less pronounced than for the Shasta River, but an undesirable degree of cool water diversion occurs through groundwater pumping, as well as from surface diversions. Other problems closely parallel those of the Shasta, but physical degradation of the main-stem channel and lower tributaries may be even more pronounced than in the Shasta.

The Salmon River drains mainly public lands, but nevertheless shows historical reduction of coho and other salmon populations. Degradation of the Salmon River is primarily physical, and is associated with inadequate forest management leading to catastrophic fires and logging practices, especially road construction and maintenance, that lead to high levels of erosion. In addition, there are some flow barriers on the Salmon River.

The Trinity River, which is much larger than the other three tributaries, shows the full complex of problems found in the Scott and Shasta rivers, but is especially affected by loss of habitat caused by installation of dams and by physical damage to channels caused by improper land-management practices. Implementation of actions called for in the Record of Decision will promote restoration and create a framework for adaptive management through a large, comprehensive effort, but this effort must be coordinated with management of the overall Klamath basin.

Small tributaries to the four large tributaries and to the Klamath main stem show a wide array of problems and will require treatment by category or individually for effective restoration. Emphasis on cold water bearing tributaries is likely to yield the most benefit for salmonid restoration.

While the Klamath main stem is less important for rearing of coho than to some other anadromous taxa on the Klamath, a number of actions on the main stem might promote the welfare of coho. Additional water during the smolt migration could enhance downstream movement, and could be tested in this respect through adaptive management procedures. In addition, removal of Iron Gate Dam and Dwinnell Dam could open new habitat, especially by making available tributaries that are now completely blocked to coho.

Application of computer modeling to habitat availability on the main stem is not likely to be relevant to coho, but would be relevant to other taxa, such as fall-run Chinook, that use the main stem extensively for rearing. In general, coho restoration requires increased attention to lands and waters beyond the Klamath Project.

Hatchery operations may have a suppressive effect on coho salmon through predation and competition; it should not be assumed that hatchery operations are beneficial to salmonids in general or to coho in particular. Hatchery operations could be viewed as adjustable rather than static and thus explored through adaptive management principles.

Because land-management practices are broadly responsible for degradation of habitat that is critical to the coho, improvement of land-management practices and restoration activities in tributary waters are the key to restoration of coho populations. Restoration will require extensive work with private parties and with agencies that are not now strongly involved in ESA actions. Restoration can succeed only through substantial technical assistance in support of the considerable private efforts that are now under way. Constant evaluation of the success of specific strategies will be important to their ultimate success.

A framework for overall management of fisheries exists already through interlocking federal statutes that require conservation and protection of habitat and fishes. The Sustainable Fisheries Act of 1996 in particular seems well suited as a model for management of environmental remediation in the Klamath basin.

9

Regulatory Context: The Endangered Species Act

Although the federal Endangered Species Act (ESA) is not alone in providing a legal framework for resolving issues related to endangered and threatened fishes in the Klamath River basin, it is the dominant legal feature now affecting federal water management in the basin. As the nation's principal federal law to protect species, the ESA's express purpose is "to provide a means whereby the ecosystems upon which endangered species and threatened species depend may be conserved" (16 U.S.C. 1531(b) [2002]). Further explanation is provided in the statute's definition of "conserve," which is "to use . . . all methods and procedures which are necessary to bring any endangered species or threatened species to the point at which the measures provided pursuant to this chapter are no longer necessary" (16 U.S.C. 1532 [2002]). It is also a policy of the ESA, however, that "Federal agencies shall cooperate with State and Local agencies to resolve water resource issues in concert with the conservation of endangered species" (16 U.S.C. 1531(c) [2002]). The difficulty of satisfying those two central objectives is well illustrated by the Klamath River basin, as attested by the U.S. Bureau of Reclamation's (USBR) Klamath Project and other public and private water-management practices. Accordingly, this chapter provides an overview of the ESA and discusses the structure and implementation of its provisions that are relevant to the Klamath River basin generally and the Klamath Project in particular. The chapter provides conclusions as to how the ESA could be implemented more productively for the benefit of species and ecosystems in the Klamath River basin.

OVERVIEW OF THE ESA IN THE KLAMATH CONTEXT

In 1988, pursuant to its authority under Section 4 of the ESA, the U.S. Fish and Wildlife Service (USFWS) listed the shortnose sucker and Lost River sucker as endangered species (53 Fed. Reg. 27130, July 18, 1988). Almost a decade after the sucker listings, in 1997, the National Marine Fisheries Service (NMFS) listed the Southern Oregon/Northern California Coast (SONCC) coho salmon, an evolutionarily significant unit (ESU) of coho salmon found in the Klamath River basin, as a threatened species (62 Fed. Reg. 24588, May 6, 1997). These listings triggered a suite of ESA regulatory responsibilities that have since had substantial influence in Klamath River basin water issues:

- Section 4 of the ESA requires the listing agency to designate "critical habitat" for endangered and threatened species unless exceptions, which are narrow, apply.
- Section 4(f) of the ESA requires the listing agency to develop and implement a "recovery plan" for endangered and threatened species.
- Section 7(a)(1) of the ESA requires all federal agencies, through consultation with the listing agency, to use their authorities to carry out programs for the "conservation" of endangered and threatened species.
- Section 7(a)(2) of the ESA requires all federal agencies, through consultation with the listing agency, to ensure that actions they carry out, fund, or authorize do not "jeopardize" the continued existence of endangered and threatened species and do not result in "adverse modification" of their critical habitat.
- Section 9(a)(1) of the ESA prohibits all persons subject to U.S. jurisdiction (including federal, state, tribal, and local governments) from "taking" endangered wildlife species—and Section 4(d) allows the listing agency to extend the same level of protection to threatened wildlife species—unless authorized by the listing agency pursuant to appropriate "incidental take authorization" provisions of the ESA.

For reasons described more fully below, some of these responsibilities have not been implemented to their full potential in the Klamath River basin. USFWS and NMFS have used ESA's authority primarily through Section 7(a)(2), which prohibits federal agencies from causing "jeopardy" to listed species. Thus, the listing agencies have focused primarily on USBR's operation of the Klamath Project.

Before proceeding to a section-by-section comparison of ESA implementation in the Klamath River basin, it is important to recognize the pervasive influence of three general principles of ESA law and policy: the "best available evidence" standard, the burden of proof applicable to the relevant decision-makers, and the species-specific orientation of the ESA.

As a package, these principles substantially affect the agencies' implementation of ESA duties and authorities under specific ESA provisions and their approach to the larger challenge of ecosystem-level management of resources in the Klamath River basin. Emphasizing the general principles also helps to clarify the distinctions between the framework within which the agencies operate under the ESA and the framework within which the NRC committee evaluated the relevant agency decisions as defined by its charge.

The "Best Available Evidence" Standard

USFWS and NMFS have ESA decision-making duties, such as listing of species under Section 4 and issuance of biological opinions under Section 7, for which they must use the "best scientific and commercial data available" as prescribed in 16 USC 1533(b) [2002] and 50 CFR 424.11(b) [2002] (listing decisions) and 16 USC 1536(b) [2002] and 50 CFR 402.14(g)(8) [2002] (consultations). Section 7 thus requires that NMFS and USFWS consult the existing body of the "best scientific and commercial data available" to determine whether USBR's proposed operation of the Klamath Project is "likely to jeopardize the continued existence of any endangered species or threatened species."

Although the statute leaves the standard for "best evidence" undefined, the courts have interpreted it to mean several things:

- The agencies may not manipulate their decisions by unreasonably relying on some sources to the exclusion of others.
- The agencies may not disregard scientifically superior evidence.
- Relatively minor flaws in scientific data do not render the data unreliable.
- The agencies must use the best data *available*, not the best data *possible.*
- The agencies must rely on even inconclusive or uncertain information if that is the best available at the time of the decision.
- The agencies cannot insist on conclusive data to make a decision.
- The agencies are not required to conduct independent research to improve the pool of available data.

A summary of the existing body of case law appears in Southwest Center for Biological Diversity v. Norton, 2002 WL 1733618 (D.D.C. 2002).

Similarly, in 1994, USFWS and NMFS issued a joint policy providing guidelines for ESA decisions (59 Fed. Reg. 34271 [1994]). The policy shows how the "best evidence" standard would apply in the context of the jeopardy consultation; it directs the agencies to follow six guidelines:

• Require that all biologists evaluate all scientific and other information that will be used to make any consultation decision.

• Gather and impartially evaluate biological, ecological, and other information that disputes official positions taken by USFWS or NMFS.

• Ensure that biologists document their evaluation of information that supports or does not support a position being proposed by the agency.

• Use primary and original sources of information, when possible, as the basis of consultation decisions or recommendations.

• Adhere to schedules established by the ESA.

• Conduct management-level review of documents developed by the agency to verify and ensure the quality of the science used to established official positions.

Appropriately, therefore, the charge of the NRC committee included a determination as to "whether the biological opinions are consistent with the *available* scientific information" (emphasis added).

The Decision-Making Burden of Proof

The NRC committee's charge to assess "whether the [agencies'] biological opinions are consistent with the available scientific information" requires the committee to adopt the burden of proof that would apply in the scientific community rather than the legal burden of proof that applies under the ESA. Scientific burden of proof may differ from legal burden of proof; this issue pervades the ESA, where science and law intersect. Keeping scientific and legal burdens of proof separate is important for proper execution of the committee's charge. The committee believes that in its interim report and in this final report it has applied an accepted scientific framework for its assessment.

Some parties to the Klamath River basin ESA actions have advocated use of a "precautionary principle," according to which a special burden of proof lies with users of resources (e.g., G. H. Spain, Pacific Coast Federation of Fisherman's Associations, personal communication, August 26, 2002). The precautionary principle, however, is a decision-making policy instrument, not a scientific standard of proof or a requirement of the ESA. Although many versions of the precautionary principle exist in the laws of many nations and in the text of many international treaties, the prototype is found in Principle 15 of the 1992 Rio Declaration of the United Nations Conference on Environment and Development (Rio Declaration on Environment and Development, UNCED, U.N. Doc. A/CONF.151/Rev. 1, 31 I.L.M. 874 [1992]):

> In order to protect the environment, the precautionary approach shall be widely applied by the States according to their capabilities. Where there

> are threats of serious or irreversible damage, lack of full scientific certainty shall not be used as a reason for postponing cost-effective measures to prevent environmental degradation.

In other words, ignorance should not justify the decision either to move forward with a proposed action that might threaten the environment or to not regulate an activity for purposes of protecting the environment.

Application of the precautionary principle in the ESA context is discussed in the National Research Council's report *Science and the Endangered Species Act* (NRC 1995), which outlines the benefits of applying such an approach to decisions about conservation of species under the ESA. As that discussion demonstrates, however, whether to apply the precautionary principle is a policy decision and as such is outside the present committee's scope of work, which pertains to "whether the biological opinions are consistent with the available scientific information."

Indeed, even when a policy decision is made to apply the precautionary principle, the question of whether the decision is consistent with the available scientific information is important. As discussed above, the ESA and the agencies' implementing regulations unequivocally require that NMFS and USFWS base their decisions, as given in their biological opinions, on the *best available* scientific evidence and that NMFS and USFWS use that evidence to decide whether Klamath Project operations are *likely* to jeopardize the listed species. These are the only explicit evidentiary standards and burdens of proof that the ESA and the agency regulations impose on the two agencies in the consultation process. In the decision-making context, relevant principles of administrative law and the ESA leave application of the precautionary principle to the discretion of USFWS and NMFS when they are confronted with substantial but inconclusive or conflicting data, especially as to whether a species deserves listing or whether a proposed action is likely to cause jeopardy (see *Conner v. Burford*, 848 F.2d 1441, 1454 9th Cir. [1988]). At some point, however, erring on the side of protection in decision-making ceases to be precautionary and becomes arbitrary. One indication that policy-based precaution has given way to bias or political forces is a major inconsistency of a presumed precautionary action with the available scientific information. Hence, the precautionary principle could not guide the NRC committee's scientific evaluation; rather, the committee evaluated the way in which NMFS and USFWS considered the best available scientific information and how they used this information to decide whether USBR's proposed operation of the Klamath Project is likely to jeopardize the continued existence of the endangered suckers and threatened coho salmon. In making this evaluation, the NRC committee recognized that scientists of federal agencies who are responsible for judging jeopardy to listed species inevitably face difficulties that derive from incomplete information even under the best of circumstances, and certainly so in the case of the Klamath basin.

The Species-Specific Orientation of the ESA

The portion of the committee's charge requiring it to evaluate "whether the biological opinions are consistent with the available scientific information" implicates one of the inherent limiting features of the ESA: it is species-specific. The biological opinions under study, therefore, are opinions about listed species and not directly about the effects of the Klamath Project on resources in the Klamath River basin that have no known linkage to listed species. Notwithstanding its stated purpose of conserving the ecosystems on which listed species depend, the ESA is strikingly short on ecosystem-focused rationale. The ESA authorizes USFWS and NMFS to list species, to designate critical habitat for species, to prepare recovery plans for species, to use authorities for conservation of species, and to issue incidental-take authorizations for species. The ESA prohibits federal agencies from jeopardizing species, and it prohibits all others (including individuals and private organizations) from taking species. Indeed, the NRC committee's charge has been conditioned by the ESA's species-specific focus, with the ultimate objective of providing "an assessment of scientific considerations relevant to strategies for promoting the recovery of listed species in the Klamath River Basin" (Appendix A).

As shown in previous chapters of this report, the listed species do not define all there is to manage in the basin; their needs encompass only a portion of the Klamath basin's combined environmental resources. In fact, a species-specific focus and an ecosystem-level focus may lead to different management policies and decisions (NRC 1995, p. 111–121). Often, actions that restore ecosystem functions are beneficial to listed species, but not always. Conversely, what is good for the listed species is not necessarily good for other ecosystem attributes or, for that matter, equally beneficial for all the listed species themselves. The dichotomy between the listed species and ecosystems limits the extent to which USFWS and NMFS can use the ESA for ecosystem management (Ruhl 2000). The ESA's species-specific focus is in itself an inadequate basis of ecosystem-wide decision-making in the Klamath River basin.

SPECIES LISTING AND DESIGNATION OF CRITICAL HABITAT

None of the conservation measures of the ESA that bear on the Klamath River basin apply unless a species is listed as endangered or threatened according to procedures specified in Section 4 of the statute. A related decision, although not necessarily made at the time of listing (or, in some cases, at all), is whether the species has "critical habitat" that should receive special protection. Listing of species and critical-habitat designations thus are the events that trigger the ESA's recovery-planning efforts and regulatory programs. A review of the background of the Klamath River basin species listings and critical-habitat determinations shows the potential and realized scope of the recovery-planning efforts and regulatory programs that have followed.

Listing of Endangered and Threatened Species

Section 4 of the ESA governs listing of species as endangered or threatened. A species is endangered if it "is in danger of extinction throughout all or a significant portion of its range" and is threatened if it "is likely to become an endangered species within the foreseeable future throughout all or a significant portion of its range" (16 U.S.C. 1532 [2002]). The agencies must consider five criteria in listing a species: the present or threatened destruction, modification, or curtailment of its habitat or range; its overuse for commercial, recreational, scientific, or educational purposes; disease or predation; the inadequacy of existing regulatory mechanisms; and other natural or anthropogenic factors affecting its continued existence (16 U.S.C. 1533(a)(1)(A)–(E) [2002]). As noted above, the agencies must evaluate these criteria for the species in question and make the listing determination "solely on the basis of the best scientific and commercial data available . . . after conducting a review of the status of the species." This limitation keeps USFWS or NMFS from considering economic factors in deciding whether to list a species.

USFWS listed the two sucker species as endangered in 1988, noting that "dams, draining of marshes, diversion of rivers and dredging of lakes have reduced the range and numbers of both species by more than 95 percent. . . . Both species are jeopardized by continued loss of habitat, hybridization with more common closely related species, competition and predation by exotic species, and insularization of remaining habitats" (53 Fed. Reg. 27130 [1988]). The agency explained some of the principal factors causing decline in amount of habitat, as given in Chapters 5 and 6.

NMFS, in listing the coho salmon as a threatened ESU in 1997, found that "threats to this ESU are numerous and varied. Several human caused factors, including habitat degradation, harvest, and artificial propagation, exacerbate the effects of natural environmental variability brought about by drought, floods, and poor ocean conditions" (62 Fed. Reg. 24588 [1997]). The agency also explained in more detail the major factors responsible for the decline of coho salmon in Oregon and California (Chapters 7 and 8).

Designation of Critical Habitat

Section 4 of the ESA also requires USFWS and NMFS, subject to specified exceptions, to designate the critical habitat of a listed species. Critical habitat consists of "the areas within the geographical area occupied by the species, at the time it is listed . . . on which are found those physical or biological features (I) essential to the conservation of the species and (II) which may require special management considerations or protection" (16 U.S.C 1531 [2002]). Areas outside the occupied area can be included if they are essential to the conservation of the species. USFWS and NMFS, in making the critical-habitat determination, consider space for individual and population growth and for normal behavior; food, water, air, light, minerals, or other nutritional or physiological requirements; cover or shelter; sites for breeding, reproduction, rearing of offspring, germination, or seed dispersal; and habitats that are protected from disturbance or are representative of the historic geographic and ecological distributions of a species (50 C.F.R. 424.12(b)(1)–(5) [2002]). In weighing these factors, the agencies focus on "primary constituent elements," which are "roost sites, nesting grounds, spawning sites, feeding sites, seasonal wetland or dryland, water quality or quantity, host species or plant pollinator, geological formation, vegetation type, tide, and specific soil types." The agencies must consider the factors "on the basis of the best scientific and commercial data available" but must also take "into consideration the economic impact, and any other relevant impact, of specifying any particular area as critical habitat" (16 U.S.C. 1533(b)(2) [2002]). Areas that otherwise satisfy the criteria for critical habitat must be excluded from designation if the costs of designation outweigh the benefits of including the area, unless failure to designate such an area would result in the extinction of the species.

The agencies are required to designate critical habitat, to the greatest extent prudent and determinable, concurrently with the listing decision (16 U.S.C. 1533(a)(3) [2002]). The time period for a designation may be extended up to 1 yr if the agency finds either that publishing the listing rule has high priority for conservation of the species or that critical habitat is not determinable at the time of listing. In such a case the agency must designate critical habitat within the 1-yr extension period "to the maximum extent prudent" (16 U.S.C. 1533(b)(6)(C) [2002]). Accordingly, the agency can find that designating critical habit is "not prudent" and thus decline to do so.

Treatment of the "economic impacts" and the "not prudent" components of critical-habitat requirement by the agencies has been the subject of intense litigation in recent years; several judicial opinions have found the agencies' approaches flawed. For the analysis of economic impacts, the agencies have taken the position that the combined legal (and thus economic) effects of jeopardy consultations under Section 7(a)(2) and of the take prohibition under Section 9(a)(1), both of which apply when a species is listed and do not require designation of critical habitat to take effect, subsume any regulatory effects that critical-habitat designation might impose. Thus the incremental economic impact of designating critical habitat is, according to the agencies, essentially nil. Adopting this position as an assumption for purposes of analyzing economic impacts has allowed the agencies to truncate the process: although economic effects were never actually quantified, the agencies took the baseline effects imposed under the jeopardy consultation and the take prohibition as the starting point for economic analysis of the effects of critical-habitat designation. The agencies thus avoided having to describe the baseline effects and routinely found—with relatively little analytic exercise, given their operating assumptions—that the incremental effects of critical-habitat designation were zero. In 2001, however, a court ruled that the agencies' approach subverted congressional intent; the court required the agency in question to quantify both the baseline effects and any incremental effects (see *New Mexico Cattle Growers Association v. USFWS*, 248 F.3d 1277, 10th Cir. [2001]).

Similarly, on the "not prudent" question, the agencies had taken the position that because designation of critical habitat triggers only the prohibition against federal agencies' adversely modifying critical habitat, it adds relatively little protection, if any, to what is already available to listed species under the jeopardy consultation and prohibition against take. Designation of critical habitat, the agencies also argued, could be detrimental to species by identifying places where unscrupulous collectors might find the species. On balance, the agencies often found that detriments associated with designation of critical habitat outweighed benefits and that a designation of critical habitat was "not prudent." This set of assumptions also has been rejected by courts in recent years on the grounds that designation of critical habitat has important educational effects at least and that Congress did not intend it to be avoided through the blanket assumptions that the agencies have adopted (see *Sierra Club v. USFWS*, 245 F.3d 434, 5th Cir. [2001]). Notwithstanding the assumptions that have prevailed in the agencies' implementation of critical-habitat rules, USFWS has proposed critical habitat for the suckers, and NMFS has done the same for the coho salmon (Chapters 5–8).

In its 1988 rule listing the suckers, USFWS declined to designate critical habitat, because "little additional benefits of notification of the species presence would be achieved through critical habitat designation" (53 Fed. Reg. 27132 [1988]). Later, however, USFWS proposed critical habitat for the species (Chapter 6; 59 Fed. Reg. 61744, December 1, 1994), but it has not promulgated a final ruling on critical habitat for the suckers, probably because of general litigation over the manner in which USFWS has implemented decisions on critical habitat. It is not clear what effect some of the recent judicial opinions on critical habitat would have on the designation of critical habitat for the listed suckers, because the analysis of economic impacts has not been developed, contrary to some judicial requirements.

In its 1997 rule listing the salmon, NMFS found that "critical habitat is not determinable at this time" and that the species should be listed before the decision on critical habitat was finalized (62 Fed. Reg. 24608 [1997]). The agency did, however, designate critical habitat for the species in 1999 (64 Fed. Reg. 24049, May 5, 1999). It adopted a watershed-based approach to the designation (64 Fed. Reg. 24052 [1999]), explaining that

> a more inclusive, watershed-based description of critical habitat is appropriate because it (1) recognizes the species' use of diverse habitats and underscores the need to account for all of the habitat types supporting the species' freshwater and estuarine life stages, from small headwater streams to migration corridors and estuarine rearing areas; (2) takes into account the natural variability in habitat use that makes precise mapping problematic (e.g., some streams may have fish present only in years of plentiful rainfall); and (3) reinforces the important linkage between aquatic areas and adjacent riparian/upland areas. While unoccupied streams are excluded from critical habitat, NMFS reiterates the proposed rule language that "it is important to note that habitat quality in this current range is intrinsically related to the quality of upland areas and of inaccessible headwater or intermittent elements (e.g., large woody debris, gravel, water quality) crucial for coho in downstream reaches."

Significantly, NMFS included riparian zones in the designation because "streams and stream functioning are inextricably linked to adjacent riparian and upland (or upslope) areas" (64 Fed. Reg. 24053 [1999]). NMFS also explained (64 Fed. Reg. 24059 [1999]) that

> activities that may require special management considerations for freshwater and estuarine life stages of listed coho salmon include, but are not limited to (1) land management; (2) timber harvest; (3) point and nonpoint water pollution; (4) livestock grazing; (5) habitat restoration; (6) beaver removal; (7) irrigation water withdrawals and returns; (8) mining; (9) road construction; (10) dam operation and maintenance; (11) diking and streambank stabilization; and (12) dredge and fill activities.

It is not clear what effect some of the recent judicial opinions on critical habitat would have on the NMFS ruling for coho salmon, because the analysis of economic impacts has not been developed.

Recovery Planning

Section 4(f) of the ESA provides that, on listing a species, USFWS or NMFS "shall develop and implement plans (hereinafter in this subsection referred to as 'recovery plans') for the conservation and survival of endangered species and threatened species listed pursuant to this section, unless [the agency] finds that such a plan will not promote the conservation of the species" (16 U.S.C. 1533(f) [2002]). Recovery plans are to include a description of site-specific management actions that may be necessary for the conservation and survival of the species and objective, measurable criteria that, when met, would result in a determination that the species be removed from the list.

Despite the requirements of Section 4(f), recovery plans do not constitute mandatory directives to USFWS, NMFS, other federal agencies, or others. USFWS and NMFS portray them as guidelines and useful menus of recovery-oriented actions that they and other parties can take voluntarily. The courts have rejected efforts to instill more legal effect into the recovery-plan program (Cheever 2001).

NMFS has prepared no formal recovery plan for the coho salmon. In contrast, USFWS finalized a formal recovery plan for the endangered sucker species on March 17, 1993. As explained in Chapter 6, the NRC committee believes that the sucker recovery plan contains many constructive recommendations but may need revision in view of extensive research efforts since 1993.

REGULATORY CONSEQUENCES

Only when a species is listed do the regulatory programs of the ESA come into play. Two of them apply directly only to federal agencies: the so-called conservation duty under Section 7(a)(1), and the duty under Section 7(a)(2) to avoid jeopardizing species or adversely modifying critical habitat. Section 7(a)(2), however, can have substantial indirect effects on state, tribal, and local governments and private entities that receive federal funding or approvals or that benefit from federal actions. The third major regulatory program, the take prohibition of Section 9(a)(1), applies directly to all entities—federal, state, tribal, and local governments and all private entities.

Federal Agency Conservation Duty

Section 7(a)(1) of the ESA states that "all federal agencies shall, in consultation and with the assistance of [USFWS and NMFS], utilize their authorities in furtherance of the purposes of this chapter by carrying out programs for the conservation of endangered species and threatened species" (16 U.S.C. 1536(a)(1) [2002]). This duty, however, is poorly defined. No procedures are specified in the ESA, nor have USFWS and NMFS provided any in their regulations.

The courts generally have construed the provision to require federal agencies to take affirmative action or to restrain from negative action to advance the purpose of conservation (Ruhl 1995). In addition, courts have confirmed that Section 7(a)(1) is a source of authority for an agency to take action in support of species conservation where no other provision of the ESA requires it, as long as the action is within the scope of and not in conflict with the agency's authority under its enabling statutes. As explained below, Sections 7(a)(2) and 9(a)(1) are prohibitions: Section 7(a)(2) prohibits federal agencies from jeopardizing species or adversely modifying critical habitat, whereas Section 9(a)(1) prohibits federal agencies from causing take (mortality or impairment). Failure of an agency to undertake actions that would promote conservation of species often would be consistent with these prohibitions. In contrast, Section 7(a)(1) is an affirmatively stated duty to promote conservation of species, and thus can serve as authority for taking actions that neither Section 7(a)(2) nor Section 9(a)(1) would require (see *Carson-Truckee Water Conservancy District v. Watt*, 549 F. Supp. 704, D. Nev. 1982, aff'd 741 F.2d 257, 9th Cir. [1984]). For example, USBR could restrict water deliveries to protect endangered fish, even though it is not required to do so under Section 7(a)(2) or 9(a)(1), because of Section 7(a)(1).

USFWS, NMFS, and other federal agencies carrying out their responsibilities in the Klamath River basin have not taken full advantage of their authority under Section 7(a)(1). For example, USBR explained in its 2002 biological assessment for the Klamath Project that Section 7(a)(1) does not expand the agency's authority beyond its enabling laws. On the basis of that principle, USBR made no additional effort to exercise its authority under Section 7(a)(1). As described above, however, Section 7(a)(1) essentially states that actions by agencies that are consistent with enabling laws and that are intended to provide for the conservation of species cannot be challenged just because they are not required by Sections 7(a)(2) or 9(a)(1). Hence, the provision creates an opportunity for conservation-promoting actions under the ESA beyond the mandates of Sections 7(a)(2) and 9(a)(1). Many of the actions outlined in this report for conservation of the listed suckers and coho salmon would be supported by Section 7(a)(1), even though they might not be required by Sections 7(a)(2) or 9(a)(1). In other words, USFWS, NMFS, and all other federal agencies carrying out actions in the Klamath River basin have substantial discretion to act on behalf of the listed species even where they do not have the duty to do so.

Section 7(a)(1) clearly does require that all relevant federal agencies at the very least consult with USFWS and NMFS about the exercise of discretionary authority (see *Sierra Club v. Glickman*, 156 F.3d 606, 5th Cir. [1998]). Unlike consultation under Section 7(a)(2), which has been the context for most ESA implementation measures in the Klamath River basin, consultation under Section 7(a)(1) is not governed by formal procedures. Working together, the agencies could establish and implement a comprehensive, flexible, multiagency consultation process that is directed specifically at the Klamath River basin and is designed to specify actions that each agency could take, under and consistent with its general authorities, to promote conservation of the listed species. In implementing such actions, agencies would be protected from legal challenge by their authority under Section 7(a)(1).

A substantial effort, justified under Section 7(a)(1), should be made to enlist all federal agencies operating in the Klamath River basin in recovery efforts. In fact, the relevant agencies—which include USFWS, NMFS, USBR, the U.S. Environmental Protection Agency, the U.S. Forest Service, and the U.S. Army Corps of Engineers—in 1994 jointly affirmed their Section 7(a)(1) authority and agreed to "identify opportunities to conserve Federally listed species and the ecosystems upon which those species depend" (Memorandum of Agreement 1994). Each of these agencies also agreed to "determine whether its respective planning processes effectively help conserve threatened and endangered species and the ecosystems upon which those species depend" and to "use existing programs, or establish a program if one does not currently exist, to evaluate, recognize, and reward the performance and achievements of personnel who are responsible for planning or implementing programs to conserve or recover listed species or the ecosystems upon which they depend." Yet there is little evidence that any federal agency operating in the Klamath River basin has been successful in fulfilling these agreements in the context of the ESA.

In summary, a multiagency consultation process under Section 7(a)(1) could expand recovery efforts beyond USBR and its Klamath Project, as needed ultimately for recovery. Section 7(a)(1) does not require any agency participating in the consultation to implement particular measures; only institutional will can bring that about. But if ever a case existed for motivating institutional will in this direction, the Klamath River basin fits the description.

Prohibition Against Jeopardy and Adverse Modification Caused by Federal Agencies

Section 7(a)(2) of the ESA requires (16 U.S.C. 1536(a)(2) 2002) that

> each federal agency shall, in consultation with and with the assistance of [USFWS and NMFS], insure that any action authorized, funded, or carried out by such agency (hereinafter in this section referred to as an "agency action") is not likely to jeopardize the continued existence of any endangered species or threatened species or result in the destruction or adverse modification of habitat of such species which is determined . . . to be critical.

The Klamath Project is subject to this requirement (see *Klamath Water Users Protection Ass'n v. Patterson*, 191 F.3d 1115, 9th Cir. 1999; *O'Neil v. United States*, 50 F.3d 667, 9th Cir. [1995]). The ESA provides an elaborate set of procedures and criteria for carrying out the jeopardy and adverse modification consultations (16 U.S.C. 1536(b)–(d) [2002]). USFWS and NMFS also have issued an extensive set of regulations covering the process (50 C.F.R. part 402 [2002]). Generally, the action agency must prepare a "biological assessment" detailing the effects that it believes its actions will have on listed species, and the consulting agency (USFWS or NMFS) must in response provide a "biological opinion" declaring whether jeopardy and adverse modification are likely to occur. If the consulting agency finds that jeopardy will occur, it must suggest "reasonable and prudent alternatives" (RPA) by which the action agency can avoid such an outcome. The RPAs, technically within the discretion of the action agency to accept or reject (see *Southwest Center for Biological Diversity v. Bureau of Reclamation*, 143 F.3d 515, 9th Cir. [1998]), carry considerable weight and are viewed as essentially mandatory in the absence of some compelling basis that the action agency might have for using different alternatives (see *Bennett v. Spear*, 520 U.S. 154 [1997]).

All agencies must fulfill all the duties by using "the best scientific and commercial data available" (50 C.F.R. 402.14(d) and 402.14(g)(8) [2002]). Action agencies also must ensure that they and their license or permit applicants "shall not make any irreversible or irretrievable commitment of resources with respect to the agency action which has the effect of foreclosing the formulation or implementation of any reasonable and prudent alternatives" (16 U.S.C. 1536(d) [2002]). A procedure established in the ESA—but rarely used, given its narrow criteria—allows an action agency to appeal a jeopardy or adverse modification finding to a committee of cabinet-level and other federal agency officials and thereby seek to carry out the action regardless of jeopardy or adverse modification (16 U.S.C. 1536(e)–(n) [2002]; 50 C.F.R. part 450 [2002]). Irrigation districts sought to initiate that procedure with respect to the 2001 jeopardy opinions that USFWS and NMFS issued for the Klamath Project, but in July 2001 the Department of the Interior declined to pursue the exemption process further. In addition to the jeopardy standards of Section 7(a)(2), the criteria for exemption involve policy matters outside the scope of this report.

The procedural details of the consultation process are not relevant to the NRC committee's charge. Rather, the key aspects of the consultation program for the committee's purposes are the meanings of *jeopardy* and *reasonable and prudent alternative*, because both USFWS and NMFS made jeopardy findings in their 2001 biological opinions and because the RPAs that they presented led USBR to suspend water deliveries in 2001. The statute defines neither term. Under USFWS and NMFS regulations, *jeopar-*

dize the continued existence means "to engage in an action that reasonably would be expected, directly or indirectly, to reduce appreciably the likelihood of both the survival and recovery of a listed species in the wild by reducing the reproduction, numbers, or distribution of that species" (50 C.F.R. 402.02 [2002]). *Reasonable and prudent alternative* means "alternative actions identified during formal consultation that can be implemented in a manner consistent with the intended purpose of the action, that can be implemented consistent with the scope of the Federal agency's legal authority, that is economically and technologically feasible, and that [USFWS or NMFS] believes would avoid the likelihood of jeopardizing the continued existence of listed species or resulting in the destruction or adverse modification of critical habitat" (50 C.F.R. 402.02 [2002]). Judgments of jeopardy are inherently difficult in a technical sense. Site-specific evidence must be used as extensively as possible in making such judgments, but use of professional judgment where site-specific evidence is inadequate or absent is inevitable and desirable for rational judgments of jeopardy (see Chapter 1).

As described in Chapter 1, USFWS has consulted with USBR regarding the Klamath Project's effects on the listed sucker species, and NMFS has done so for the coho salmon. The history of the consultations is long and has at times been controversial (see, e.g., *Bennett v. Spear*, 5 F.Supp.2d 882 D. Or. [1998]; *Pacific Coast Federation of Fishermen v. Bureau of Reclamation*, 138 F.Supp.2d 1228 N.D. Cal. [2001]). For consistency with its charge, the NRC committee's principal focus has been on the 2001 and 2002 consultation documents.

In addition to the Klamath Project, numerous other actions in the Klamath River basin are carried out, funded, or authorized by federal agencies (Chapter 2). USFWS and NMFS do not appear to maintain comprehensive inventories of actions for which consultation is necessary and for which each action agency's consultation is satisfied or deficient, nor is there any basinwide strategy for conservation of the species through coordinated Section 7(a)(2) consultations. The agencies should prepare and implement such an inventory and strategy.

The Authorities to Prohibit Take and Incidental Take

Section 9(a)(1) of the ESA provides that "with respect to any endangered species of fish or wildlife . . . it is unlawful for any person subject to the jurisdiction of the United States to . . . take any such species within the United States or the territorial sea of the United States" (16 U.S.C. 1538(a)(1) [2002]). Although threatened species, such as the coho salmon, are not covered directly in this provision, Section 4(d) of the ESA provides that USFWS and NMFS "may by regulation prohibit with respect to any threatened species any act prohibited under section 1538(a)(1) of this title . . . with respect to endangered species" (16 U.S.C. 1533(d) [2002]).

Under the statute, to *take* is to "harass, harm, pursue, hunt, shoot, wound, trap, capture, or collect, or to attempt to engage in any such action" (16 U.S.C. 1532 [2002]). USFWS and NMFS have further defined *harm* to mean "an act which actually kills or injures wildlife. Such an act may include significant habitat modification or degradation where it actually kills or injures wildlife by significantly impairing essential behavioral patterns, including breeding, feeding or sheltering" (50 C.F.R. 17.3 [2002]). The U.S. Supreme Court has upheld the latter definition as consistent with the congressional intent of the ESA but in so ruling construed the regulation to limit findings of harm to cases in which actual death or injury to identifiable members of a protected species is the proximate and foreseeable result of a habitat modification (*Babbitt v. Sweet Home Chapter of Communities for a Great Oregon*, 515 U.S. 687 [1995]).

When USFWS and NMFS prepare biological opinions in connection with consultations under Section 7(a)(2) of the ESA, they most often find that no jeopardy or adverse modification will occur. Even in such cases, however, incidental take of a species might be a foreseeable consequence of the action. In such instances the consulting agency must (16 U.S.C. 1536(b)(4) [2002])

> provide the Federal agency and the applicant concerned, if any, with a written statement that (i) specifies the impact of such incidental take on the species, (ii) specifies those reasonable and prudent measures that the Secretary considers necessary or appropriate to minimize such impact, (iii) in the case of marine mammals, specifies those measures that are necessary to comply with section 1371(a)(5) of this title with regard to such taking, and (iv) sets forth the terms and conditions (including, but not limited to, reporting requirements) that must be complied with by the Federal agency or applicant (if any), or both, to implement the measures specified under clauses (ii) and (iii).

A similar procedure for authorization of incidental take is available under Section 10(a)(1)(B) of the ESA for projects and actions not carried out, funded, or authorized by a federal agency and thus not subject to the consultation requirement of Section 7(a)(2). Under the procedure, the entity carrying out an action that will cause take of a listed species must submit a habitat conservation plan (HCP) to USFWS or NMFS on which the agency bases its decision of whether to grant a permit for the incidental take (16 U.S.C. 1539(a)(1)(B) [2002]).

USFWS and NMFS consistently have found in their biological opinions for the Klamath Project that USBR's actions will result in take of the species in question and have prepared incidental-take statements with reasonable and prudent measures and terms and conditions for implementing them.

Some take of the listed species is an undisputed consequence of USBR's operation of the Klamath Project (see Chapters 6 and 8), and the reasonable and prudent measures for avoiding this take are well founded. One concern of the NRC committee, however, is the lack of attention that USFWS and NMFS appear to have given to take of the listed species by actions other than USBR's operation of the Klamath Project. Throughout the Klamath River basin, actions by public and private entities are causing take of the listed species. Many of these actions are outside the control of the USBR and thus not susceptible to correction though the Klamath Project consultations. Such sources of take, which the committee believes may be substantial, should not be ignored simply because USBR and the Klamath Project present a bigger and easier target for consultation. Indeed, doing so leads inevitably to the potential for overregulation of the Klamath Project and, indirectly, its beneficiaries and thus an inequitable distribution of the social and economic costs of the conservation of species. The Klamath Project is a valid target for scrutiny and regulation, but not the only one.

Examples of take outside the reach of the Klamath Project are given in the listing documents and in Chapters 5–8 of this report. For example, Chiloquin Dam causes take of endangered suckers but is not under the control of USBR in connection with the Klamath Project. Even so, there is no organized effort by USFWS to enforce the take prohibition at Chiloquin Dam or elsewhere outside the Klamath Project where persons causing take must modify their behavior so as to avoid take or submit HCPs under Section 10(a)(1)(B). NMFS has a similar record in relation to coho salmon. In other parts of the nation, however, such as Austin, Portland, Tucson, and southern California, USFWS and NMFS have expended considerable resources to limit incidental take of listed species caused by dispersed actions, including those of private parties. It is not clear why the agencies have not initiated similar enforcement actions in the Klamath River basin.

There is ample basis for each agency to extend its authority to prohibit take, and doing so is likely to benefit the listed species. For example, NMFS listed the coho salmon ESU as threatened, requiring the agency to adopt conservation regulations under Section 4(d) of the ESA, and thus to regulate take of the species. In July 1997, the agency published an interim Section 4(d) rule extending the full extent of Section 9(a) take prohibitions to the species, except for specified benign and beneficial actions, including aspects of habitat restoration programs that the states had initiated (62 Fed. Reg. 38479 July 18, 1997). In July 2000, the agency included the coho salmon in a rule establishing general take authorizations for specified activities, subject to limits, covering 14 salmonid ESUs (65 Fed. Reg. 42421 July 10, 2000).

When describing the activities that would be affected by the take prohibition in its July 1999 interim Section 4(d) rule for the coho salmon ESU,

NMFS explained that "agricultural activities that might result in take of SONCC coho are . . . sediment from cultivation or livestock movements on the banks or in the beds of streams; unscreened water diversions and reductions of flow through irrigation could also result in take; and NMFS . . . would expect the 4(d) rule to result in some curtailment of [timber] harvest on lands owned by small entities over and above the impacts of state regulation" (62 Fed. Reg. 38481-38483 [1997]). In the July 2000 rule, NMFS explained that the general take authorizations cover "properly screened water diversion devices" (62 Fed. Reg. 42423 [1997]). Other agricultural, logging, and land-use activities were not covered in any general or specific way by the general take authorizations.

As is the case for the listed sucker species, there clearly are numerous common activities outside the control of USBR that are recognized by NMFS as causing unauthorized take of coho salmon in the Klamath basin. NMFS recognized this in its 2002 biological opinion on the Klamath Project, for example, when it acknowledged that USBR accounts for 57% rather than 100% of the total irrigation-related depletions of flow at Iron Gate Dam. If, as NMFS has concluded, USBR's flow-depletion component has triggered jeopardy of the species, the other irrigation flow depletions most likely are also causing take of the species. Yet there is little evidence that NMFS has actively enforced the take prohibition in these contexts in the Klamath River basin (as it recently did, in contrast, against the Grants Pass Irrigation District for its take of salmon at its Savage Rapids Dam diversion structure). The NRC committee has not examined the full extent of the potential measures that NMFS might take in enforcing the take prohibition in the Klamath basin beyond USBR's operation of the Klamath Project, but there is ample basis for the agency to do so, and doing so is likely to benefit the species (Chapter 8).

For take caused by the Klamath Project, USBR obtains approval through the procedure on incidental-take statements given in Section 7(b)(4). If USFWS and NMFS were faithfully to enforce the take prohibition, there would be many more additional nonproject actions that, if not modified to discontinue the take, also would require authorization of incidental take. Take of listed suckers and salmon caused by actions other than the Klamath Project may even be associated with some federal agency funding or approval, in which case Section 7(b)(4) also would apply. For take caused by actions not carried out, funded, or authorized by a federal agency, Section 10(a)(1) supplies the applicable procedure.

Given the multiplicity of actions that may be causing take of the listed suckers and salmon, it may be productive for representatives of various interests to consider organizing an effort to explore a regional HCP that would form the basis for USFWS and NMFS to issue an "umbrella" authorization of incidental take for several actions in the Klamath River basin.

Regional HCPs have been used or explored in a number of urban and rural settings as a means of avoiding piecemeal administration of incidental take permitting and to enhance opportunities for more efficient and effective habitat conservation and mitigation measures (Thornton 2001). Moreover, like the Section 7(a)(1) multiagency consultation proposed above, the regional HCP process involves coordination of numerous diverse interests and thus has the potential to produce more sustainable decisions than would incremental, action-specific permitting. As an earlier National Research Council committee found (NRC 1995, p. 92, 198–199), "habitat conservation planning . . . has the potential to be effective in protecting ecosystems and has realized that potential in a few cases," leading it to "endorse regionally based, negotiated approaches to the development of habitat conservation plans."

NMFS appears to have recognized the benefits of such interest planning processes in its 2002 biological opinion when it recommended creation of a "task force" to address the 43% of irrigation-related flow depletion at Iron Gate Dam that is not attributable to USBR. The NRC committee sees no reason why a Section 7(a)(1) process and a regional HCP process cannot be undertaken simultaneously and in coordination to fulfill the objectives of such a task force and of related species-conservation goals in the Klamath River basin.

CONCLUSIONS

The ESA is not a panacea for the challenges of ecosystem management and species conservation posed in the Klamath River basin. However, ESA authorities could be implemented more effectively, more extensively, and more creatively than they are now. Specifically, the relevant federal agencies have failed in several ways to exercise their full ESA authorities.

- USFWS and NMFS recovery planning for listed species under Section 4(f) has stalled.
- Federal agencies operating in the Klamath River basin have not been successful in the full use of discretionary conservation authority given in Section 7(a)(1).
- USFWS and NMFS appear to have focused jeopardy consultation under Section 7(a)(2) narrowly on USBR's operation of the Klamath Project, notwithstanding the many other federal agency actions carried out, funded, or authorized in or affecting the Klamath River basin. Neither agency has made any basinwide inventory of or strategy for federal actions and consultations a prominent part of its public discourse on the Klamath basin.
- USFWS and NMFS have not actively enforced the ESA Section 9 take prohibition outside the context of the Klamath Project itself, notwith-

standing ample evidence that numerous other actions are causing take of the species.

Those problems in large part could be remedied as follows:

- NMFS could prepare and promulgate a recovery plan for the coho salmon, and USFWS could revise, update, and repromulgate the sucker recovery plan. In each case, the recovery plan could be designed with the specific purpose of enabling federal agency consultations under Sections 7(a)(1) and 7(a)(2) and individual or regional habitat-conservation planning under Section 10(a)(1), and it ideally would be capable of being carried out more comprehensively—that is, across the full spectrum of issues in the Klamath River basin and not just the Klamath Project—and through adaptive-management principles.
- NMFS and USFWS could inventory all federal agencies that are exercising any authority in or affecting the Klamath River basin and could initiate a multiagency consultation process with them under Section 7(a)(1). The consultation process would be most effective if centered on adaptive-management principles. Each federal agency engaging in the process could direct its institutional will toward fulfilling the agreements it made in the 1994 interagency agreement regarding the exercise of discretionary authority under Section 7(a)(1), with the Klamath River basin specifically in mind.
- If they have not already done so, NMFS and USFWS could inventory all active and potential federal agency consultations that are or could be carried out in the Klamath River basin under Section 7(a)(2), and develop a more coordinated basin-wide approach to the entire package of consultations. If these instruments already exist, the agencies could use them more overtly and provide the public more information about them.
- NMFS and USFWS could identify the inventory of federal, state, local, tribal, and private actions that are causing unauthorized take of the suckers and coho salmon. NMFS and USFWS could work with the agencies and persons causing the takes to help them either to modify their behavior to avoid the takes or to obtain incidental-take authorization under Sections 7(b)(4) or 10(a)(1). NMFS and USFWS could explore with those interests, which include private-sector and government actors, the possibility of a regional habitat-conservation planning approach.

10

Adaptive Management for Ecosystem Restoration in the Klamath Basin

This report has described many ways in which the status of Klamath basin ecosystems can be improved for the benefit of endangered or threatened species and other fish and wildlife resources. The report also shows that geographic expansion of restoration efforts beyond the lakes and the main stem of the Klamath River is necessary for recovery of listed species. Recovery efforts will require adjustments in policies of agencies, in cooperation between institutions, and human use of resources in the basin.

Ecosystem management in the Klamath basin today is disjointed, occasionally dysfunctional, and commonly adversarial. Thus, it often is inefficient or ineffective in dealing with issues related to restoration of listed species in the basin. Cooperation among agencies has been poor; potential restoration activities have been generally restricted to actions or operations of the Klamath Project; and local communities, stakeholders, and individuals that control resources critical to long-term solutions often have been alienated, uninterested, or simply left out. Changes that occurred during consultations leading to the biological assessment and opinions of 2002 appear to show some movement toward remedies for these deficiencies, but much remains to be done, and an overall integrated strategy still is missing.

This chapter discusses alternative or modified management frameworks that might allow resources for recovery to be used more effectively than in the past. First, the potential value of adaptive management is explored. The chapter then presents specific examples of policy instruments, approaches, and activities that may facilitate environmental restoration. The last section suggests specific changes in management that probably would improve the efficacy of public and private investments in habitat or minimize the costs to private landowners as they adjust to the needs of listed species.

ADAPTIVE MANAGEMENT AS AN ORGANIZING FRAMEWORK

Regional restoration programs—which typically are large, complex, and fraught with uncertainties and competing interests—must include a process for implementing restoration activities and a means of measuring their effectiveness. The concepts of adaptive assessment (analysis leading to adaptations) and adaptive management (adjustment of management in light of new information) are often suitable for those purposes; for brevity, they are referred to here collectively as adaptive management.

Adaptive management is a formal, systematic, and rigorous program of learning from the outcomes of management actions, accommodating change, and improving management (Holling 1978). Its primary purpose is to establish a continuous, iterative process for increasing the probability that a plan for environmental restoration will be successful. In practice, adaptive management uses conceptual and numerical models and the scientific method to develop and test management options. It requires the explicit recognition that management policies can, with appropriate precautions, be applied as experimental treatments (Walters 1997). Decision makers use the results as a basis for improving knowledge of the system and adjusting management accordingly (Haley 1990, McLain and Lee 1996).

Adaptive management is being applied to major ecosystem restoration projects in the Florida Everglades, Chesapeake Bay, and California's Sacramento and San Joaquin River system (CERP 2002, CALFED 2002), and it recently has been used in an evaluation of flow regimes for the Grand Canyon (NRC 1999) and the Trinity River component of the Klamath River system (USFWS/HVT 1999). The following description of the adaptive management process is drawn from the CALFED Sacramento-San Joaquin Comprehensive Study working paper (2002), the appendix to USFWS/HVT (1999), Nagle and Ruhl (2002), and other sources. Not all features of adaptive management will be applicable to the Klamath basin, given legal constraints arising from the federal Endangered Species Act (ESA; Chapter 9). The general principles of adaptive management do, however, provide useful guidance as managers consider development and implementation of recovery plans. Adaptive management on the Trinity River could serve as a useful model for the rest of the basin.

Ecosystem Management and Adaptive Management

Ecosystem management refers to policy goals directed at ensuring the sustainability of natural resources in ecologically functional units (Grumbine

1994). Grumbine defines *adaptive management* as a set of policy tools intended to move decision-making from a process of incremental trial and error to one of experimentation that uses continuous monitoring, assessment, and recalibration. Ecosystem management and adaptive management are not interchangeable, but they are nearly inseparable (Nagle and Ruhl 2002). Successful ecosystem management usually requires some form of adaptive management, and use of adaptive management in the context of natural-resources conservation generally requires that goals be expressed in terms of ecosystem management.

Through research already completed, scientists and managers have come to understand much about Klamath basin ecosystems and the species that depend on them, but many of the important ecological and human processes and interactions that animate the ecosystem remain unknown. Furthermore, ecosystem processes, habitats, and species are modified continually by changing environmental conditions and human activities. Presently and in the future, uncertainty is inevitable. Adaptive management provides an iterative process for continually reducing uncertainty by refining the implementation of environmental restoration projects in response to information from monitoring and scientific analysis.

Extreme events such as drought, flood, and unexpected human actions are anticipated by a properly designed adaptive-management program. Adaptive management incorporates processes for early detection and interpretation of the unexpected and for maximizing the learning opportunities associated with these events. Adaptive management is valuable in that it treats all responses, expected or not, as learning opportunities.

An example of an incidental experiment from the Klamath basin is the variation of water levels of Upper Klamath Lake over the last 15 yr. Drought and human management have caused the water level of the lake to fluctuate over a range of about 6 ft (Chapter 3). Changes in water levels now can be compared with changes in water quality (Chapter 3) or in sucker populations (Chapter 6). A number of other experiments, planned or inadvertent, have occurred in the basin, such as changes in seasonal and annual flows at Iron Gate Dam; they provide useful information about recovery, but in many cases monitoring programs have been inadequate to support analysis and interpretation that would lead to adaptation of management based on the new information.

Key Components of Adaptive Management

The key components of adaptive management are as follows:

- *Definition of the problem.* Examples are loss of critical habitat for species and the need for protection and restoration of habitat for species, such as those listed under the ESA.

- *Determination of goals and objectives for management of ecosystems.* Examples are restoration of habitat protection and recovery of endangered species and other fish and wildlife resources at minimum social or economic cost.
- *Determination of the ecosystem baseline.* The ecosystem baseline includes all relevant information, past and present, such as physical, chemical, and biological features and benchmark indicators of the abundance of critical species. The baseline is the reference condition against which progress toward management goals is measured.
- *Development of conceptual models.* The analytical basis of adaptive management typically is a set of conceptual and numerical models. For example, conceptual ecological models convert broad, policy-level objectives into specific, measurable indicators of the status of natural and human systems. Conceptual modeling requires knowledge of ecosystem functions, of alteration or degradation, and of potential improvements. This information is framed in terms of major stressors and indicators (ecological attributes) that provide the most useful measures of ecological and social response to change. The conceptual model can be used to identify a small number of representative biological, chemical, and physical indicators of system-wide responses to restoration on various spatial and temporal scales. The indicators then can be used in developing models or protocols for monitoring and testing the efficiency of the restoration efforts. Performance measures are developed for each of the elements (ideally for both stressors and indicators) and are used as the standards for evaluating the restoration program.
- *Selection of future restoration actions.* The conceptual models shape the character of restoration actions by identifying key kinds of uncertainty or by revealing the extent of confidence that a particular action will achieve a given objective. On the basis of past and current conditions of the ecosystem, and insights from the conceptual models about the ecological and social consequences of management actions, managers apply two processes for changing management activities: identification of alternative-management procedures to achieve objectives and selection of alternatives that appear to move the system toward management objectives. One aspect of the selection process should be the social and economic costs of achieving an objective. When two alternatives are effective, lower cost is preferred. If alternative actions are proposed for the same purpose, comparison (perhaps in consecutive years) leads to selection of the action that most efficiently achieves the objectives.
- *Implementation of management actions.* A group of scientists and agency managers collectively is responsible for determining the criteria and procedures for management actions. This work requires coordination, organization, and accountability among the agencies, which can be difficult if

the agencies have conflicting missions, as is the case in the Klamath basin. Experts in modeling, simulation, experimental design, and prediction forecast responses to managerial actions. Each iteration of simulation is tested through post-audit comparisons of observed and expected results. As part of an evaluation program, agency managers may support short-term and long-term experiments, such as alternative water levels or stream flows, habitat restoration efforts in selected areas, or other ecosystem changes. Experiments often involve major change, as would be the case for closure of the Iron Gate or Trinity Hatchery or removal of major dams (Chapter 8).

• *Monitoring of the ecosystem response.* "It is critical to monitor the implementation of restoration actions to gage how the ecosystem responds to management interventions. Monitoring provides the information necessary for tracking ecosystem health, for evaluating progress toward restoration goals and objectives, and for evaluating and updating problems, goals and objectives, conceptual models, and restoration actions. Monitoring requires measuring the baseline condition, abundance, distribution, change or status of ecological indicators" (CALFED Bay-Delta Program 2000).

• *Evaluation of restoration efforts and proposals for remedial actions.* After implementation of specific restoration activities and procedures, the status of the ecosystem is regularly and systematically reassessed and described. Comparison of the new state with the baseline state is a measure of progress toward objectives. The evaluation process feeds directly into adaptive management by informing the implementation team and leading to testing of management hypotheses, new simulations, and proposals for adjustments in management experiments or development of wholly new experiments or management strategies.

Status of Adaptive Management in the Klamath Basin

There has been little effort to implement adaptive-management strategies in the Klamath basin, except through the Trinity River Restoration Program, which deals only with the Trinity River. Even the 2002 biological assessment of the U.S. Bureau of Reclamation (USBR), which prescribes Klamath Project operations for the next 10 yr, gives only weak indications of mechanisms for adapting to new information. One exception is the proposed water bank, which if properly structured will provide annual information on the quantities of water available for voluntary transfer across uses and locations, and on the economic and social value of such water. This information can then be used by USBR to manage the water bank and to develop more accurate estimates of water availability for both agricultural and environmental uses in the basin, and to establish a long-term mechanism to address demands for water.

Ecosystem management in the Klamath basin typically has pursued the widely recognized alternatives to adaptive management: deferred action and trial and error involving crisis management In the deferred-action approach, management methods are not changed until ecosystems are fully understood (Walters and Hillborn 1978, Walters and Holling 1990, Wilhere 2002). This approach is cautious but has two notable drawbacks: deferral of management changes may magnify losses, and knowledge acquired by deferred action may reveal little about the response of ecosystems to changes in management. Stakeholder groups or agencies that are opposed to changes in management often are strong proponents of deferred action.

Crisis management is common throughout the Klamath basin and permeates most restoration efforts, particularly on the tributaries. The approach often involves restoration actions, but neglects assessment (Wilhere 2002). Thus, management becomes based principally on casual observations and anecdotal reports. Trial and error without assessment and adaptation undervalues information, which is the most critical need in restoration, and overvalues action for its own sake. The trial and error without assessment and adaptation may cause more harm than good, but its benefits typically cannot be determined.

The legislative potential for watershed planning and restoration based on an adaptive-management framework already exists through the Klamath Act (Public Law 99-552), which was passed by Congress in 1986. The act led to formation of the Klamath River Basin Conservation Area Restoration Program, which includes the Klamath Basin Restoration Task Force. The task force is comprised of federal, state, and local officials and representatives of several tribes and other stakeholders, including the private sector. In addition, other committees, organizations and ad hoc working groups, such as the Upper Klamath Basin Working Group, the Klamath Basin Ecosystem Foundation, and several watershed councils have been created for improvement of dialogue among parties in the basin, and for development of solutions to water issues within the basin. The task force and other groups have facilitated discussion, but it is not clear that any group has contemplated extensive use of adaptive management. Considerable public and private funds have been invested in restoration and management of the ecosystems of the Klamath basin. It is not clear what benefits have been derived from the investments, or how management will be improved as a result. Adaptive management as applied to the Klamath basin would need to function within the legal framework of the ESA (Chapter 9), but the key point of the process is to set goals, develop a plan, determine whether it is achieving specific goals, and make adjustments as needed to be effective. This approach is both ecologically and socially responsible, given that ultimately all agencies and other stakeholders have limited resources with which to operate. As specific goals are achieved, resources become available for other socially desirable purposes within and outside the basin.

POLICY OPTIONS AND RESTORATION ACTIVITIES

Federal legislation and regulations, including the ESA and Federal Tribal Trust responsibilities, supercede state laws, including state water law. Thus, water demands for ESA purposes or to meet treaty obligations to Indian tribes have generally been upheld by federal courts (see, for example, the Winters Doctrine). Since such federal rulings reinforcing the ESA or tribal water needs typically do not apply to all waters in a basin or watershed, the Prior Appropriation Doctrine is still the major allocation device for waters in much of the West, including the Klamath basin. The prior appropriation system requires that the first individual to divert water for a beneficial use shall have the right to do so into perpetuity ("first in time, first in right"). The right of use generally is defined in terms of a given amount of water at a particular point of diversion. The rights of later diverters are junior (subordinate) to the right of the first diverter (senior right); in times of shortage, those holding water rights with earlier diversion dates are the last to be denied water. These water rights are established and protected by the states in which the diversions occur, usually by a state department of water resources.

The prior appropriation system of rights provides an efficient mechanism for allocating water during times of shortage, but has many limitations (Getches 2003). One is that the use of water by the holders of senior rights (seniors) may in some cases be of lower economic or social value than that of holders of junior rights (juniors). For example, a senior may divert water onto pastureland, of low economic value while a junior has the opportunity to use water to produce crops of high value. In a time of drought, there may not be sufficient water for both users, and only the crop of lower value would receive water. A related issue is that the most senior water rights are for diversions, primarily to agriculture. Values of flow in the stream itself have only recently been recognized as beneficial. As a result, seniors have the potential to divert all usable flow, thus dewatering portions of streams, even if the marginal value of water in the stream could produce substantially higher benefits than the diversion.

Another shortcoming of the prior-appropriation doctrine as applied by most states is that water rights are defined for a specific location. Thus, water rights are tied to a particular parcel of land unless a change is approved by a state authority. Defining water rights as appurtenant to land creates inflexibilities in the use of the water (for example, it restricts water trading), which leads to substantial economic and social costs with respect to maximizing the value of water to society, as demonstrated in the example of the preceding paragraph.

Because of problems with the prior-appropriation doctrine, states began using water markets about 30 yr ago (Colby Saliba and Bush 1987, National Research Council 1992, Getches 2003). The idea of a water market is that willing buyers and sellers should engage in transfers of water, thereby increasing the value of water to society. To use the preceding example, the junior may be willing to pay more for water than the senior can realize from using it. In such a case, both parties would gain and society would have realized greater value through the transfer.

To facilitate creation of a water market, states have changed laws and rules to allow a water right to be separated from the land to which it was originally applied. In such cases, the right is redefined as a particular flow or volume of water instead of a diversion at a particular location. Thus, a downstream user can purchase water from an upstream user. The magnitude of the gain from such a transaction is determined by the seller's increase in returns (over the value of the water on site) plus the additional increase in income or averted loss realized by the downstream purchaser. Obviously, trades will not occur unless they are of mutual benefit to buyer and seller. The existence of a market also allows other prospective water users to obtain water that was previously unavailable. For example, conservation groups or fisheries agencies may purchase water for maintenance of stream flows that benefit fish and wildlife (Colby 1990, Adams et al. 1993). In some western states (such as, Colorado and Arizona) municipalities purchase agricultural water rights through water markets to meet rising water demand due to residential growth.

Water markets create their own problems. They include so-called third-party effects by which someone who is not party to the sale may be harmed. For example, harm could come to an irrigator who has been using return flows from an upstream irrigator. If the upstream irrigator ceases irrigation, there would be no return flows for the neighboring irrigator. In addition, some return flows create wetlands or supplement groundwater supplies; if the water is moved to a new location as a result of a water transfer, these local benefits may be lost.

Water markets also may affect rural communities. If large amounts of water are diverted from agriculture to other uses, rural communities, including Indian tribal groups, who depend on the economic activity generated by irrigated agriculture will suffer. Thus, although the traders gain from the existence of markets and society gains from water transfers to use of higher value, rural communities may lose economic viability.

Despite the problems created by water markets, their use is increasing throughout the West. Many of the western states allow water to be sold or leased. Permanent transfer of water rights occurs in the case of a true water market, but a water bank typically involves the temporary transfer (lease) of a water right. Water banks are particularly useful during drought. Water banks also reduce some of the adverse effects of a permanent transfer of a water right. Farmers and rural communities often are more supportive of the water-bank concept than of sales of water rights (Keenan et al. 1999).

Water banks hold promise for water problems such as those of the Klamath basin. As noted earlier, Indian tribal claims to waters of the upper Klamath basin must be addressed as part of the adjudication process. Remaining water rights then will be assigned based on demonstrated proof of the initiation of beneficial use. Indeed, the recent USBR biological assessment (2002) contains a 10-yr plan that calls for creation of a water bank of 100,000 acre-ft of water per yr (see Table 1-1 for comparison with total annual flows). The water would come from groundwater and from surface water within and outside the Klamath Project. USBR would purchase the water, which would be used for environmental purposes.

The Klamath basin shows one of the necessary conditions for a water market or bank to be successful: a pronounced difference in the value of water across crops and other uses. For example, crops of both low and high value are grown in the Klamath Project and in the basin. In addition to providing a mechanism by which USBR could purchase water for environmental uses, a properly structured water bank would allow irrigators to trade among themselves. In a hypothetical analysis of the events of the 2001 water year in the Klamath basin, Jaeger (2002) has shown that a fully functioning water bank would have reduced losses to agriculture by over 50%. A water bank also could allow irrigation water to be shifted to nonagricultural uses. For example, the California water bank, which is administered by the California Department of Water Resources, reserves a small portion of each exchange between farmers to be used for environmental purposes in the Sacramento-San Joaquin delta.

The necessary economic conditions exist for a water bank in the Klamath basin, but institutional conditions do not. Specifically, before water can be traded, water rights must be clearly defined. In California, such rights have been established by the state. Oregon, however, has not finished the adjudication process for water rights in its portion of the Klamath basin. In the short term, water banking will need to rely on water sales from the California portion of the basin or among farmers in the Klamath Project who have water available for transfer, such as from wells. Even a limited water bank that is based on adjudicated surface water in California or from groundwater in the Klamath Project has the potential to improve water allocation in the basin.

IMPROVEMENT OF RESOURCE MANAGEMENT IN THE KLAMATH BASIN

The present management structure for restoring the two sucker species and coho salmon in the Klamath basin consists of the federal agencies involved in the ESA Section 7 (a)(2) consultations—USBR, the U.S. Fish and Wildlife Service (USFWS), and the National Marine Fisheries Service (NMFS)—and, less directly, a number of other federal and state agencies, such as the U.S. Forest Service, Bureau of Land Management, U.S. Fish and Wildlife Service National Wildlife Refuges, the Natural Resource Conservation Service, the Bureau of Indian Affairs, the Environmental Protection Agency, the Army Corps of Engineers, the California and Oregon Departments of Environmental Quality, and state water-resources departments. Because the ESA is federal legislation, USBR, USFWS, and NMFS are the primary agencies that respond to ESA rules and procedures for the Klamath Project. Given the conflicting objectives and missions of these agencies, however, tensions among them are inevitable. ESA processes also have been joined by a number of advocacy groups that oppose or support actions of the various federal agencies. For example, the National Research Council staff has identified at least 29 environmental advocacy groups that have joined in litigation or taken positions against USBR and at least seven water-user advocacy groups that have brought suit against or opposed actions of USFWS. In addition, stakeholders in and outside the Klamath Project and local communities have not been adequately included in actions implemented under the ESA (Chapter 9). Entities outside the federal agencies feel disempowered by the present process (Lach et al. 2002). Their sense of powerlessness may contribute to the litigious nature of interaction among parties in the Klamath basin.

The current management structure includes the Klamath Basin Ecosystem Restoration Office (ERO), which fills two important functions in implementing the ESA in the Klamath basin: it provides money for research on the status of suckers in the upper basin, and it reviews USBR's biological assessments and prepares the USFWS biological opinions for the Section 7(a)(2) consultations. In fulfilling these functions, it operates essentially as a regulatory agency and could be viewed as an adversary to regulated parties (in this case, USBR and the irrigators in the Klamath Project). It also funds "restoration activities and practices" as part of the recovery program for the listed species. The activities and practices may lead to changes in land-use patterns on private lands in support of the sucker-recovery efforts.

The ERO serves as both a regulator and a funding agency; it is staffed primarily by USFWS personnel. It apparently does not effectively monitor and evaluate the success of its restoration actions. As noted earlier in this chapter, monitoring and evaluation are the most critical components of adaptive management for measuring the success of any ecosystem-restoration effort and incorporating new knowledge into the management process. In fact, USFWS and the ERO do not appear to have an operational recovery plan for the two sucker species (Chapter 9).

The underlying presumption of ERO managers appears to be that expenditure of money by the ERO on selected restoration actions is an acceptable measure of performance. In this regard, the ERO functions in a manner similar to that of many federal and state agencies in the basin that mistake input for output when evaluating their performance.

Federal and state emergency funding to assist farmers and agencies in 2001 was well intended but only exacerbated the problem of accountability. Similarly, the recent farm bill legislation that earmarked $50 million specifically for the Environmental Quality Incentives Program (NRCS 2003) and similar U.S. Department of Agriculture programs in the basin raises the issue of accountability in the absence of any central plan for recovery of the suckers. It is clear that the present level of emergency and supplemental funding for the basin may not be sustainable. Managers, therefore, need to have mechanisms in place to ensure that such funds, when available, are achieving the goals of the ESA recovery plans or, where appropriate, are being spent effectively in assisting stakeholders as they adjust to the consequences of the ESA.

Management of species in the Klamath basin should have two goals: maintenance and recovery of listed species and, among the actions that meet this objective, minimization of cost to society. The first goal is mandated by the ESA; the second is not the main objective of the ESA but is consistent with it (Chapter 9). The present management system in the Klamath basin is not ideal for reaching either goal.

If institutional deficiencies in the Klamath basin could be remedied, the likelihood of achieving the recovery of species and minimizing costs would increase. The design of research should begin with a broad set of objectives and scientific hypotheses; such breadth may require information from sources beyond local agencies and their supporting scientists and staff. The strong focus on water levels in Upper Klamath Lake and flows in the main stem, although driven by a desire to deal with issues over which the federal agencies have immediate control (through USBR operations), is indicative of an excessively narrow consideration of possibilities for restoration at the expense of other activities and solutions that may be effective over the long term. Furthermore, locations of restoration activities and their effects on water quality and habitat should be considered in the acquisition of land or other major investments. The assignment of priorities should recognize budgetary limitations of the agencies and others. Estimation of the cost effectiveness of restoration efforts is needed, as are the integrated monitoring and assessment programs to evaluate them.

Management requires external oversight by a committee or group capable of resolving conflicts between federal agencies. There appears to have been closer collaboration between USBR, USFWS, and NMFS in developing their most recent biological assessment and biological opinions than in previous years (Chapter 1), perhaps in response to external review. There is no guarantee, however, that such collaboration will continue and some mechanism should be in place for coordination of federal management efforts. For example, such a management role could be played by the Committee on Environment and Natural Resources of the National Science and Technology Council; this committee of the executive branch was founded for such purposes. At the same time that there is need for oversight of federal agencies, the management structure for ecosystem restoration needs to involve local groups and private landowners as well in the design of restoration activities and investments. As a part of these efforts, federal management agencies should recognize the nature of incentives in the ESA for private landowners to participate in ecosystem recovery. Specifically, the ESA may prohibit taking of endangered species by private landowners; it does not contain provisions that encourage landowners to increase the abundance of fish populations. Indeed, landowners who increase populations of endangered species on their land may face increased government regulation. Thus, although the ESA does not prohibit the use of incentives that would encourage landowners to promote the welfare of endangered species, it is often viewed by landowners as more stick than carrot. This perception could be changed by cooperative arrangements that promote the welfare of the listed species without threatening landowners.

Third, the management structure should, through monitoring and evaluation, improve the efficiency of expenditures for both research and restoration activities. That requires better mechanisms for setting spending priorities. Research demonstrates that cumulative effects are typical of stream restoration and that thresholds for recovery require implementation of corrective measures on a geographically broad scale (Adams et al. 1993, Li et al. 1994, Wu et al. 2000). The present pattern of federal, state, and private land acquisition for restoration in the upper Klamath basin shows little evidence of being guided by any systematic plan.

The process-oriented issues described above can be addressed by use of the adaptive-management framework, subject to the limitations imposed by the ESA. The Klamath River Basin Restoration Task Force (KRBRTF) or some other broadly constituted group may be a logical starting point in developing and implementing a set of basinwide restoration activities. At a minimum, an adaptive management approach, whether through an existing group or through a new entity, would address current shortcomings that arise from a lack of clearly defined benchmarks and a failure to monitor the biological and economic efficiency of current expenditures. The use of external advisory groups or panels for oversight would also provide fresh perspective and perhaps reduce some of the tensions and distrust inherent in the current system.

CONCLUSIONS

The listing agencies for endangered and threatened species in the Klamath basin accept adaptive management as a principle for pursuing restoration of these species, as does USBR. Even so, working examples of adaptive management in the upper Klamath basin are virtually absent. Erratic funding, lack of recovery plans, absence of systematic external review of research, and other deficiencies having to do with lack of continuity have been the direct cause of deficiencies in adaptive management.

Adaptive management is an ideal approach for the Klamath basin insofar as the effects of specific actions intended to benefit the endangered and threatened species cannot be evaluated fully except on a conditional trial basis. Conditional trials require thoughtful design and organized monitoring that will reveal responses to management actions. Efforts to implement ESA requirements for the benefit of fishes in the Klamath basin cannot succeed without aggressive pursuit of adaptive management principles, which in turn require continuity, master planning, flexibility, and conscientious evaluation of the outcomes of management.

11

Recommendations

BASINWIDE ISSUES

Scope of ESA Actions

Recovery of endangered suckers and threatened coho salmon in the Klamath basin cannot be achieved by actions that are exclusively or primarily focused on operation of USBR's Klamath Project. While continuing consultation between the listing agencies and USBR is important, distribution of the listed species well beyond the boundaries of the Klamath Project and the impairment of these species through land- and water-management practices that are not under control of USBR require that the agencies use their authority under the ESA much more broadly than they have in the past.

Recommendation 1. The scope of ESA actions by NMFS and USFWS should be expanded in several ways, as follows (Chapters 6, 8, 9).

- NMFS and USFWS should inventory all governmental, tribal, and private actions that are causing unauthorized take of endangered suckers and threatened coho salmon in the Klamath basin and seek either to authorize this take with appropriate mitigative measures or to eliminate it.
- NMFS and USFWS should consult not only with USBR, but also with other federal agencies (e.g., U.S. Forest Service) under Section 7(a)(1); the federal agencies collectively should show a will to fulfill the interagency agreements that were made in 1994.

• NMFS and USFWS should use their full authority to control the actions of federal agencies that impair habitat on federally managed lands, not only within but also beyond the Klamath Project.

• Within 2 yr, NMFS should prepare and promulgate a recovery plan for coho salmon, and USFWS should do the same for shortnose and Lost River suckers. The new recovery plans should facilitate consultations under ESA Sections 7(a)(1), 7(a)(2), and 10(a)(1) across the entire geographic ranges of the listed species.

• NMFS and USFWS should more aggressively pursue opportunities for non-regulatory stimulation of recovery actions through the creation of demonstration projects, technical guidance, and extension activities that are intended to encourage and maximize the effectiveness of non-governmental recovery efforts.

Planning and External Review

For all three of the listed fish species, monitoring, research, and remediation have been handicapped by lack of effective central planning, by insufficient external review, and by poor connections between research and remediation (Chapters 6, 8, 10).

Recommendation 2. Planning and organization of research and monitoring for listed species should be implemented as follows.

• Research and monitoring programs for endangered suckers should be guided by a master plan for collection of information in direct support of the recovery plan; the same should be true of coho salmon.

• A recovery team for suckers and a second recovery team for coho salmon should administer research and monitoring on the listed species. The recovery team should use an adaptive management framework that serves as a direct link between research and remediation by testing the effectiveness and feasibility of specific remediation strategies.

• Research and monitoring should be reviewed comprehensively by an external panel of experts every 3 yr.

• Scientists participating in research should be required to publish key findings in peer-reviewed journals or in synthesis volumes subjected to external review; administrators should allow researchers sufficient time to do this important aspect of their work.

• Separately or jointly for the upper and lower basins, a broadly based, diverse committee of cooperators should be established for the purpose of pursuing ecosystem-based environmental improvements throughout the basin for the benefit of all fish species as a means of preventing future listings while also preserving economically beneficial uses of water that are compatible with high environmental quality. Where possible, existing federal and state legislation should be used as a framework for organization of this effort.

ENDANGERED LOST RIVER AND SHORTNOSE SUCKERS

Needs for New Information

The endangered suckers have been extensively studied, particularly in Upper Klamath Lake, in ways that have proven very useful to the diagnosis of causes for decline in the abundance of suckers. Research and monitoring programs will continue to be valuable in revealing mechanisms that cause decline of the listed species, in developing a scientific basis for recovery actions, and in evaluating trial remediation measures through adaptive management. Research that is focused on gaps in knowledge or on mechanisms that appear to be particularly important to the recovery of the suckers will be most useful in support of the recovery effort.

Recommendation 3. Research and monitoring on the endangered suckers should be continued. Topics for research should be adjusted annually to reflect recent findings and to address questions for which lack of knowledge is a handicap to the development or implementation of the recovery plan. Gaps in knowledge that require research in the near future are as follows (Chapters 5, 6).

- Efforts should be expanded to estimate annually the abundance or relative abundance of all life stages of the two endangered sucker species in Upper Klamath Lake.
- At intervals of 3 yr, biotic as well as physical and chemical surveys should be conducted throughout the geographic range of the endangered suckers. Suckers should be sampled for indications of age distribution, qualitative measures of abundance, and condition factors. Sampling should include fish other than suckers on grounds that the presence of other fish is an indicator of the spread of nonnative species, of changing environmental conditions, or of changes in abundance of other endemic species that may be approaching the status at which listing is needed. Habitat conditions and water-quality information potentially relevant to the welfare of the suckers should be recorded in a manner that allows comparison across years. The resulting survey information, along with the more detailed information available from annual monitoring of populations in Upper Klamath Lake, should be synthesized as an overview of status.
- Detailed comparisons of the Upper Klamath Lake populations (which are suppressed) and the Clear Lake and Gerber Reservoir populations (which are apparently stable), in combination with studies of the environmental factors that may affect welfare of the fish, should be con-

ducted as a means of diagnosing specific life-history bottlenecks that are affecting the Upper Klamath Lake populations.

- Multifactorial studies under conditions as realistic as practicable should be made of tolerance and stress for the listed suckers relevant to poor water-quality conditions in Upper Klamath Lake and elsewhere.
- Factors affecting spawning success and larval survival in the Williamson River system should be studied more intensively in support of recovery efforts that are focused on improvements in physical habitat protection for spawners and larvae in rivers.
- An analysis should be conducted of the hydraulic transport of larvae in Upper Klamath Lake.
- Relevant to the water quality of Upper Klamath Lake, more intensive studies should be made of water-column stability and mixing, especially in relation to physiological status of *Aphanizomenon* and the occurrence of mass mortality; of mechanisms for internal loading of phosphorus; of winter oxygen concentrations; and of the effects of limnohumic acids on *Aphanizomenon*.
- A demographic model of the populations in Upper Klamath Lake should be prepared and used in integrating information on factors that affect individual life-history stages.
- Studies should be done on the degree and importance of predation on young fish by nonnative species.
- Additional studies should be done on the genetic identities of subpopulations.

Remedial Actions

Because the suckers currently are not showing evidence of recovery, new types of actions intended to promote recovery are essential. The main focus of action in the recent past has been maintenance of specific minimum water levels in Upper Klamath Lake. Current evidence suggests that these manipulations will not be effective in causing restoration of suckers in Upper Klamath Lake, despite evidence that higher water levels maximize certain habitat features that are known to be important to the suckers. Additional harm to the suckers might result, however, from changes in the Klamath Project operations that would allow greater degrees of mean or maximum drawdown than those observed in the 1990s. USFWS may continue to investigate the effects of lake level in a more directed way by collaborating with USBR in experiments involving water-level manipulations. Some new types of manipulations not produced by past operating procedures might be especially informative. In planning experiments USFWS should consider the possibility that sustained high water levels could be detrimental to the suckers by increasing the severity of mass mortality through maintenance of high water-column stability, thus exacerbating surface oxygen depletion at times of mixing during the late growing season. Water levels in Clear Lake and Gerber Reservoir appear to have been adequate to sustain stable populations except at extreme drawdown, the occurrence of which is a risk to the suckers.

Current evidence indicates that attempts to intercept nutrients from the watershed will not improve the quality of water of Upper Klamath Lake, and thus cannot be taken as a likely way to achieve recovery of suckers.

Recovery actions for suckers of Upper Klamath Lake at present should emphasize measures that maximize production and survival of young fish on the basis that additional recruitment into the subadult and adult stages could partially or fully offset mass mortality of adults. In addition, experiments should be done on artificial oxygenated refugia that may be used by large fish. Recovery planning should assume that, because mass mortality of adults will likely continue in Upper Klamath Lake, significant efforts should be made to establish self-sustaining populations elsewhere in the Klamath basin.

Recommendation 4. Recovery actions of highest priority based on current knowledge of endangered suckers are as follows (Chapter 6):

- Removal of Chiloquin Dam to increase the extent of spawning habitat in the upper Sprague River and expand the range of and conditions under which larvae enter Upper Klamath Lake.
- Removal or facilitation of passage at all small blockages, dams, diversions, and tributaries where suckers are or could be present.
- Screening of water intakes at Link River Dam.
- Modification of screening and intake procedures at the A Canal as recommended by USFWS (2002).
- Protection of known spawning areas within Upper Klamath Lake from disturbance (including hydrologic manipulation, in the case of springs), except for restoration activities.
- For river spawning suckers of Upper Klamath Lake, protection and restoration of riparian conditions, channel geomorphology, and sediment transport; elimination of disturbance at locations where suckers do spawn or could spawn. These actions will require changes in grazing and agricultural practices, land management, riparian corridors, and public education.
- Seeding of abandoned spawning areas in Upper Klamath Lake with new spawners and physical improvement of selected spawning areas.
- Restoration of wetland vegetation in the Williamson River estuary and northern portions of Upper Klamath Lake.
- Use of oxygenation on a trial basis to provide refugia for large suckers in Upper Klamath Lake.

• Rigorous protection of tributary spawning areas on Clear Lake and Gerber Reservoir, where populations are apparently stable.

• Reintroduction of endangered suckers to Lake of the Woods after elimination of its nonnative fish populations.

• Reestablishment of spawning and recruitment capability for endangered suckers in Tule Lake and Lower Klamath Lake, even if the attempts require alterations in water management, provided that preliminary studies indicate feasibility; increased control of sedimentation in Tule Lake.

• All proposed changes in Klamath Project operations should be reviewed for potential adverse effects on suckers; water level limits for the near future should be maintained as proposed by USBR in 2002 but with modifications as required by USFWS in its most recent biological opinion (2002).

THREATENED COHO SALMON

Needs for New Information

While the biology of coho salmon is well known in general, studies of coho salmon specific to the Klamath River basin have been few and do not provide the requisite amount of information to support quantitative assessments of population strength and distribution, environmental correlates of successful spawning and rearing, overwintering losses and associated habitat deficiencies, water temperatures at critical points in tributary waters, and effects of hatchery-reared fish on wild coho. Main-stem conditions are primarily of interest with respect to the spawning run and the downstream migration of smolts. Tributary conditions, which have been much less studied than main-stem conditions, are critical to both spawning and rearing; habitat includes but extends beyond the main stems of the large tributaries and into the small tributaries and headwaters that strongly favor spawning and rearing of coho.

Recommendation 5. Needs for new information on coho salmon are as follows (Chapters 7, 8).

• Annual monitoring of adults and juveniles should be conducted at the mouths of major tributaries and the main stem as a means of establishing a record of year-class strength for coho. Every 3 yr, synoptic studies of the presence and status of coho should be made of coho in the Klamath basin. Physical and chemical conditions should be documented in a manner that allows interannual comparisons. Not only coho but other fish species present in coho habitats should be sampled simultaneously on grounds that changes in the relative abundance of species are relevant to the welfare of coho and may serve as an early warning of declines in the abundance of other species. Results of synoptic studies, along with the annual monitoring at tributary mouths, should be synthesized as an overview of population status at 3-yr intervals.

- Detailed comparisons should be made of the success of coho in specific small tributaries that are chosen so as to represent gradients in potential stressors. The objective of the study should be to identify thresholds for specific stressors or combinations of stressors and thus to establish more specifically the tolerance thresholds for coho salmon in the Klamath basin.
- The effect on wild coho of fish released in quantity from hatcheries should be determined by manipulation of hatchery operations according to adaptive-management principles. As an initial step, release of hatchery fish from Iron Gate Hatchery (all species) should be eliminated for 3 yr, and indicators of coho response should be devised. Complementary manipulations at the Trinity River Hatchery would be desirable as well.
- Selected small tributaries that have been impaired should be experimentally restored, and the success of various restoration strategies should be determined.
- Success of specific livestock-management practices in improving channel conditions and promoting development of riparian vegetation should be evaluated systematically.
- Relationships between flow and temperature at the junctions of tributaries with the main stem and the estuary should be quantified; possible benefits of coordinating flow management in the Trinity and Klamath main stem should be studied.

Remediation

Actions intended to improve environmental conditions for the threatened coho salmon to date primarily have involved hydrologic manipulation of the main stem at Iron Gate Dam. Continual focus on hydrologic conditions in the main stem is an excessively narrow basis for recovery actions or for a recovery plan in that coho salmon are strongly oriented toward tributaries for all phases of the freshwater phase of their life cycle except migration at the adult and smolt stages. Changes required by NMFS in the flow of the main stem include additional water specifically for smolt migration; it is unknown whether this will be a major benefit to coho, but in the absence of information to the contrary it is a reasonable requirement. Establishment of more stringent minimum flows for the other parts of the year, as compared to the operations during the 1990s, are of uncertain benefit to coho salmon, although they may be of substantial benefit to other species that use the main stem more extensively. In apportioning responsibility to USBR for providing minimum flows according to its proportional use of water, NMFS is recognizing in a realistic way the need for all consumptive uses to be factored into any minimum-flow regime.

Major tributaries as well as small tributaries must benefit from remediation if recovery is to occur. Although more detailed information would be desirable as a basis for remediation, beginning points for remediation are obvious in locations where tributaries have been critically dewatered or warmed to the lethal threshold for coho salmon (a problem that could be exacerbated by climate change), or where appropriate substrate has been eliminated and cover is absent. Thus, there is ample justification for beginning remediation immediately. This will require extensive work on private lands, and also the establishment of improved management practices for mining and forestry, some of which is under the direct control of other agencies that are subject to ESA authority through NMFS. Blockage of coho migration, which occurs in dozens of locations at various scales within the Klamath basin, is inconsistent with ESA regulations on take and must be dealt with by NMFS.

Recommendation 6. Remediation measures that can be justified from current knowledge include the following (Chapter 8).

- Reestablishment of cool summer flows in the Shasta and Scott rivers in particular but also in small tributaries that reach the Klamath main stem or the Trinity main stem where water has been anthropogenically warmed. Reestablishment of cool flows should be pursued through purchase, trading, or leasing of groundwater flows (including springs) for direct delivery to streams; by extensive restoration of woody riparian vegetation capable of providing shade; and by increase of annual or seasonal low flows.
- Removal or provision for effective passage at all small dams and diversions throughout the distribution of the coho salmon, to be completed within 3 yr. In addition, serious evaluation should be made of the benefits to coho salmon from elimination of Dwinnell Dam and Iron Gate Dam on grounds that these structures block substantial amounts of coho habitat and, in the case of Dwinnell Dam, degrade downstream habitat as well.
- Prescription of land-use practices for timber management, road construction, and grazing that are sufficiently stringent to prevent physical degradation of tributary habitat for coho, especially in the Scott, Salmon, and Trinity river basins as well as small tributaries affected by erosion.
- Facilitation through cooperative efforts or, if necessary, use of ESA authority to reduce impairment of spawning gravels and other critical habitat features by livestock, fine sediments derived from agricultural practice, timber management, or other human activities.
- Changes in hatchery operations to the extent necessary, including possible closure of hatcheries, for the benefit of coho salmon as determined through research by way of adaptive management of the hatcheries.

COSTS

The costs of remediation actions are difficult to estimate without more detail on their mode of implementation by the agencies. Based on general knowledge of costs of research and monitoring at other locations, an approximate figure for the recommendations on endangered suckers over a 5-yr period is $15–20 million, including research, monitoring, and remedial actions of minor scope. Excluded are administrative costs and the costs of remedial actions of major scope (e.g., removal of Chiloquin Dam), which would need to be evaluated individually for cost. For coho salmon, research, monitoring, and remedial projects of small scope over 5 yr is estimated at $10–15 million. Thus, the total for all three species over 5 yr is $25–35 million, excluding major projects such as removal of dams. These costs are high relative to past expenditures on research and remediation in the basin, but the costs of further deterioration of sucker and coho populations, along with crisis management and disruptions of human activities, may be far more costly. A hopeful vision is that increased knowledge, improved management, and cohesive community action will promote recovery of the fishes. This outcome, which would be of great benefit to the Klamath basin, could provide a model for the nation.

References

Abbott, A. 2002. Effect of Riparian Vegetation on Stream Temperature in the Shasta River. M.S. Thesis, University of California, Davis, CA. 144 pp.

Adams, R.M., P.R. Berrens, A. Cerda, H. Li, and C.P. Klingeman. 1993. Developing a bioeconomic model for riverine management: Case of the John Day River, Oregon. Rivers 4(3):213–226.

Amory, C. 1926. Agriculture and wild fowl conservation at Lower Klamath Lake. New Reclamation Era (May)17:80–82.

Andreasen, J.K. 1975. Systematics and Status of the Family Catostomidae in Southern Oregon. Ph.D. Dissertation, Oregon State University, Corvallis, OR. 76 pp.

Anglin, D.R. 1994. Lower Klamath River Instream Flow Study. Lower Columbia River Fisheries Office, U.S. Fish and Wildlife Service, Dept. of Interior, Vancouver, WA. 46 pp.

Argus, D.F., and R.G. Gordon. 1990. Pacific-North American plate motion from very long baseline interferometry compared with motion inferred from magnetic anomalies, transform faults, and earthquake slip vectors. J. Geophys. Res. 95(11): 17315–17324.

Baker, P.F., T.P. Speed, and F.K. Ligon. 1995. Estimating the influence of temperature on the survival of chinook salmon smolts (Oncorhynchus tshawytscha) migrating through the Sacramento-San Joaquin River Delta of California. Can. J. Fish. Aquat. Sci. 52(4):855–863.

Balance Hydrologics, Inc. 1996. Initial Assessment of Pre- and Post-Klamath Project Hydrology on the Klamath River and Impacts of the Project on In-Stream Flows and Fisheries Habitat. Balance Hydrologics, Inc., Berkeley, CA. 39 pp.

Baltz, D.M., J.W. Fleeger, C.F. Rakocinski, and J.N. McCall. 1998. Food, density, and microhabitat: Factors affecting growth and recruitment potential of juvenile saltmarsh fishes. Environ. Biol. Fish. 53(1):89–103.

Barnhart, R.A. 1994. Salmon and steelhead populations of the Klamath-Trinity Basin, California. Pp. 73–97 in Klamath Basin Fisheries Symposium: Proceedings of a Symposium held in Eureka, California, 23–24 March 1994, T.J. Hassler, ed. Arcata, CA: California Cooperative Fishery Research Unit, Humboldt State University.

Bartholow, J.M. 1995. Review and Analysis of Klamath River Basin Water Temperatures as a Factor in the Decline of Anadromous Salmonids with Recommendations for Mitigation. Final Draft. River Systems Management Section, Midcontinent Ecological Science Center, Fort Collins, CO. May 11, 1995. 53 pp.

Beckman, T. 1998. The view from native California: Lifeways of California's indigenous people. In The Yurok and Hupa of the Northern Coast, Chapter 10. [Online]. Available: http://www4.hmc.edu:8001/Humanities/indian/ca/ch10.htm [accessed Feb. 4, 2003].

Behnke, R.J. 1992. Native Trout of Western North America. American Fisheries Society Monograph No. 6. Bethesda, MD: American Fisheries Society. 275 pp.

Bell, E., W.G. Duffy, and T.D. Roelofs. 2001. Fidelity and survival of juvenile coho salmon in response to a flood. Trans. Am. Fish. Soc. 130(3):450–458.

Bennett, R.A., B.P. Wernicke, and J.L. Davis. 1998. Continuous GPS measurements of contemporary deformation across the northern Basin and Range province. Geophys. Res. Lett. 25(4):563–566.

Bennett, R.A., J.L. Davis, and B.P. Wernicke. 1999. Present-day pattern of Cordilleran deformation in the western United States. Geology 27(4):371–374.

Bentivoglio, A.A. 1998. Investigations into the Endemic Sculpins (Cottus princeps, Cottus tenuis), in Oregon's Upper Klamath Lake Watershed, with Information on Other Sculpins of Interest (C. evermanni, C. spp., and C. "pretendor"). U.S. Geological Survey, Reston, VA. 29 pp.

Bienz, C.S., and J.S. Ziller. 1987. Status of Three Lacustrine Sucker Species (Catostomidae). Report by the Klamath Tribe, Klamath OR, and Oregon Dept. of Fish and Wildlife to the U.S. Fish and Wildlife Service, Sacramento, CA. 39 pp.

Bisson, P.A., J.L. Nielsen, and J.W. Ward. 1988. Summer production of coho salmon stocked in Mount St. Helens streams 3–6 years after the 1980 eruption. Trans. Am. Fish. Soc. 117:322–335.

Blake, T.A., M.G. Blake, and W. Kittredge. 2000. Balancing Water: Restoring the Klamath Basin. Berkeley, CA: University of California Press.

Bohnsack, J.A. 1994. Marine reserves: They enhance fisheries, reduce conflicts, and protect resources. NAGA, ICLARM Quarterly (July):4–7.

Boles, G. L. 1990. Food Habits of Juvenile Wild and Hatchery Steelhead Trout, Oncorhynchus mykiss, in the Trinity River, California. Inland Fisheries Administrative Report 90-10. Sacramento, CA: California Department of Fish and Game. 8 pp.

Bond, C.E., E. Rexstad, and R.M. Hughes. 1988. Habitat use of twenty-five common species of Oregon freshwater fishes. Northwest Sci. 62(5):223–232.

Bortleson, G.C., and M.O. Fretwell. 1993. A Review of Possible Causes of Nutrient Enrichment and Decline of Endangered Sucker Populations in Upper Klamath Lake, Oregon. Water-Resources Investigations Report 93-4087. Portland, OR: U.S. Geological Survey. 24 pp. [Online]. Available: http://www.krisweb.com/krisweb_kt/biblio/uperklam/ usgs/usgs93/usgs9301.htm [accessed Oct. 9, 2003].

Bowers, W., R. Smith., R. Messmer, C. Edwards, and R. Perkins. 1999. Conservation Status of Oregon Basin Redband Trout. Oregon Department of Fish and Wildlife. [Online]. Available: http://www.dfw.state.or.us/ODFWhtml/redband.pdf. [accessed June 29, 2003].

Boyd, M., S. Kirk, M. Wiltsey, B. Kasper, J. Wilson, and P. Leinenbach. 2001. Upper Klamath Lake Drainage Total Maximum Daily Load (TMDL), Draft. Oregon Department of Environmental Quality, Portland, OR. November 2001.

Boyd, M., S. Kirk, M. Wiltsey, and B. Kasper. 2002. Upper Klamath Lake Drainage Total Maximum Daily Load (TMDL) and Water Quality Management Plan (WQMP). State of Oregon Department of Environmental Quality. May 2002 [Online]. Available: http://www.deq.state.or.us/wq/TMDLs/UprKlamath/UprKlamathTMDL.pdf [accessed Feb. 4, 2003].

Boydstun, L.B., M. Palmer-Zwahlen, and D. Viele. 2001. Pacific salmon. Pp. 407–417 in California's Living Marine Resources: A Status Report, W.S. Leet, C.M. Dewees, R. Klingbeil, and E.J. Larson, eds. Sacramento, CA: California Department of Fish and Game.

Brett, J.R. 1952. Temperature tolerances of young Pacific salmon, Oncorhynchus. J. Fish. Res. Board Can. 9(6):264–323.

Brown, L.R., P.B. Moyle, and R.M. Yoshiyama. 1994. Historical decline and current status of coho salmon (Oncorhynchus kisutch) in California. N. Am. J. Fish. Manage. 14(2):237–261.

Buchanan, D.V., M.L. Hanson, and R.M. Hooton. 1997. Status of Oregon's Bull Trout. Portland, OR: Dept. of Fish and Wildlife.

Buer, K. 1981. Klamath and Shasta Rivers Spawning Gravel Enhancement Study. Northern District, California Dept. of Water Resources, Red Bluff, CA. 178 pp.

Buettner, M. 1992. Potential Refugial Habitat for Suckers, in Upper Klamath and Agency Lakes, at Different Lake Elevations (Habitat). Memo KO-750, ENV-4.00 to Files. August 27, 1992.

Buettner, M., and G. Scoppettone. 1990. Life History and Status of Catastomids in Upper Klamath Lake, Oregon: Completion Report (1990). Corvallis, OR: Oregon Dept. of Fish and Wildlife Service, Fishery Research Division.

Buettner, M.E., and G.G. Scoppettone. 1991. Distribution and Information on the Taxonomic Status of the Shortnose Sucker, Chasmistes brevirostris, and Lost River Sucker, Deltistes luxatus, in the Klamath River Basin, California. Completion Report. Contract FG-8304. California Department of Fish and Game, Seattle National Fisheries Research Center, Reno Substation. 101 pp.

Buhl, K.J., and S.J. Hamilton. 1990. Comparative toxicology of inorganic contaminants released by placer mining to early life stages of salmonids. Ecotoxicol. Environ. Saf. 20:325–342.

Buhl, K.J., and S.J. Hamilton. 1991. Relative sensitivity of early life stages of Arctic grayling, coho salmon, and rainbow trout to nine inorganics. Ecotoxicol. Environ. Saf. 22:184–197.

Burke, S. 2002. The effects of water allocation decisions on crop revenue in the Klamath Reclamation Project. Pp. 231–250 in Water Allocation in the Klamath Reclamation Project, 2001: An Assessment of Natural Resource, Economic, Social, and Institutional Issues in the Upper Klamath Basin, W.S. Braunworth, Jr., T. Welch, and R. Hathaway, eds. Corvallis, OR: Oregon State University Extension Service. [Online]. Available: http://eesc.orst.edu/agcomwebfile/edmat/html/sr/sr1037/sr1037.html [accessed March 10, 2003].

Busby, P.J., T.C. Wainwright, and R.S. Waples. 1994. Status Review for Klamath Mountains Province Steelhead. NOAA Technical Memorandum NMFS-NWFSC-19. National Marine Fisheries Service, Seattle, WA. 130 pp. [Online]. Available: http://www.nwfsc.noaa.gov/pubs/tm/tm19/tm19.html [accessed Feb. 6, 2003].

CALFED. 2002. Evaluating and Comparing Proposed Water Management Actions, Draft. CALFED Bay-Delta Program, Sacramento, CA. February 2002. [Online]. Available: http://www.calfed.water.ca.gov/BDPAC/Subcommittees/WMSCEFullReport_020202.pdf [accessed Feb. 6, 2003].

CALFED Bay-Delta Program. 2000. Adaptive management process. Pp. 17–22 in Strategic Plan for Ecosystem Restoration. CALFED Bay-Delta Program, Sacramento, CA. [Online]. Available: http://calfed.ca.gov/Programs/EcosystemRestoration/adobe_pdf/ 304c_TOC.pdf [accessed Feb. 6, 2003].

California Agricultural Statistics Service. 2003. Summary of Annual Production by County. California Agricultural Statistics Service, State of California, Sacramento, CA.

Campbell, E.A., and P.B. Moyle. 1991. Historical and recent population sizes of spring-run chinook salmon in California. Pp. 155–216 in Proceedings of the 1990 Northeast Pacific Chinook and Coho Salmon Workshop: Humboldt State University, Arcata, CA, September 18–22, 1990, T.J. Hassler, ed. Arcata, CA: California Cooperative Fishery Research Unit, Humboldt State University.

Campbell, S.G. 1995. Klamath River Basin Flow-Related Scoping Study – Phase I, Water Quality. In Compilation of Phase I Reports for the Klamath River Basin, May 1995. Prepared for the Technical Work Group of the Klamath River Basin Fisheries Task Force by River Systems Management Section, National Biological Service, Midcontinent Ecological Science Center, Fort Collins, CO.

Campbell, S.G. 2001. Water Quality and Nutrient Loading in the Klamath River Between Keno, Oregon and Seiad Valley, California from 1996–1998. U.S. Geological Survey Open File Report 01-301. Ft. Collins, CO: U.S. Dept. of the Interior, U.S. Geological Survey. 55 pp.

Carlson, L., D. Simon, B. Shields, and D. Markle. 2002. Interannual Patterns of Ectoparasite Infestation in Age 0 Klamath Suckers, Poster. Oregon Chapter American Fisheries Society Annual Meeting, February 2002, Sunriver, OR.

Castleberry, D.T., and J.J. Cech. 1992. Critical thermal maxima and oxygen minima of five fishes from the Upper Klamath Basin. Cal. Fish Game Bull. 78:145–152.

CDFG (California Department of Fish and Game). 1934. Stream Survey: Scott River, Tributary to Klamath River. California Department of Fish and Game, Sacramento, CA. 3 pp.

CDFG (California Department of Fish and Game). 1965. Klamath River Hydrographic Unit. Pp. 249–271 in North Coastal Area Investigation, Appendix C: Fish and Wildlife. California Department of Water Resources Bulletin 136. Sacramento, CA: California Department of Fish and Game.

CDFG (California Department of Fish and Game). 1979a. Salmon River Waterway Management Plan. Sacramento, CA: California Department of Fish and Game. 92 pp.

CDFG (California Department of Fish and Game). 1979b. Scott River Waterway Management Plan. Sacramento, CA: California Department of Fish and Game. 121 pp.

CDFG (California Department of Fish and Game). 1999. Klamath River Basin Fall Chinook Salmon Spawner Escapement, In-River Harvest and Run-Size Estimates 1978–1999. California Department of Fish and Game, Trinity River Project Office, Arcata, CA. [Online]. Available: www.krisweb.com/krisweb_kt/trtour [accessed August 26, 2003].

CDFG (California Department of Fish and Game). 2002. Status Review of California Coho Salmon North of San Francisco. Report to the California Fish and Game Commission. Candidate Species Status Review Report 2002–3. 232 pp. [Online]. Available: http://www.dfg.ca.gov/nafwb/pubs/2002/2002_04_coho_status.pdf [accessed Feb. 6, 2003].

CDFG (California Department of Fish and Game). 2003. September 2002 Klamath River Fishkill—Preliminary Analysis of Contributing Factors. State of California, the Resources Agency, California Department of Fish and Game. 63 pp. [Online]. Available: www.dfg.ca.gov/html/krfishkill-rpt.pdf [accessed Oct. 7, 2003]

CDWR (California Department of Water Resources). 1964. Shasta Valley Investigation. Bulletin No. 87. Sacramento, CA: Dept. of Water Resources.

CDWR (California Department of Water Resources). 1965. North Coastal Area Investigation. Bulletin No. 136. Sacramento, CA: Dept. of Water Resources.

CDWR (California Department of Water Resources). 1986. Shasta/Klamath Rivers Water Quality Study. Northern District, Dept. of Water Resources, Red Bluff, CA. [Online]. Available: http://www.krisweb.com/biblio/regional/klamath_trinity/dwr/dwr1.htm [accessed Feb. 12, 2003].

CDWR (California Department of Water Resources). 2002. California Data Exchange Center. California Department of Water Resources. [Online]. Available: http://cdec.water.ca.gov [accessed Feb. 12, 2003].

CERP. 2002. The Comprehensive Everglades Restoration Plan (CERP). [Online]. Available: http://www.evergladesplan.org/about/rest_plan.cfm

Chan, F. 2001. Ecological Controls on Estuarine Planktonic Nitrogen-Fixation: The Roles of Grazing and Cross-Ecosystem Patterns in Phytoplankton Mortality. Ph.D. Thesis, Cornell University, Ithaca, NY. 288 pp.

Chapin, D.M. 1997. Aerial Photographic Analysis of Exposure and Inundation of Marsh Vegetation Bordering Upper Klamath and Agency Lakes. Prepared for EA Engineering, Science, and Technology, Lafayette, CA. March 1997. (as cited in Reiser et al. 2001)

CH2M HILL. 1985. Klamath River Basin Fisheries Resource Plan. Prepared for the Bureau of Indian Affairs, U.S. Department of Interior, Redding, CA.

CH2M HILL. 1996. Chiloquin Dam Conceptual Design and Cost Estimate for Upstream Fish Passage Facilities. Prepared for The Nature Conservancy, Portland, OR. March 1996.

Chavez, F.P., J. Ryan, S.E. Lluch-Cota, and C.M. Ñiquen. 2003. From anchovies to sardines and back: Multidecadal change in the Pacific Ocean. Science 299(5604):217–221.

Cheever, F. 2001. Recovery planning, the courts and the Endangered Species Act. Nat. Resour. Environ. 16(2):106–111, 135.

Chesney, W.R. 2000. Shasta and Scott River Juvenile Steelhead Trapping 2000, Annual Report, Study 3a1. California Department of Fish and Game, Steelhead Research and Monitoring Program, Yreka, CA. December. 37 pp.

Chesney, W.R. 2002. Shasta and Scott River Juvenile Outmigrant Study, 2000–2001, Annual Report: Project 2a1. Steelhead Research and Monitoring Program, California Department of Fish and Game, Yreka, CA.

Close, D.A., M.S. Fitzpatrick, and H.W. Li. 2002. The ecological and cultural importance of a species at risk of extinction, Pacific lamprey. Fisheries 27(7):19–25.

Cobb, J.N. 1930. Pacific Salmon Fisheries. Document No. 1092. U.S. Bureau of Fisheries, U.S. Dept. of Commerce. Washington, DC: U.S. Government Printing Office.

Coen, M.A., and R.S. Shively. 2001. Sampling of suckers in Upper Klamath Lake, OR to identify shoreline spawning sites. In Monitoring of Lost River and Shortnose Suckers in the Upper Klamath Basin, 2000. Prepared by U.S. Geological Survey, Western Fisheries Research Center, Klamath Falls Field, Klamath Falls, OR, for the U. S. Bureau of Reclamation, Mid-Pacific Region, Klamath Area Office, Klamath Falls, OR. Contract #00AA200049.

Coen, M.A., E.C. Janney, R.S. Shively, B.S. Hayes, and G.N. Blackwood. 2002. Monitoring of Lost River and shortnose suckers in Upper Klamath Lake, Oregon, Annual Report 2001. Pp. 106–135 in Monitoring of Lost River and Shortnose Suckers in the Upper Klamath Basin, 2001. Prepared by U.S. Geological Survey, Western Fisheries Research Center, Klamath Falls Field, Klamath Falls, OR, for the U.S. Bureau of Reclamation, Mid-Pacific Region, Klamath Area Office, Klamath Falls, OR. Contract #00AA200049.

Colby, B.G. 1990. Enhancing instream flow benefits in an era of water marketing. Water Resour. Res. 26(6):1113–1120.

Colby-Saliba, B., and D. Bush. 1987. Water Markets in Theory and Practice: Market Transfer, Water Values, and Public Policy. Boulder, CO: Westview Press.

Coleman, M.E., J. Kann, and G.G. Scoppettone. 1988. Life History and Ecological Investigations of Catostomids from the Upper Klamath Lake Basin, Oregon. Draft Annual Report. National Fisheries Research Center, U.S. Fish and Wildlife Service, Seattle, WA.

Conover, D.O., and S.B. Munch. 2002. Sustaining fisheries yields over evolutionary time scales. Science 297(5578):94–95.

Contreras, G.P. 1973. Distribution of Fishes of the Lost River System, California-Oregon, With a Key to the Species Present. M.S. Thesis, University of Nevada, Reno, NV.

Cooperman, M. 2002. Clarification of the Role of Submergent Macrophytes to Larval Shortnose and Lost River Suckers in the Williamson River and Upper Klamath Lake. Comments sent by Department of Fisheries and Wildlife, Oregon State University, Corvallis, OR to NRC, Washington, DC. March 7, 2002. 3 pp.

Cooperman, M., and D.F. Markle. 2000. Ecology of Upper Klamath Lake Shortnose and Lost River suckers. 2. Larval Ecology of Shortnose and Lost River Suckers in the Lower Williamson River and Upper Klamath Lake. Oregon State University, Department of Fisheries and Wildlife, Corvallis, OR. 27pp. [Online]. Available: http://www.mp.usbr.gov/kbao/esa/99report2.PDF [accessed March 5, 2003].

Cooperman, M.S., and D.F. Markle. 2003. Endangered species, the Endangered Species Act and the National Research Council's interim judgment in Klamath basin. Fisheries 28(3):10–19.

Coots, M. 1965. Occurrences of the Lost River sucker, Deltistes luxatus (Cope), and shortnose sucker, Chasmistes brevirostris (Cope), in northern California. Calif. Fish Game Bull. 51:68–73.

Cope, E.D. 1879. The fishes of Klamath Lake, Oregon. Am. Nat. 13:784–785.

Crandall, D.R. 1989. Gigantic Debris Avalanche of Pleistocene Age from Ancestral Mount Shasta Volcano, California, and Debris-Avalanche Hazard Zonation. U.S. Geological Survey Bulletin 1861. Denver, CO: U.S. G. P.O. 32 pp. [Online]. Available: http://www.krisweb.com/krisweb_kt/biblio/biblio.htm#Shasta River Sub-basin [accessed Oct 9, 2003].

Cressman, L.S. 1956. Klamath Prehistory: The Prehistory of the Klamath Lake Area, Oregon. Lancaster, PA: Lancaster Press, Inc.

CSWRCB (California State Water Resources Control Board). 1975. Report on Hydrogeologic Conditions, Scott River Valley, Scott River Adjudication, Siskiyou County. California Division of Water Rights, Sacramento, CA. 18 pp.

Cunningham, M.E., and R.S. Shively. 2001. Monitoring report of Lost River and shortnose suckers in the Lower Williamson River, Oregon, 2000. In Monitoring of Lost River and Shortnose Suckers in the Upper Klamath Basin, 2000, Annual Report. Prepared by U.S. Geological Survey, Western Fisheries Research Center, Klamath Falls Field, Klamath Falls, OR, for the U.S. Bureau of Reclamation, Mid-Pacific Region, Klamath Area Office, Klamath Falls, OR. Contract #00AA200049.

Cunningham, M.E., R.S. Shively, E.C. Janney, and G.N. Blackwood. 2002. Monitoring of Lost River and shortnose suckers in the lower Williamson River, Oregon, 2001. Pp. 1–36 in Monitoring of Lost River and Shortnose Suckers in the Upper Klamath Basin, 2001. Prepared by U.S. Geological Survey, Western Fisheries Research Center, Klamath Falls Field, Klamath Falls, OR, for the U.S. Bureau of Reclamation, Mid-Pacific Region, Klamath Area Office, Klamath Falls, OR. Contract #00AA200049.

Davison, W. 1993. Iron and manganese in lakes. Earth Sci. Rev. 34(2):119–163.

Deas, M.L. 2000. Application of Numerical Water Quality Models in Ecological Assessment. Ph.D. Dissertation, University of California. Davis, CA. 408 pp.

Desjardins, M., and D.F. Markle. 2000. Distribution and Biology of Suckers in the Lower Klamath Reservoirs, 1999. Final Report. Prepared for PacifiCorp., Portland, OR, by Department of Fisheries and Wildlife, Oregon State University, Corvallis, OR. March 28, 2000. [Online]. Available: http://www.mp.usbr.gov/kbao/esa/Finalreport_17.PDF [accessed March 5, 2003].

Devlin, E.W., and N.K. Mottet. 1992. Embryotoxic action of methyl mercury on coho salmon embryos. Bull. Environ. Contam. Toxicol. 49(3):449–454.

Dileanis, P.D., S.E. Schwarzbach, J. Bennett, J. Sefchick, R. Boyer, J. Henderson, T. Littleton, D.E. MacCoy, E. Materna, T. Maurer, D. Mead, and M. Moore. 1996. Detailed Study of Water Quality, Bottom Sediment, and Biota Associated with Irrigation Drainage in the Klamath Basin, California and Oregon, 1990–92. Water Resources Investigations Report 95-4232. U.S. Geological Survey, Sacramento, CA. 68 pp. [Online]. Available: http://www.krisweb.com/krisweb_kt/biblio/uperklam/usgs/usgs96/usgs96.pdf [accessed Oct. 3, 2003].

Docker, M.F., J.H. Youson, R.J. Beamish, and R.H. Devlin. 1999. Phylogeny of the lamprey genus Lampetra inferred from mitochondrial cytochrome b and ND3 gene sequences. Can. J. Fish. Aquat. Sci. 56(12):2340–2349.

Dong, A.E., K.W. Beatty, and R.C. Averett. 1974. Limnological Study of Lake Shastina, Siskiyou County, California. U.S. Geological Survey, Water-Resources Investigation Report No. 19-74. Washington, DC: U.S. Government Printing Office.

Dowling, T.E. 2000. Conservation Genetics of Endangered Suckers of the Klamath Region: Mitochondrial DNA. Interim Report. Amendment to Cooperative Agreement 9-FC-20-17650. Prepared by Department of Biology, Arizona State University, Tempe, AZ, to U.S. Bureau of Reclamation, Klamath Falls, OR. [Online]. Available: http://www.mp.usbr.gov/kbao/esa/99nwsuckreport.pdf [accessed March 5, 2003].

Downton, M.W., and K.A. Miller. 1998. Relationships between Alaskan salmon catch and North Pacific climate on interannual and interdecadal time scales. Can. J. Fish. Aquat. Sci. 55(10):2255–2265.

Drake, D., K.W. Tate, and H. Carlson. 2000. Analysis shows climate-caused decreases in Scott River fall flows. California Agriculture 54(6):46–49.

Dunsmoor, L. 1993. Laboratory Studies of Fathead Minnow Predation on Catostomid Larvae. Klamath Tribes Research Report KT-93-01. The Klamath Tribes, Chiloquin, OR. 16 pp.

Dunsmoor, L., L. Basdekas, B. Wood, and B. Peck. 2000. Quantity, Composition and Distribution of Emergent Vegetation along the Lower River and Upper Klamath Lake Shorelines of the Williamson River Delta, Oregon. Completion Report. Klamath Tribes, Chiloquin, OR, and U.S. Bureau of Reclamation, Klamath Basin Area Office, Klamath Falls, OR.

Earthinfo, Inc. 1995. U.S. Geological Survey Quality of Water Surface, West 1, 1996. [Online]. Available: http://www.earthinfo.com/.

Echelle, A.A. 1991. Conservation genetics and genic diversity in freshwater fishes of western North America. Pp. 141–153 in Battle Against Extinction: Native Fish Management in the American West, W.L. Minckley and J.E. Deacon, eds. Tucson, AZ: University of Arizona Press.

Eilers, J., J. Kann, J. Cornett, K. Moser, A. St. Amand, and C. Gubala. 2001. Recent Paleolimnology of Upper Klamath Lake, Oregon. Prepared by J.C. Headwaters, Inc., Rosenburg, OR, for U.S. Bureau of Reclamation, Klamath Falls, OR. March 16, 2001. [Online]. Available: http://www.mp.usbr.gov/kbao/esa/FinalReport23_3-29-01.pdf [accessed Feb. 25, 2003].

Elder, D., B. Olson, A. Olson, J. Villeponteaux, and P. Brucker. 2002. Salmon River Subbasin Restoration Strategy: Steps to Recovery and Conservation of Aquatic Resources. Prepared for The Klamath River Basin Fisheries Restoration Task Force, Yreka Fish and Wildlife Office, Yreka, CA. 53pp. [Online]. Available: http://www.r5.fs.fed.us/klamath/mgmt/analysis.html [accessed March 6, 2003].

Elmore, W., and R.L. Beschta. 1987. Riparian areas: Perceptions in management. Rangelands 9:260–265.

EPA (U.S. Environmental Protection Agency). 1998. South Fork Trinity River and Hayfork Creek Sediment Total Maximum Daily Loads. Region 9, U.S. Environmental Protection Agency, San Francisco, CA. [Online]. Available: www.epa.gov/region9/water/tmdl/fsftmdl.pdf [accessed Oct. 3, 2003].

EPA (U.S. Environmental Protection Agency). 2001. Trinity River Total Maximum Daily Load for Sediment. Region 9, U.S. Environmental Protection Agency, San Francisco, CA. [Online]. Available: http://www.epa.gov/Region9/water/tmdl/trinity/finaltrinitytmdl.pdf [accessed Oct. 3, 2003].

Falter, M.A., and J. Cech. 1991. Maximum pH tolerance of three Klamath Basin fishes. Copeia 4:1109–1111.

Fausch, K.D. 1988. Tests of competition between native and introduced salmonids in streams: What have we learned? Can. J. Fish. Aquat. Sci. 45:2238–2246.

Fausch, K.D. 1998. Interspecific competition and juvenile Atlantic salmon (Salmo salar): On testing effects and evaluating the evidence across scales. Can. J. Fish. Aquat. Sci. 55 (suppl. 2):218–231.

Fausch, K.D., C.E. Torgersen, C.V. Baxter, and H.W. Li. 2002. Landscapes to riverscapes: Bridging the gap between research and conservation of stream fishes. BioScience 52 (6):483–498.

FEMAT (Forest Ecosystem Management Assessment Team). 1993. Forest Ecosystem Management: An Ecological, Economic, and Social Assessment. U.S. Department of Agriculture, Forest Service, and other agencies, Portland, OR.

Fletcher, K.M., and S.E. O'Shea. 2000. Essential fish habitat: Does calling it essential make it so? Environ. Law 30(1):51–98.

Foott, J.S. 1997. Results of Histological Examination of Sucker Tissues. Memorandum 1997. U.S. Fish and Wildlife Service, Fish Health Center, Anderson, CA.

FOTR (Friends of the Trinity River). 2003. Trinity Division History, Restoration Program Projects and Costs. Friends of the Trinity River, Mill Valley, CA. [Online]. Available: www.fotr.org/history6.html [accessed August 26, 2003].

Franzius, A. 1997. California Gold: The Diary and Letters of James M. Burr, 1850–1853. The Digital Scriptorium. Special Collections Library, Duke University. [Online]. Available: http://www.duke.edu/~agf2/history391/index.html [accessed Feb.13, 2003].

Fuller, P.L., L.G. Nico, and J.D. Williams. 1999. Nonindigenous Fishes Introduced Into Inland Waters of the United States. Special Pub. No. 27. Bethesda, MD: American Fisheries Society.

Gabrielson, I.N. 1943. Wildlife Refuges. New York: Macmillan.

Gahler, A.R., and W.D. Sanville. 1971. Lake sediments: Characterization of Lake Sediments and Evaluation of Sediment-Water Nutrient Interchange Mechanisms in the Upper Klamath Lake System: Lake Sediments. Corvallis, OR: Pacific Northwest Water Laboratory. 40 pp.

Gale, D.B., T.R. Hayden, L.S. Harris, and H.N. Voight. 1998. Assessment of Anadromous Fish Stocks in Blue Creek, Lower Klamath River, California, 1994–1996. Yurok Tribal Fisheries Program, Habitat Assessment and Biological Monitoring Division Technical Report 4. April 1998. 101 pp. [Online]. Available: http://www.krisweb.com/krisweb_kt/biblio/biblio.htm#General Methods [accessed Oct. 9, 2003].

Ganf, G.G. 1983. An ecological relationship between Aphanizomenon and Daphnia pulex. Aust. J. Mar. Freshwater Res. 34(5):755–773.

Ganf, G.G., and R.L. Oliver. 1982. Vertical separation of light and available nutrients as a factor causing replacement of green algae by blue-green algae in the plankton of a stratified lake. J. Ecol. 70(3):829–844.

Gearheart, R.A., J.K. Anderson, M.G. Forbes, M. Osburn, and D. Oros. 1995. Watershed Strategies for Improving Water Quality in Upper Klamath Lake, Oregon, Vol. 2. Prepared for Klamath Basin Area Office, U.S. Bureau of Reclamation, by Humboldt State University, Arcata, CA. 270 pp.

Geiger, N.S. 2001. Reassociating Wetlands with Upper Klamath Lake to Improve Water Quality. Presented at Klamath Fish and Water Management Symposium, May 22–25, 2001, Arcata, CA.

Getches, D.H. 2003. Constrains of law and policy on the management of western water. Pp. 183–234 in Water and Climate in the Western United States, W.M. Lewis, ed. Boulder, CO: University Press of Colorado.

Gilbert, C.H. 1898. The Fishes of the Klamath Basin. Bulletin of the United States Fisheries Commission 17:1–13. [Online]. Available: http://www.krisweb.com/krisweb_kt/ biblio/ biblio.htm#General Methods [accessed Oct. 9, 2003].

Gillilan, D.M., and T.C. Brown. 1997. Instream Flow Protection: Seeking Balance under Western Water Use. Covelo, CA: Island Press. 417 pp.

Golden, M.P. 1969. The Lost River Sucker. Oregon Game Commission Administrative Report. No. 1–69. 11 pp.

Gordon, C. 1883. Report on Cattle, Sheep, and Swine, Supplementary to Enumeration of Live Stock on Farms in 1880. In Report on the Production of Agriculture as Returned at the Tenth Census (June 1, 1880): Embracing General Statistics and Monographs on Cereal Production, Flour-Milling, Tobacco Culture, Manufacture and Movement of Tobacco, Meat Production. Tenth Census of the United States, 1880, Vol. 3. Washington, DC: U.S. Government Printing Office.

Griffiths, D. 1902. Forage Conditions on the Northern Border of the Great Basin, Being a Report Upon Investigations Made During July and August, 1901, in the Region Between Winnemucca, Nevada, and Ontario, Oregon. U.S. Department of Agriculture, Bureau of Plant Industry Bulletin No. 15. Washington, DC: U.S. Government Printing Office.

Grumbine, R.E. 1994. What is ecosystem management? Conserv. Biol. 8(1):27–38.

Gutermuth, B., D. Beckstrand, and C. Watson. 1998. New Earth Harvest Site Monitoring, 1996–1997. Final Report. New Earth/Cell Technology, Klamath Falls, OR.

Gutermuth, B., E. Pinkston, and D. Vogel. 2000. A-Canal Fish Entrainment During 1997 and 1998 with Emphasis on Endangered Suckers. Prepared for U.S. Bureau of Reclamation, by New Earth/Cell Tech, Klamath Falls, OR, and Natural Resource Scientists, Inc., Red Bluff, CA.

Gwynne, B.A. 1993. Investigation of Water Quality Conditions in the Shasta River, Siskiyou County. Interim Report. California Regional Water Quality Control Board, Santa Rosa, CA. September 20, 1993. 28 pp. [Online]. Available: http://www.krisweb.com/ krisweb_kt/biblio/biblio.htm#Shasta River Sub-basin [accessed Oct. 9, 2003].

Haley, K.B. 1990. Operational research and management in fishing. Pp. 3–7 in Operations Research and Management in Fishing, A.G. Rodrigues, ed. NATO ASI Series F: Applied Sciences Vol. 189. Dordrecht: Kluwer.

Hanna, R.B., and S.G. Campbell. 2000. Water Quality Modeling in the System Impact Assessment Model for the Klamath River Basin – Keno, Oregon to Seiad Valley, California. USGS Open File Report 99-113. U.S. Geological Survey, Fort Collins, CO. 84 pp.

Hardy, T.B., and R.C. Addley. 2001. Evaluation of Interim Instream Flow Needs in the Klamath River: Phase 2. Final Report. Institute for Natural Systems Engineering, Utah Water Research Laboratory, Utah State University, Logan, UT.

Hare, S.R., N.J. Mantua, and R.C. Francis. 1999. Fisheries habitat-inverse production regimes: Alaska and west coast Pacific salmon. Fisheries 24(1):6–14.

Harris, G.P. 1986. Phytoplankton Ecology: Structure, Function and Fluctuation. London: Chapman and Hall Ltd.

Hart, D.D., and N.L. Poff. 2002. A special section on dam removal and river restoration. BioScience 52(8):653–655.

Harvey, B.C., and R.J. Nakamoto. 1996. Effects of steelhead density on growth of coho salmon in a small coastal stream. Trans. Am. Fish. Soc. 125(2):237–243.

Havens, K.E., E.J. Phlips, M.F. Cichra, and B.L. Li. 1998. Light availability as a possible regulator of cyanobacteria species composition in a shallow subtropical lake. Freshwater Biol. 39(3):547–556.

Hayden, T.R., and D.B. Gale. 1999. Juvenile Salmonid Emigration Monitoring from the Hunter Creek Basin, Lower Klamath River, California, 1996–1997. Yurok Tribal Fisheries Program, Habitat Assessment. Biological Monitoring Division Technical Report 6. 63 pp.

Hayes, B.S., R.S. Shively, E.C. Janney, and G.N. Blackwood. 2002. Monitoring of Lost River and shortnose suckers at shoreline spawning areas in Upper Klamath Lake, Oregon. Pp. 37–78 in Monitoring of Lost River and Shortnose Suckers in the Upper Klamath Basin, 2001. Prepared by U.S. Geological Survey, Western Fisheries Research Center, Klamath Falls Field, Klamath Falls, OR, for the U. S. Bureau of Reclamation, Mid-Pacific Region, Klamath Area Office, Klamath Falls, OR. Contract #00AA200049

Haynal, P. 2000. Termination and tribal survival: The Klamath Tribes of Oregon. Oreg. Hist. Quart. 101(3):271–301.

Healey, M.H. 1991. Life history of chinook salmon (Oncorhynchus tshawytscha). Pp. 311–393 in Pacific Salmon Life Histories, C. Groot and L. Margolis, eds. Vancouver, BC, Canada: University of British Columbia Press.

Heinz (H. John Heinz III Center for Science, Economics, and the Environment). 2002. Dam Removal: Science and Decision Making. Washington, DC: The H. J. Heinz III Center for Science, Economics, and the Environment. [Online]. Available: http://www.heinzctr. org/publications.htm [accessed Feb.14, 2003].

Heizer, R.F. 1978. Handbook of North American Indians, Vol. 8. California. Washington, DC: Smithsonian Institution.

Helfman, G.S., and E.T. Schultz. 1984. Social transmission of behavioral traditions in a coral reef fish. Anim. Behav. 32:379–384.

Henshaw, F.F., and H.J. Dean. 1915. Surface Water Supply of Oregon: 1878–1910. Water Supply Paper 370. Washington, DC: U.S. Government Printing Office.

Hicks, M. 2002. Evaluating Standards for Protecting Aquatic Life in Washington's Surface Water Quality Standards Temperature Criteria. Draft Discussion Paper and Literature Summary. Pub. No. 00-10-070. Washington State Department of Ecology, Water Quality Program, Watershed Management Section, Olympia, WA. [Online]. Available: http://www.ecy.wa.gov/biblio/0010070.html [accessed Oct. 7, 2003].

Hilborn, R., and J. Winton. 1993. Learning to enhance salmon production: Lessons from the salmonid enhancement program. Can. J. Fish. Aquat. Sci. 50:2043–2056.

Hobday, A.J., and G.W. Boehlert. 2001. The role of coastal ocean variation in spatial and temporal patterns in survival and size of coho salmon (Oncorhynchus kisutch). Can. J. Fish. Aquat. Sci. 58(10):2021–2036.

Hoglund, P. 2003. Keep it real for Klamath Basin trout, Oregon fishing map feature. Fishing and Hunting News, ESPN Outdoors. [Online]. Available: http://www.espn.go.com/outdoors/fishing/s/f_map_OR_Klamath_basin.html [accessed June 29, 2003].

Holling, C.S., ed. 1978. Adaptive Environmental Assessment and Management. New York: Wiley.

Holt, R. 1997. Upper Klamath Lake Fish Disease Exam Report. RH-96-126. Oregon Department of Fish and Wildlife, Corvallis, OR.

Hopelain, J.S. 1998. Age, Growth, and Life History of Klamath River Basin Steelhead Trout (Oncorhynchus mykiss irideus) as Determined from Scale Analysis. Inland Fisheries Administration Report 98-3. California Department of Fish and Game, Sacramento, CA. 19 pp. [Online]. Available: http://www.krisweb.com/krisweb_kt/biblio/biblio.htm#General Methods [accessed Oct. 9, 2003].

Hopelain, J.S. 2001. Lower Klamath River Angler Creel Census with Emphasis on Upstream Migrating Fall Chinook Salmon, Coho Salmon, and Steelhead Trout during July through

October, 1983 through 1987. Inland Fisheries Administration Report 2001-1. California Department of Fish and Game, Sacramento, CA. 79 pp.

Horne, A.J. 2002. 2001 Data Summary Report. Limnological Survey During August 2001 in Support of the Pilot Oxygenation Project Upper Klamath Lake, Oregon. Findings and Recommendations for the Prevention of Fish Kills and the Restoration of the Lake. Prepared for the U.S. Department of the Interior, Bureau of Reclamation, Mid-Pacific Region, Klamath, OR, in association with Burleson Consulting, Inc., El Cerrito, CA.

Houde, E.D. 1987. Fish early life dynamics and recruitment variability. Pp. 17–29 in 10th Annual Larval Fish Conference: Proceedings, May 18–23, 1986, Miami, FL, R.D. Hoyt, ed. American Fisheries Symposium 2. Bethesda, MD: American Fisheries Society.

Houde, E.D. 1997. Patterns and consequences of selective processes in teleost early life histories. Pp. 173–196 in Early Life History and Recruitment in Fish Populations, R.C. Chambers and E.A. Trippel, eds. London: Chapman and Hall.

Howarth, R.W., F. Chan, and R. Marino. 1999. Do top-down and bottom-up controls interact to exclude nitrogen-fixing cyanobacteria from the plankton of estuaries? An exploration with a simulation model. Biogeochemistry 46(1/3):203–231.

Hughes, R.M., and J.R. Gammon. 1987. Longitudinal changes in fish assemblages and water quality in the Willamette River, Oregon. Trans. Am. Fish. Soc. 116:196–209.

Hughes, R.M., P.R. Kaufman, A.T. Herlihy, T.M. Kincaid, L. Reynolds, and D.P. Larsen. 1998. A process for developing and evaluating indices of fish assemblage integrity. Can. J. Fish. Aquat. Sci. 55(7):1618–1631.

Hydrosphere Data Products, Inc. 1993. Hydro data Regional CD-ROMS: U.S. Geological Survey Daily Values, Vols. West 1, West 2. Hydrosphere Data Products, Inc., Boulder, CO.

ILM (Integrated Land Management Working Group). 2000. Integrated Land Management on Lake National Wildlife Refuge: An Alternate Strategy Developed by the Integrated Land Management Working Group. October 30, 2000. 30 pp.

IPCC (Intergovernmental Panel on Climate Change). 2001. Working Group I. Climate Change 2001: The Scientific Basis. [Online]. Available: http://www.unep.ch/ipcc/. [accessed Feb. 14, 2003].

Ishida, Y., T. Hariu, J. Yamashiro, S. McKinnell, T. Matsuda, and H. Kaneko. 2001. Archaeological evidence of Pacific salmon distribution in northern Japan and implications for future global warming. Progr. Oceanogr. 49(1):539–550.

Jaeger, W.K. 2002. Water allocation alternatives for the upper Klamath basin. Pp. 365–391 in Water Allocation in the Klamath Reclamation Project, 2001: An Assessment of Natural Resource, Economic, Social, and Institutional Issues in the Upper Klamath Basin, W.S. Braunworth, Jr., T. Welch, and R. Hathaway, eds. Corvallis, OR: Oregon State University Extension Service. [Online]. Available: http://eesc.oregonstate.edu/agcomwebfile/edmat/html/sr/sr1037/alternatives.pdf [accessed March 10, 2003].

Janney, E.C., R. Shively, and G.N. Blackwood. 2002. Monitoring of Lost River and shortnose suckers at the Sprague River Dam Fish Ladder, Oregon. Pp. 79–105 in Monitoring of Lost River and Shortnose Suckers in the Upper Klamath Basin, 2001. Prepared by U.S. Geological Survey, Western Fisheries Research Center, Klamath Falls Field, Klamath Falls, OR, for the U. S. Bureau of Reclamation, Mid-Pacific Region, Klamath Area Office, Klamath Falls, OR. Contract #00AA200049.

Jennings, C.A., J.L. Shelton, B.J. Freeman, and G.L. Looney. 1998. Culture Techniques and Ecological Studies of the Robust Redhorse, Moxostoma robustum: Assessment of Reproductive and Recruitment Success and Incubation Temperatures and Flows, 1996 Annual Report. Prepared for Georgia Power Co., Environmental Affairs Division, Atlanta, GA.

Jessup, L.T. 1927. Report on Proposed Reflooding of a Portion of Lower Klamath Lake California. Oregon State University Archives. William L. Finley Papers, Box 7, Klamath-Reclamation 1934–1935 folder.

Jones, R.I. 1998. Phytoplankton, primary production and nutrient cycling. Pp. 145–175 in Aquatic Humic Substances: Ecology and Biogeochemistry, D.O. Hessen and L.J. Tranvik, eds. Berlin: Springer.

Jong, H.W. 1997. Evaluation of Chinook Salmon Spawning Habitat Quality in the Shasta and South Fork Trinity Rivers, 1994. Inland Fisheries Administrative Report No. 97-5. California Department of Fish and Game. 23 pp.

Kahler, T.H., P. Romi, and T.P. Quinn. 2001. Summer movement and growth of juvenile anadromous fishes in small western Washington streams. Can. J. Fish. Aquat. Sci. 58(10):1947–1956.

Kann, J. 1998. Ecology and Water Quality Dynamics of a Shallow Hypereutrophic Lake Dominated by Cyanobacteria (Aphanizomenon flos-aquae). Ph.D. Dissertation, University of North Carolina, Chapel Hill, NC.

Kann, J., and V.H. Smith. 1999. Estimating the probability of exceeding elevated pH values critical to fish populations in a hypereutrophic lake. Can. J. Fish. Aquat. Sci. 56(12): 2262–2270.

Kann, J., and W.W. Walker. 2001. Nutrient and Hydrologic Loading to Upper Klamath Lake, Oregon, 1991–1998. Prepared for the U.S. Bureau of Reclamation, Klamath Falls, OR.

Karr, J.R., K.D. Fausch, P.L. Angermeier, P.R. Yant, and I.J. Schlosser. 1986. Assessing Biological Integrity in Running Waters: A Method and Its Rationale. Special Pub. No. 5. Illinois Natural History Survey.

Keenan, S.P., R.S. Krannich, and M.S. Walker. 1999. Public perceptions of water transfers and markets: Describing differences in water use communities. Soc. Nat. Resour. 12(4): 279–292.

Kelsey, D.A., C.B. Schreck, J.L. Congleton, and L.E. Davis. 2002. Effects of juvenile steelhead on juvenile chinook salmon behavior and physiology. Trans. Am. Fish. Soc. 131(4):676–689.

Kemper, S. 2001. A flap over water. Smithsonian Magazine. September 2001. [Online]. Available: http://www.smithsonianmag.si.edu/smithsonian/issues01/sep01/poi.html [accessed Feb. 20, 2003].

Kesner, W.D., and R.A. Barnhart. 1972. Characteristics of fall-run steelhead trout (Salmo gairdneri) of the Klamath River system, with emphasis on the half-pounder. Calif. Fish Game Bull. 58(3):204–220. [Online]. Available: http://www.krisweb.com/biblio/regional/klamath_trinity/kesner.pdf [accessed Feb. 20, 2003].

Kier (William A.) Associates. 1998. Mid-Term Evaluation of the Klamath River Basin Fisheries Restoration Program. Kier Associates, Sausalito, CA.

Kier (William A.) Associates. 1999. Mid-Term Evaluation of the Klamath River Basin Fisheries Restoration Program. Prepared for The Klamath River Basin Fisheries Task Force, by Kier Associates, Sausalito and Arcata, CA. 303 pp. [Online]. Available: http://www.krisweb.com/krisweb_kt/biblio/biblio.htm#General Methods [accessed Oct. 9, 2003].

Kim, B., and R.G. Wetzel. 1993. The effect of dissolved humic substances on the alkaline phosphatase and the growth of microalgae. Verh. Int. Verein. Limnol. 25(1):129–132.

Kim, J. 2001. A nested modeling study of elevation-dependent climate change signals in California induced by increased atmospheric CO_2. Geophys. Res. Lett. 28(15): 2951–2954.

Klamath Tribe. 1991. Effects of Water Management in Upper Klamath Lake on Habitats Important to Endangered Catostomids. Internal Report. Klamath Tribe, Chiloquin, OR.

Klamath Water Users Association. 2001. Protecting the Beneficial Uses of Waters of Upper Klamath Lake: A Plan to Accelerate Recovery of the Lost River and Shortnose Suckers. Klamath Water Users Association, Klamath Falls, OR.

Knudsen, E.E., C.R. Steward, D.D. McDonald, J.E. Williams, and D.W. Reiser, eds. 2000. Sustainable Fisheries Management: Pacific Salmon. Boca Raton, FL: Lewis Publishers.

Koch, D.L., J.J. Cooper, G.P. Contreras, and V. King. 1975. Survey of Fishes of the Clear Lake Reservoir Drainage. Report No. 37. Reno, NV: Center for Water Resources Research, Desert Research Institute. 38 pp.

Kostow, K.E., A.R. Marshall, and S.R. Phelps. 2003. Naturally spawning hatchery steelhead contribute to smolt production but experience low reproductive success. Trans. Amer. Fish. Soc. 132(4):780–790.

Krist, J. 2001. Mercury Contamination Rising. Ventura County Star Special Report. May 12, 2001. [Online]. Available: http://www.insidevc.com/vcs/news/article/0,1375,VCS_ 121_ 465450,00.html [accessed Feb. 20, 2003].

Lach, D., L. Richards, C. Corson, and P. Case. 2002. Effects of the 2001 water allocation decisions on project area communities. Pp. 177–207 in Water Allocation in the Klamath Reclamation Project, 2001: An Assessment of Natural Resource, Economic, Social, and Institutional Issues in the Upper Klamath Basin, W.S. Braunworth, Jr., T. Welch, and R. Hathaway, eds. Corvallis, OR: Oregon State University Extension Service. [Online]. Available: http://eesc.orst.edu/agcomwebfile/edmat/html/sr/sr1037/sr1037.html [accessed March 10, 2003].

Langston, N.E. 1995. Forest Dreams, Forest Nightmares: The Paradox of Old Growth in the Inland West. Seattle, WA: University of Washington Press.

Langston, N.E. 2003. Where Land and Water Meet: A Western Landscape Transformed. Seattle, WA: University of Washington Press.

Lawson, P.W. 1993. Cycles in ocean productivity, trends in habitat quality, and the restoration of salmon runs in Oregon. Fisheries 18(8):6–10.

Lease, H.M. 2000. Histopathological Changes in Gills of Lost River Suckers (Deltistes luxatus) Exposed to Elevated Ammonia and Elevated pH. M.S. Thesis, University of Wyoming, Laramie, WY.

Lettenmaier, D.P., and A.F. Hamlet. 2003. Improving water-resource system performance through long-range climate forecasts: The Pacific Northwest experience. Pp. 107–122 in Water and Climate in the Western United States, W.M. Lewis, ed. Boulder, CO: University Press of Colorado.

Lettenmaier, D.P., A.W. Wood, R.N. Palmer, E.F. Woof, and E.A. Stakhiv. 1999. Water resources implications of global warming: A U.S. regional perspective. Climate Change 43(3):537–579.

Levin, P.S., R.W. Zabel, and J.G. Williams. 2001. The road to extinction is paved with good intentions: Negative association of fish hatcheries with threatened salmon. Proc. R. Soc. London Bull. Biol. Sci. 268(1472):1153–1158.

Lewis, A.G. 1992. Scott River Riparian Zone Inventory and Evaluation. Etna, CA: Siskiyou Resource Conservation District.

Lewis Jr., W.M. 1983. A revised classification of lakes based on mixing. Can. J. Fish. Aquat. Sci. 40(10):1779–1787.

Lewis Jr., W.M. 1986. Nitrogen and phosphorus runoff losses from a nutrient-poor tropical moist forest. Ecology 67(5):1275–1282.

Lewis Jr., W.M., and S.N. Levine. 1984. The light response of nitrogen fixation in Lake Valencia, Venezuela. Limnol. Oceanogr. 29(4):894–900.

Li, H.W., G.A. Lamberti, T.N. Pearson, C.K. Tait, J.L. Li, and J.C. Buckhouse. 1994. Cumulative effects of riparian disturbances along high desert trout streams of the John Day basin, Oregon. Trans. Am. Fish. Soc. 123(4):627–640.

Loftus, M. 2001. Assessment of Potential Water Quality Stress to Fish. Supplement to "Effects of Water Quality and Lake Level on the Biology and Habitat of Selected Fish Species in Upper Klamath Lake." Prepared by R2 Resources Consultants, Redmond, WA, for the Bureau of Indian Affairs, U.S. Dept. of the Interior, Portland, OR.

Lo Piccollo, M. 1962. Some Aspects of the Range Cattle Industry of Harney County, Oregon, 1870–1900. M.S. Thesis, University of Oregon, Eugene, OR.

Lorion, C.M., D.F. Markle, S.B. Reid, and M.E. Docker. 2000. Redescription of the presumed-extinct Miller Lake lamprey, Lampetra minima. Copeia 2000(4):1019–1028.

Lowney, C.L. 2000. Stream temperature variation in regulated rivers: Evidence for a spatial pattern in daily minimum and maximum magnitudes. Water Resour. Res. 36(10):2947–2955.

Lufkin, A. 2000. History: Indians Saw the Best Years. Eel River Salmon Restoration Project. The Humboldt Institute for Technological Studies, Garberville, CA. [Online]. Available: http://www.hits.org/salmon98/history.html [accessed March 20, 2003].

Lynch, D.D., and J.C. Risley. 2003. Klamath River Basin Hydrologic Conditions Prior to the September 2002 Die-Off of Salmon and Steelhead. Water-Resources Investigations Report 03-4099. U.S. Geological Survey, Department of the Interior, Portland, OR. 17 pp. [Online]. Available: oregon.usgs.gov/pubs_dir/WRIR03-4099/wri034099.pdf [accessed Oct. 7, 2003].

Mack, S. 1958. Geology and Ground-Water Features of Scott Valley, Siskiyou County, California. U.S. Geological Survey Water-Supply Paper 1462. Washington, DC: U.S. Government Printing Office. 98 pp.

Mack, S. 1960. Geology and Ground-Water Features of Shasta Valley, Siskiyou County, California. U.S. Geological Survey Water-Supply Paper 1484. Washington, DC: U.S. Government Printing Office. 115 pp.

MacIntyre, S., and J.M. Melack. 1984. Vertical mixing in Amazon floodplain lakes. Verh. Int. Verein. Limnol. 22:1283–1287.

Magill, J.R., R.E. Wells, R.W. Simpson, and A.V. Cox. 1982. Post-12 m.y. rotation of southwest Washington. J. Geophys. Res. 87:3761–3776.

Malouf, C.I., and J.M. Findlay. 1986. Euro-American impact before 1870. Pp. 499–516 in Handbook of North American Indians, Vol. 11. Great Basin, W.C. Sturtevant, and W.L. D'Azevedo, eds. Washington, DC: Smithsonian Institution.

Marino, R., F. Chan, R.W. Howarth, M. Pace, and G.E. Likens. 2002. Ecological and biogeochemical interactions constrain planktonic nitrogen fixation in estuaries. Ecosystems 5(7):719–725.

Markle, D.F. 1992. Evidence of bull trout x brook trout hybrids in Oregon. Pp. 58–67 in Proceedings of the Gearhart Mountain Bull Trout Workshop, P.J. Howell and D.V. Buchanan, eds. Corvallis, OR: American Fisheries Society.

Markle, D.F., and M.S. Cooperman. 2002. Relationships between Lost River and shortnose sucker biology and management of Upper Klamath Lake. Pp. 93–117 in Water Allocation in the Klamath Reclamation Project, 2001: An Assessment of Natural Resource, Economic, Social, and Institutional Issues in the Upper Klamath Basin, W.S. Braunworth, Jr., T. Welch, and R. Hathaway, eds. Corvallis, OR: Oregon State University Extension Service. [Online]. Available: http://eesc.orst.edu/agcomwebfile/edmat/html/sr/sr1037/sr1037.html [accessed March 10, 2003].

Markle, D.F., and D.C. Simon. 1994. Larval and Juvenile Ecology of Upper Klamath Lake Suckers, Annual Report. Department of Fisheries and Wildlife, Oregon State University, Corvallis, OR.

Markle, D.F., D.L. Hill, and C.E. Bond. 1996. Sculpin Identification Workshop and Working Guide to Freshwater Sculpins of Oregon and Adjacent Areas, Revision 1.1. Oregon State University, Corvallis, OR. 50 pp.

Marsden, W.W. 1989. Lake restoration by reducing external phosphorous loading: The influence of sediment phosphorous release. Freshwater Biol. 21(2):139–162.

Martin, B.A. 1997. Effects of Ambient Water Quality on the Endangered Lost River Sucker (Deltistes luxatus) in Upper Klamath Lake, Oregon. M.S. Thesis, Humboldt State University, Arcata, CA.

Martin, B.A., and M.K. Saiki. 1999. Effects of ambient water quality on the endangered Lost River sucker in Upper Klamath Lake, Oregon. Trans. Am. Fish. Soc. 128(5):953–961.

Mayfield, R.B. 2002. Temperature Effects on Green Sturgeon (Acipenser medirostris Ayres) Bioenergetics: An Experimental Lab Study. M.S. Thesis, University of California, Davis, CA. 41 pp.

McBain and Trush. 1997. Trinity River Flow Maintenance Study. Contract #30255. Prepared for the Hoopa Valley Tribe Fisheries Department, Hoopa, CA. [Online]. Available: http://www.krisweb.com/krisweb_kt/biblio/biblio.htm [accessed Oct. 2, 2003].

McCaffrey, R., M.D. Long, C. Goldfinger, P.C. Zwick, J.L. Nabelek, C.K. Johnson, and C. Smith. 2000. Rotation and plate locking at the southern Cascadia subduction zone. Geophys. Res. Lett. 27(19):3117–3120.

McCullough, D.A. 1999. A Review and Synthesis of Effects of Alterations of the Water Temperature Regime on Freshwater Life Stages of Salmonids, With Special Reference to Chinook Salmon. EPA910-R-99-010. Region 10, U.S. Environmental Protection Agency, Seattle, WA. 279 pp. [Online]. Available: http://www.krisweb.com/krisweb_ kt/biblio/biblio.htm#General Methods Documents [accessed Oct. 9, 2003].

McLain, R.J., and R.G. Lee. 1996. Adaptive management: Promises and pitfalls. Environ. Manage. 20(4): 437–448.

McMichael, G.A., T.N. Pearsons, and S.A. Leider. 1999. Behavioral interactions among hatchery-reared steelhead smolts and wild Oncorhynchus mykiss in natural streams. N. Am. J. Fish. Manage. 19(4):948–956.

Miller, B.A., and S. Sadro. 2003. Residence time and seasonal movements of juvenile coho salmon in the ecotone and lower estuary of Winchester Creek, South Slough, Oregon. Trans. Am. Fish. Soc. 132(3):546–559.

Miller, R.R., and G.R. Smith. 1981. Distribution and evolution of Chasmistes (Pisces: Catostomidae) in western North America. Occasional Papers of the Museum of Zoology 696:1–46.

Minckley, W.L., D.A. Hendrickson, and C.E. Bond. 1986. Geography of western North American freshwater fishes: Description and relationships to intercontinental tectonism. Pp. 519–614 in Zoogeography of North American Freshwater Fishes, C.H. Hocutt and E.O. Wiley, eds. New York: Wiley.

Minckley, W.L., P.C. Marsh, J.E. Brooks, J.E. Johnson, and B.L. Jensen. 1991. Management toward recovery of the razorback sucker. Pp. 303–357 in Battle against Extinction: Native Fish Management in the American West, W.L. Minckley and J.E. Deacon, eds. Tucson, AZ: University of Arizona Press.

Moffett, J.W., and S.H. Smith. 1950. Biological Investigations of the Fishery Resources of the Trinity River, California. Special Scientific Report No. 12. Washington, DC: U.S. Fish and Wildlife Service. 71 pp. [Online]. Available: http://www.krisweb.com/krisweb_kt/biblio/biblio.htm#Trinity River Basin [accessed Oct. 9, 2003].

Morris, D.P., and W.M. Lewis, Jr. 1988. Phytoplankton nutrient limitation in Colorado mountain lakes. Freshwater Biol. 20:315–327.

Morris, D.P., H. Zagarese, C.E. Williamson, E.G. Balseiro, B.R. Hargreaves, B. Modenutti, R. Moeller, and C. Queimalinos. 1995. The attenuation of solar UV radiation in lakes and the role of dissolved organic carbon. Limnol. Oceanogr. 40(8):1381–1391.

Morris, J.M., M.J. Suedkamp, E. Snyder-Conn, R.A. Holt, H.M. Lease, S.J. Clearwater, and J.S. Meyer. 2000. Survival and Growth of Juvenile Lost River Suckers (Deltistes luxatus) Challenged with a Bacterial Pathogen (Flavobacterium columnare) During Exposure to Sublethal Ammonia Concentrations at pH 9.5. Abstract of Paper Presented at Annual Meeting of Rocky Mountain Chapter of Society of Environmental Toxicology and Chemistry, April 7, 2000, Laramie, WY.

Mortimer, C.H. 1941. The exchange of dissolved substances between mud and water in lakes. J. Ecol. 29(2):280–329.

Mortimer, C.H. 1942. The exchange of dissolved substances between mud and water in lakes. J. Ecol. 30(1):147–201.

Mount, J.F. 1995. California Rivers and Streams: The Conflict between Fluvial Processes and Land Use. Berkeley, CA: University of California Press.

Moyle, P.B. 2002. Inland Fishes of California. Berkeley, CA: University of California Press. 502 pp.

Moyle, P.B., and R.A. Leidy. 1992. Loss of biodiversity in aquatic ecosystems: Evidence from fish faunas. Pp. 127–169 in Conservation Biology: The Theory and Practice of Nature Conservation, Preservation and Management, P.L. Fiedler and S.K. Jain, eds. New York: Chapman and Hall.

Moyle, P.B., and T. Light. 1996. Biological invasions of fresh water: Empirical rules and assembly theory. Biol. Conserv. 78(1):149–161.

Moyle, P.B., R.M. Yoshiyama, J.E. Williams, and E.D. Wikramanayake. 1995. Fish Species of Special Concern of California, 2nd Ed. California Department of Fish and Game, Sacramento, CA.

Mueller, G., and P. Marsh. 1995. Bonytail and razorback sucker in the Colorado River Basin. In Our Living Resources, a Report to the Nation on the Distribution, Abundance, and Health of U.S. Plants, Animals, and Ecosystems. U.S. Department of the Interior, National Biological Service. [Online]. Available: http://biology.usgs.gov/s+t/noframe/r166.htm [accessed Oct. 3, 2003].

Myers, J.M., R.G. Kope, G.J. Bryant, D. Teel, L.J. Lierheimer, T.C. Wainwright, W. S. Grant, F.W. Waknitz, K. Neely, S.T. Lindley, and R.S. Waples. 1998. Status Review of Chinook Salmon from Washington, Idaho, Oregon, and California. NOAA Technical Memorandum NMFS-NWFSC-35. National Marine Fisheries Service, National Oceanic and Atmospheric Administration, U.S. Dept. of Commerce, Seattle, WA. 443 pp. [Online]. Available: http://www.nwfsc.noaa.gov/publications/techmemos/tm35/index.htm [accessed Oct. 9, 2003].

Myrick, C.A., and J.J. Cech, Jr. 2001. Temperature Effects on Chinook Salmon and Steelhead: A Review Focusing on California's Central Valley Populations. Technical Publication 01-1. Sacramento, CA: Bay-Delta Modeling Forum. [Online]. Available: http://www.cnr.colostate.edu/~camyrick/Publications.html [accessed Feb. 28, 2003].

Nagle, J.C., and J.B. Ruhl. 2002. The Law of Biodiversity and Ecosystem Management. Foundation Press. New York: Foundation Press.

NAST (National Assessment Synthesis Team). 2001. Climate Change Impacts on the United States: Foundation Report: The Potential Consequences of Climate Variability and Change. Cambridge: Cambridge University Press. 618 pp.

Nelson, B. 1988. Our Home Forever: The Hupa Indians of Northern California. Salt Lake City, UT: Howe Brothers Press.

Nickelson, T.E., M.F. Solazzi, and S.L. Johnson. 1986. Use of hatchery coho salmon (Oncorhynchus kisutch) presmolts to rebuild wild populations in Oregon coastal streams. Can. J. Fish. Aquat. Sci. 43:2443–2449.

Nickelson, T.E., M.F. Solazzi, J.D. Rodgers, and S.L. Johnson. 1992a. Effectiveness of selected stream improvement techniques to create suitable summer and winter rearing habitat for juvenile coho salmon (Oncorhynchus kisutch) in Oregon coastal streams. Can. J. Fish. Aquat. Sci. 49(4):790–794.

Nickelson, T.E., J.D. Rodgers, S.L. Johnson, and M.F. Solazzi. 1992b. Seasonal changes in habitat use by juvenile coho salmon (Oncorhynchus kisutch) in Oregon coastal streams. Can. J. Fish. Aquat. Sci. 49(4):783–789.

NMFS (National Marine Fisheries Service). 2001. Biological Opinion. Ongoing Klamath Project Operations. National Marine Fisheries Service, Southwest Region, National Oceanic and Atmospheric Administration, Long Beach, CA. April 6, 2001 [Online]. Available: http://www.mp.usbr.gov/kbao/esa/38_cohobo_4_6-01.pdf [accessed March 5, 2003].

NMFS (National Marine Fisheries Service). 2002. Biological Opinion. Klamath Project Operations. National Marine Fisheries Service, Southwest Region, National Oceanic and Atmospheric Administration, Long Beach, CA. May 31, 2002. [Online]. Available: http://swr.ucsd.edu/psd/kbo.pdf; also available through the National Research Councils Public Access File

NRC (National Research Council). 1992. Water Transfers in the West: Efficiency, Equity, and the Environment. Washington, DC: National Academy Press.

NRC (National Research Council). 1995. Science and the Endangered Species Act. Washington, DC: National Academy Press.

NRC (National Research Council). 1996. Upstream: Salmon and Society in the Pacific Northwest. Washington, DC: National Academy Press. 242 pp.

NRC (National Research Council). 1999. Downstream: Adaptive Management of Glen Canyon Dam and the Colorado River Ecosystem. Washington, DC: National Academy Press.

NRC (National Research Council). 2001. Climate Change Science: An Analysis of Some Key Questions. Washington, DC: National Academy Press.

NRC (National Research Council). 2002. Scientific Evaluation of Biological Opinions on Endangered and Threatened Fishes in the Klamath River Basin. Interim Report. Washington, DC: National Academy Press.

NRCS (Natural Resources Conservation Service). 2003. Environmental Quality Incentives Program. Natural Resources Conservation Service, U.S. Department of Agriculture. [Online]. Available: http://www.nrcs.usda.gov/programs/eqip/ [accessed June 24, 2003].

OECD (Organization for Economic Co-operation and Development). 1982. Eutrophication of Waters: Monitoring, Assessment and Control. Paris, France: OECD.

Ogden, P.S. 1971. Snake Country Journals, 1827–28 and 1828–29, G. Williams, ed. Publications of the Hudson's Bay Record Society 28. London: Hudson's Bay Record Society.

Olson, A.D., and O.J. Dix. 1993. Lower Salmon River Sub-Basin Fish Habitat Condition and Utilization Assessment 1990/1991. Final Report. U.S. Dept. of Agriculture Forest Service, Klamath National Forest, Yreka, CA. 41 pp.

O'Neal, K. 2002. Effects of Global Warming on Trout and Salmon in U. S. Streams. Defenders of Wildlife and Natural Resources Defense Council, Washington, DC. [Online]. Available: http://www.defenders.org/publications/fishreport.pdf [accessed Feb. 24, 2003].

OWRD (Oregon Water Resources Department. 2000. Resolving the Klamath: Klamath Basin Adjudication. Oregon Water Resources Department, Salem, OR. [Online]. Available: http://www.wrd.state.or.us/programs/klamath/summary/textonly.shtml [accessed Feb. 24, 2003].

PCFFA (Pacific Coast Federation of Fisherman's Associations). 2002. Myths and Facts about the Klamath Water Issues. Klamath Basin Water Issues, Economic Impacts and Salmon Recovery. Pacific Coast Federation of Fisherman's Associations. [Online]. Available: http://www.pcffa.org/kl-myths.htm [accessed July 1, 2003].

PacifiCorp. 2000. Klamath Hydroelectric Project. First Stage Consultation Document. FERC Project No. 2082. PacifiCorp, Portland, OR. December 15, 2000.

Pacific Watershed Associates. 1994. Action Plan for Restoration of the South Fork Trinity River Watershed and its Fisheries. Prepared for U.S. Bureau of Reclamation and the Trinity River Task Force. Contract No. 2-CS-20-01100. Pacific Watershed Associates, Arcata, CA. [Online]. Available: http://www.krisweb.com/biblio/general/pwa/pwa23.htm [accessed Feb. 25, 2003].

Parson, E.A., P.W. Mote, A. Hamlet, N. Mantua, A. Snover, W. Keeton, E. Miles, D. Canning, and K.G. Ideker. 2001. Potential consequences of climate variability and change for the Pacific Northwest. Pp. 247–280 in Climate Change Impacts on the U.S., the Potential Consequences of Climate Variability and Change. National Assessment Synthesis Team, U.S. Global Change Research Program, Washington, DC. [Online]. Available: http://www.usgcrp.gov/usgcrp/Library/nationalassessment/09PNW.pdf [accessed March 8, 2003].

Pearcy, W.G. 1992. Ocean Ecology of North Pacific Salmonids. Washington Sea Grant Program, University of Washington Press, Seattle. 179 pp.

Pearsons, T.N., and C.W. Hopley. 1999. Fisheries management: A practical approach for assessing ecological risks associated with fish stocking programs. Fisheries 24(9):16–23.

Perkins, D.L., and G.G. Scoppettone. 1996. Spawning and Migration of Lost River Suckers (Deltistes luxatus) and Shortnose Suckers (Chasmistes brevirostris) in the Clear Lake Drainage, Modoc County, California. National Biological Service—California Science Center, Reno Field Station, Reno, NV. 52 pp.

Perkins, D., G.G. Scoppettone, and M. Buettner. 2000a. Reproductive Biology and Demographics of Endangered Lost River and Shortnose Suckers in Upper Klamath Lake, Oregon. Report to the Bureau of Reclamation. Reno, NV, by U.S. Geological Survey-Biological Resources Division, Western Fisheries Science Center, Reno Field Station, Reno, NV. 52 pp. [Online]. Available: http://www.mp.usbr.gov/kbao/esa/UKL%20 SPAWNING-REPORT%20VERSION.pdf [accessed March 5, 2003].

Perkins, D.J., J. Kann, and G.G. Scoppettone. 2000b. The Role of Poor Water Quality and Fish Kills in the Decline of Endangered Lost River and Shortnose Suckers in the Upper Klamath Lake. Final Report. Contract 4-AA-29-12160. Prepared for the U.S. Bureau of Reclamation, Klamath Falls Project Office, Klamath Falls, OR, by U.S. Geological Survey, Biological Resources Division, Western Fisheries Science Center, Reno Field Station, Reno, NV. [Online]. Available: http://www.mp.usbr.gov/kbao/esa/Fish_Kill.pdf [accessed Feb. 25, 2003].

Pierce, R. 1991. The Lower Klamath fishery: Recent times. Pp. 142–150 in California's Salmon and Steelhead: A Struggle to Restore an Imperiled Resource, A. Lufkin, ed. Berkeley, CA: University of California Press.

Plunkett, S.R., and E. Snyder-Conn. 2000. Anomalies of Larval and Juvenile Shortnose and Lost River Suckers in the Upper Klamath Lake, Oregon. U.S. Fish and Wildlife Service, Klamath Falls, OR. 26 pp.

Poff, N.L., J.D. Allan, M.B. Bain, J.R. Karr, K.L. Prestegaard, B.D. Richter, R.E. Sparks, and J.C. Stromberg. 1997. The natural flow regime: A paradigm for river conservation and restoration. BioScience 47(11):769–784.

Poff, N.L., J.D. Allan, M.A. Palmer, D.D. Hart, B.D. Richter, A.H. Arthington, K.H. Rogers, J.L. Meyer, and J.A. Stanford. 2003. River flows and water wars: Emerging science for environmental decision making. Front. Ecol. Environ. 1(6): 298–306.

Power, J.H. 2001. Letters: Scott River flows. California Agriculture 55(2):4.

Prager, M.H., and M.S. Mohr. 2001. The harvest rate model for Klamath River fall chinook salmon, with management applications and comments on model development and documentation. N. Am. J. Fish. Manage. 21(3):533–547.

PWA (Philip Williams & Associates). 2001. Evaluation of Proposed Lake Management on Hydrodynamics, Water Quality and Eutrophication in Upper Klamath Lake. Report PWA #1412. Prepared by Philip Williams & Associates, Portland, OR, and Corte Madera, CA, for U.S. Bureau of Reclamation, Klamath Falls, OR. [Online]. Available: http://www.mp.usbr.gov/kbao/esa/UKL-Report26_3-29-01.pdf [accessed March 5, 2003].

Railsback, S.F., and B.C. Harvey. 2001. Individual-Based Model Formulation for Cutthroat Trout, Little Jones Creek, California. General Technical Report PSW-GTR-182. Pacific Southwest Research Station, Forest Service, U.S. Dept. of Agriculture, Albany, CA. 80 pp.

Rantz, S.E. 1972. Runoff Characteristics of California Streams. U.S. Geological Survey Water-Supply Paper No. 2009-A. Washington, DC: U.S. Goernment Printing Office.

Ratliff, D.E., and P.J. Howell. 1992. The status of bull trout populations in Oregon. Pp. 10–17 in Proceedings of the Gearhart Mountain Bull Trout Workshop, P.J. Howell, and D.V. Buchanan, eds. Corvallis, OR: American Fisheries Society.

Redmond, K.T. 2003. Climate variability in the west: Complex spatial structure associated with topography, and observational issues. Pp. 29–49 in Water and Climate in the Western United States, W.M. Lewis, ed. Boulder, CO: University Press of Colorado.

Reiser, D.W., M. Loftus, D. Chapman, E. Jeanes, and K. Oliver. 2001. Effects of Water Quality and Lake Level on the Biology and Habitat of Selected Fish Species in Upper Klamath Lake. Prepared for the Bureau of Indian Affairs.

Reynolds, C.S. 1971. The ecology of the planktonic blue-green algae in the north Shropshire meres. Field Studies 3:409–432.

Reynolds, C.S. 1984. The Ecology of Freshwater Phytoplankton. Cambridge: Cambridge University Press.

Reynolds, C.S. 1993. Scales of disturbance and their role in plankton ecology. Hydrobiologia 249(1/3):157–171.

Rhodes, J.S., and T.P. Quinn. 1998. Factors affecting the outcome of territorial contests between hatchery and naturally reared coho salmon parr in the laboratory. J. Fish Biol. 53(6):1220–1230.

Ricker, S.J. 1997. Evaluation of Salmon and Steelhead Spawning Habitat Quality in the Shasta River Basin, 1997. Inland Fisheries Administrative Report No. 97-9. Sacramento, CA: California Dept. of Fish and Game. 13 pp.

Risley, J.C., and A. Laenen. 1999. Upper Klamath Lake Basin Nutrient Loading Study—Assessment of Historic Flows in the Williamson and Sprague Rivers. Survey Water Resources Investigation Report 98-4198. Portland, OR: U.S. Geological Survey, U.S. Dept. of Interior. 22 pp.

Robins, C.R., and R.R. Miller. 1957. Classification, variation, and distribution of the sculpins, genus Cottus, inhabiting Pacific slope waters in California and southern Oregon, with a key to the species. Calif. Fish Game Bull. 43(3):213–233.

Ruhl, J.B. 1995. Section 7(a)(1) of the "new" Endangered Species Act: Rediscovering and redefining the untapped power of federal agencies' duty to conserve species. Environ. Law 25(4):1107–1163.

Ruhl, J.B. 2000. Ecosystem management, the ESA, and the seven degrees of relevance. Nat. Resour. Environ. 14:156–161.

Ruppert, J.B., R.T. Muth, and T.P. Nesler. 1993. Predation on fish larvae by adult red shiner, Yampa and Green rivers, Colorado. The Southwestern Naturalist 38(4):397–399.

Russell, I.C. 1903. Notes on the Geology of Southwestern Idaho and Southeastern Oregon. Bulletin No. 217. Washington, DC: U.S. Government Printing Office.

Ruzycki, J.R., D.A. Beauchamp, and D.L. Yule. 2003. Effects of introduced lake trout on native cutthroat trout in Yellowstone Lake. Ecol. Appl. 13(1):23–37.

Saiki, M.K., D.P. Monda, and B.L. Bellerud. 1999. Lethal levels of selected water quality variables to larval and juvenile Lost River and shortnose suckers. Environ. Pollut. 105(1):37–44.

Salmon River Restoration Council. 2002. Salmon River Restoration Council, Sawyers Bar, CA. [Online]. Available: http://www.srrc.org/ [accessed Feb. 26, 2003].

Sandercock, F.K. 1991. Life history of coho salmon. Pp. 395–446 in Pacific Salmon Life Histories, C. Groot, and L. Margolis, eds. Vancouver: University of British Columbia Press.

Schindler, D.W., R.W. Newbury, K.G. Beatty, and P. Campbell. 1976. Natural water and chemical budgets for a small Precambrian lake basin in central Canada. J. Fish. Res. Board Can. 33:2526–2543.

Scoppettone, G.G. 1986. Upper Klamath Lake, Oregon, Catostomid Research. Completion report. U.S. Fish and Wildlife Service, National Fisheries Research Center, Seattle, WA.

Scoppettone, G.G. 1988. Growth and longevity of the Cui-ui and longevity of other catostomids and cyprinids in western North America. Trans. Am. Fish. Soc. 117: 301–307.

Scoppettone, G.G., and G.L. Vinyard. 1991. Life history and management of four endangered lacustrine suckers. Pp. 359–377 in Battle Against Extinction: Native Fish Management in the American West, W.L. Minckley and J.E. Deacon, eds. Tucson, AZ: University of Arizona Press.

Scoppettone, G.G., S. Shea, and M.E. Buettner. 1995. Information on Population Dynamics and Life History of Shortnose Suckers (Chasmistes brevirostris) and Lost River Suckers (Deltistes luxatus) in Tule and Clear Lakes. National Biological Service, Reno Field Station, Reno, NV.

Scott River Watershed Coordinated Resource Management Planning Council. 1997. Scott River Watershed Fish Population and Habitat Plan. Scott River Watershed Coordinated Resource Management Planning Council, Etna, CA. 25 pp.

Shively, R.S., A.E. Kohler, B.J. Peck, M.A. Coen, and B.S. Hayes. 2000a. Water Quality, Benthic Macroinvertebrates, and Fish Community Monitoring in the Lost River Sub-basin, Oregon and California, 1999. Annual Report 1999. U.S. Geological Survey, Biological Resources Division, Western Fisheries Research Center, Klamath Falls Duty Station, Klamath Falls, OR, Johnson Controls Wold Services Inc, NERC Operation, Fort Collins, CO, and U.S. Bureau of Reclamation, Mid-Pacific Region, Klamath Falls Area Office, Klamath Falls, OR. [Online]. Available: http://www.mp.usbr.gov/kbao/esa/Lostriverrpt.pdf [accessed March 5, 2003].

Shively, R.S., E.B. Neuman, A.E. Kohler, and B.J. Peck. 2000b. Species composition and distribution of fishes in the Lost River, Oregon. Pp. 72–92 in Water Quality, Benthic Macroinvertebrates, and Fish Community Monitoring in the Lost River Sub-basin, Oregon and California, 1999. Annual Report 1999, R.S. Shively, A.E. Kohler, B.J. Peck, M.A. Coen, and B.S. Hayes, eds. U.S. Geological Survey, Biological Resources Division, Western Fisheries Research Center, Klamath Falls Duty Station, Klamath Falls, OR, Johnson Controls Wold Services Inc, NERC Operation, Fort Collins, CO, and U.S. Bureau of Reclamation, Mid-Pacific Region, Klamath Falls Area Office, Klamath Falls, OR. [Online]. Available: http://www.mp.usbr.gov/kbao/esa/Lostriverrpt.pdf [accessed March 5, 2003].

Simon, D.C., and D.F. Markle. 1997a. Inter-annual abundance of non-native fathead minnows (Pimephales promelas) in Upper Klamath Lake, Oregon. Great Basin Nat. 57:142–148.

Simon, D.C., and D.F. Markle. 1997b. Larval and Juvenile Ecology of Upper Klamath Lake Suckers, Annual Report: 1996. Department of Fisheries and Wildlife, Oregon State University, Corvallis, OR.

Simon, D.C., and D.F. Markle. 2001. Ecology of Upper Klamath Lake Shortnose and Lost River suckers. Annual Survey of Abundance and Distribution of Age 0 Shortnose and Lost River Suckers in Upper Klamath Lake, Annual Report: 2000. Oregon Cooperative Research Unit, Department of Fisheries and Wildlife, Corvallis, OR.

Simon, D.C., G.R. Hoff, D.J. Logan, and D.F. Markle. 1996. Larval and Juvenile Ecology of Upper Klamath Lake Suckers. Annual Report: 1995. Department of Fisheries and Wildlife, Oregon State University, Corvallis, OR. 60 pp.

Simon, D.C., M.R. Terwilliger, P. Murtaugh, and D.F. Markle. 2000. Larval and Juvenile Ecology of Upper Klamath Lake Suckers: 1995–1998. Final Report. Prepared by Oregon State University, Corvallis, OR, for U.S. Bureau of Reclamation, Klamath Falls, OR.

January 7, 2000. 108 pp. [Online]. Available: http://www.mp.usbr.gov/kbao/esa/95-98report_4.pdf [accessed March 5, 2003].

Simpson, P.K. 1987. The Community of Cattlemen: A Social History of the Cattle Industry in Southeastern Oregon 1869–1912. Moscow, ID: University of Idaho Press.

Smith, J.J., and H.W. Li. 1983. Energetic factors influencing foraging tactics of juvenile steelhead trout, Salmo gairdneri. Pp. 173–180 in Predators and Prey in Fishes, D.L.G. Noakes, D.G. Lindquist, G.S. Helfman, and J.A. Ward, eds. The Hague: W. Junk.

Snyder, D.T., and J.L. Morace. 1997. Nitrogen and Phosphorus Loading from Drained Wetlands Adjacent to Upper Klamath and Agency Lakes, Oregon. Water-Resources Investigations Report 97-4059. Portland, OR: U.S. Dept. of Interior, U.S. Geological Survey. 67 pp. [Online]. Available: http://www.krisweb.com/krisweb_kt/biblio/biblio.htm# Upper Klamath Basin [accessed Oct. 9, 2003].

Snyder, J.O. 1931. Salmon of the Klamath River, California. California Department of Fish and Game Bulletin 34. Sacramento, CA: California State Print. Off. 130 pp. [Online]. Available: http://www.krisweb.com/krisweb_kt/biblio/biblio.htm#General Methods [accessed Oct. 9, 2003].

Snyder, J.O. 1933. A steelhead migration in the Shasta River. Calif. Fish Game Bull. 19(4): 252–254.

Snyder, M.A., J.L. Bell, L.C. Sloan, P.B. Duffy, and B. Govindasamy. 2002. Climate responses to a doubling of atmospheric carbon dioxide for a climatically vulnerable region. Geophys. Res. Lett. 29(11):9.

Sommerstram, S., E. Kellogg, and J. Kellogg. 1990. Scott River Watershed Granitic Sediment Study. Prepared by Tierra Data Systems, for Siskiyou Resource Conservation District, Etna, CA. [Online]. Available: http://www.krisweb.com/krisweb_kt/biblio/biblio.htm# Scott River Sub-basin [accessed Oct. 9, 2003].

Sorte, B., and B. Wyse. In press. The Lower Klamath Basin Economy and the Roles of Agriculture and Commercial Fishing. Special Report. Oregon State University Extension Service, Corvallis, OR.

SRCD (Siskiyou Resource Conservation District). 2001. Water Temperatures in the Scott River Watershed in Northern California. Prepared for the U.S. Fish and Wildlife Service, Etna, CA. 49 pp.

Stern, M.A. 1990. Strategies for Improving Fish Passage for the Lost River and Shortnose Sucker at the Chiloquin Dam. Report to U.S. Fish and Wildlife Service, Sacramento Field Station, Sacramento, CA, by Oregon Natural Heritage Program, The Nature Conservancy, Portland, OR. November.

Stern, T. 1965. The Klamath Tribe: A People and Their Reservation. Seattle, WA: University of Washington Press. 356 pp.

Strzepek, K.M., and D.N. Yates. 2003. Assessing effects of climate change on the water resources of the western United States. Pp. 93–106 in Water and Climate in the Western United States, W.M. Lewis, ed. Boulder, CO: University Press of Colorado.

Stubbs, K., and R. White. 1993. Lost River (Deltistes luxatus) and Shortnose (Chasmistes brevirostris) Sucker Recovery Plan. U.S. Fish and Wildlife Service, Portland, OR.

Taft, A.C., and L. Shapovalov. 1935. A Biological Survey of Streams and Lakes in the Klamath and Shasta National Forests in California. Washington, DC: U.S. Department of Commerce, Bureau of Fisheries.

Thomas, J.W., and the Scientific Analysis Team. 1993. Viability Assessments and Management Considerations for Species Associated with Late-Successional and Old-Growth Forests of the Pacific Northwest. Portland, OR: U.S. Dept. of Agriculture, National Forest System, Forest Service Research. 530 pp.

Thompson, D.E., J.E. Lauman, J.D. Fortune, and R. Todd. 1989. Controlled Grazing for Riparian Zone Rehabilitation. Oregon State University Extension Office, Klamath Falls, OR. 99 pp.

Thornton, R.D. 2001. Habitat conservation plans: Frayed safety nets or creative partnerships? Nat. Resour. Environ. 16(2):94–101.

Thurman, E.M. 1985. Organic Geochemistry of Natural Waters. Dordrecht: M. Nijhoff.

Tranah, G.J. 2001. Molecular Genetic Analysis of Hybridization and Population Structure in Endangered Sturgeon and Sucker Species. Ph.D. Dissertation, University of California, Davis, CA.

Trihey and Associates, Inc. 1996. Instream Flow Requirements for Tribal Trust Species in the Klamath River. Prepared by Trihey and Associates, Inc., Concord, CA, for the Yurok Tribe, Eureka, CA. 43 pp. [Online]. Available: http://www.krisweb.com/krisweb_kt/biblio/biblio.htm#General Methods [accessed Oct. 9, 2003].

Trippel, E.A. 1995. Age at maturity as a stress indicator in fisheries. BioScience 45(11):759–771.

USBR (U.S. Bureau of Reclamation). 1994. Biological Assessment on Long-Term Operations of the Klamath Project, with Special Emphasis on Clear Lake Operations. U.S. Bureau of Reclamation, Klamath Basin Area Office, Klamath Falls, OR.

USBR (U.S. Bureau of Reclamation). 1999. Bathymetry of Upper Klamath and Agency Lake provided in GIS format. U.S. Bureau of Reclamation, Mid-Pacific Region, Klamath Basin Area Office, OR. (as cited in Welch and Burke 2001).

USBR (U.S. Bureau of Reclamation). 2000a. Inventory of Water Diversions in the Klamath Project Service Area that Potentially Entrain Endangered Lost River and Shortnose Suckers, Draft. U.S. Bureau of Reclamation, Mid-Pacific Region, Klamath Basin Area Office, OR. January 19, 2000.

USBR (U.S. Bureau of Reclamation). 2000b. Klamath Project Historical Operation. U.S. Bureau of Reclamation, Mid-Pacific Region, Klamath Basin Area Office, Klamath Falls, OR.

USBR (U.S. Bureau of Reclamation). 2001a. Biological Assessment of Klamath Project's Continuing Operations on the Endangered Lost River Sucker and Shortnose Sucker. U.S. Department of the Interior, U.S. Bureau of Reclamation, Mid-Pacific Region, Klamath Basin Area Office, Klamath Falls, OR. February 13, 2001. [Online]. Available: http://www.mp.usbr.gov/kbao/esa/34_final_sucker_bo_4_06_01.pdf [accessed August 18, 2001].

USBR (U.S. Bureau of Reclamation). 2001b. Biological Assessment of the Klamath Project's Continuing Operations on Southern Oregon/Northern California ESU Coho Salmon and Critical Habitat for Southern Oregon/Northern California ESU Coho Salmon. U.S. Department of the Interior, Bureau of Reclamation, Mid-Pacific Region, Klamath Basin Area Office, OR. 53 pp. [Online]. Available: http://www.mp.usbr.gov/kbao [accessed August 18, 2001]

USBR (U.S. Bureau of Reclamation). 2002a. Final Biological Assessment: The Effects of Proposed Actions Related to Klamath Project Operation (April 1, 2002 – March 31, 2012) on Federally-Listed Threatened and Endangered Species. U.S. Bureau of Reclamation, Mid-Pacific Region, Klamath Basin Area Office, Klamath Falls, OR.

USBR (U.S. Bureau of Reclamation). 2002b. Summary of Water Quality and Fisheries Sampling Conducted in Gerber Reservoir, OR. U.S. Department of the Interior, Bureau of Reclamation, Mid-Pacific Region. Klamath Basin Area Office, Klamath Falls, OR. October, 2002. 70 pp.

USBR (U.S. Bureau of Reclamation). 2003. Klamath River Water Quality 2000 Monitoring Program: Project Report. U.S. Bureau of Reclamation, Klamath Falls Area Office, with support from PacifiCorp., Watercourse Engineering, Inc., Napa, CA. January 25, 2003.

USDA (U.S. Department of Agriculture). 1999. Census of Agriculture: 1997. U.S. Summary and State Data, Vol.1. Geographic Area Series Part 51, Introduction. AC97-A-51. National Agricultural Statistics Service, U.S. Department of Agriculture. [Online]. Available: http://www.nass.usda.gov/census/census97/volume1/us1into.pdf [accessed Feb.12, 2003].

USFS (U.S. Forest Service). 2002. Scott River Watershed Adult Coho Spawning Survey, December 2001 – January 2002. Klamath Forest Service, Fort Jones, CA. 27 pp.

USFWS (U.S. Fish and Wildlife Service). 1991. Young Salmon are Sensitive to Arsenite and Mercury Released by Placer Mining Activities. Research Information Bulletin No. 30. Columbia Environmental Research Center, Columbia, MO.

USFWS (U.S. Fish and Wildlife Service). 1992a. Biological Opinion on Effects of Long-term Operation of the Klamath Project. U.S. Fish and Wildlife Service, Klamath Falls, OR.

USFWS (U.S. Fish and Wildlife Service). 1992b. Cui-ui (Chasmistes cujus) Recovery Plan, Second Revision. U.S. Fish and Wildlife Service, Region 1, Portland, OR. 47 pp.

USFWS (U.S. Fish and Wildlife Service). 1994. Federal Interagency Memorandum of Understanding for Implementation of the Endangered Species Act. September 28, 1994.

USFWS (U.S. Fish and Wildlife Service). 1999. Environmental Impact Statement/Report. Trinity River Mainstem Fishery Restoration, Draft Environmental Impact Report. U.S. Fish and Wildlife Service, Arcata, CA. [Online]. Available: http://www.ccfwo.r1.fws.gov/fisheries/treis.html [accessed Oct. 6, 2003].

USFWS (U.S. Fish and Wildlife Service). 2000. Record of Decision, Trinity River Mainstem Fishery Restoration, Final Environmental Impact Statement/Environmental Impact Report, U.S. Department of the Interior, Washington, DC. [Online]. Available: http://www.ccfwo.r1.fws.gov/fisheries/reports/treis/ROD12-19-00(b).pdf [accessed August 22, 2003].

USFWS (U.S. Fish and Wildlife Service). 2001. Biological/Conference Opinion Regarding the Effects of Operation of the Bureau of Reclamation's Klamath Project on the Endangered Lost River Sucker (Deltistes luxatus), Endangered Shortnose Sucker (Chasmistes brevirostris), Threatened Bald Eagle (Haliaeetus leucocephalus), and Proposed Critical Habitat for the Lost River/Shortnose Suckers. U.S. Fish and Wildlife Service, Klamath Falls Fish and Wildlife Office, Klamath Falls, OR.

USFWS (U.S. Fish and Wildlife Service). 2002. Biological/Conference Opinion Regarding the Effects of Operation of the U.S. Bureau of Reclamation's Proposed 10-Year Operation Plan for the Klamath Project and its Effect on the Endangered Lost River Sucker (Deltistes luxatus), Endangered Shortnose Sucker (Chasmistes brevirostris), Threatened Bald Eagle (Haliaeetus leucocephalus), and Proposed Critical Habitat for the Lost River and Shortnose Suckers. U.S. Fish and Wildlife Service, Klamath Falls Fish and Wildlife Office, Klamath Falls, OR.

USFWS/HVT (U.S. Fish and Wildlife Service and Hoopa Valley Tribe). 1999. Trinity River Flow Evaluation. Final Report. U.S. Fish and Wildlife Service, Arcata Fish and Wildlife Office, and Hoopa Valley Tribe, U.S. Department of the Interior, Washington, DC.

USGS (U.S. Geological Survey). 1995. Klamath River Basin Characterization of Hydrology Data from U.S.G.S. Records. In Compilation of Phase I Reports for the Klamath River Basin, May 1995. River Systems Management Section, National Biological Service, Mid-Continent Ecological Science Center, Fort Collins, CO.

USGS (U.S. Geological Survey). 1998. Status of Listed Species and Recovery Plan Development June Sucker Chasmistes liorus-Endangered, Utah. Northern Prairie Wildlife Research Center, U.S. Geological Survey. [Online]. Available: http://www.npwrc.usgs.gov/resource/distr/others/recoprog/states/species/chaslior.htm [accessed June 23, 2003].

USGS (U.S. Geological Survey). 2002. Monitoring of Lost River and Shortnose Suckers in the Upper Klamath Basin, 2001. Prepared by U.S. Geological Survey, Western Fisheries Research Center, Klamath Falls Field, Klamath Falls, OR, for the U. S. Bureau of Reclamation, Mid-Pacific Region, Klamath Area Office, Klamath Falls, OR. Contract #00AA200049.

Viola, A.E., and M.L. Schuck. 1995. A method to reduce the abundance of residual hatchery steelhead in rivers. N. Am. J. Fish. Manage. 15(2):488–493.

Voight, H.N., and D.B. Gale. 1998. Distribution of Fish Species in Tributaries of the Lower Klamath River: An Interim Report for 1996. Yurok Tribal Fisheries Program. Technical Report 3. Habitat Assessment and Biological Monitoring Division, Klamath, CA. 71 pp.

Wales, J.H. 1951. The Decline of the Shasta River King Salmon Run. Inland Fisheries Administrative Report 51-18. California Department of Fish and Game. 82 pp.

Walker, W.W. 2001. Development of a Phosphorus TMDL for Upper Klamath Lake, Oregon. Oregon Department of Environmental Quality. March 7, 2001.

Wallace, M. 2000. Length of Residency of Juvenile Chinook Salmon in the Klamath River Estuary. Final Performance Report. Federal Aid in Sport Fish Restoration Act. Project F-51-R; Subproject 32. California Department of Fish and Game, Arcata, CA. [Online]. Available: http://www.krisweb.com/krisweb_kt/biblio/biblio.htm#General Methods [accessed Oct. 9, 2003].

Wallace, M., and B.W. Collins. 1997. Variation in use of the Klamath River estuary by juvenile chinook salmon. Calif. Fish Game Bull. 83(4):132–143.

Walters, C.J. 1997. Adaptive policy design: Thinking at larger spatial scales. Pp. 386–394 in Wildlife and Landscape Ecology: Effects of Pattern and Scale, J.A. Bissonette, ed. New York: Springer.

Walters, C.J., and R. Hilborn. 1978. Ecological optimization and adaptive management. Ann. Rev. Ecol. Syst. 8:157–188.

Walters, C.J., and C.S. Holling. 1990. Large-scale management experiments and learning by doing. Ecology 71(6):2060–2068.

Warner, R.R. 1988. Traditionality of mating site preferences in a coral reef fish. Nature (Lond.) 335:719–721.

Warner, R.R. 1990. Resource assessment versus traditionality in mating site determination. Am. Nat. 135(2):205–217.

Warren, Jr., M.L., and B.M. Burr. 1994. Status of freshwater fishes of the United States: Overview of an imperiled fauna. Fisheries 19(1):6–18.

Waterman, T.T. 1920. Yurok Geography. Berkeley: University of California Press. (Reprinted by Trinidad Museum Society, Trinidad, CA, 1993).

Waters, E.C., B.A. Weber, and D.W. Holland. 1999. The role of agriculture in Oregon's economic base: Findings from a social accounting matrix. J. Agric. Resour. Econ. 24(1): 266–280.

Weber, B., and B. Sorte. 2002. The Upper Klamath basin economy and the role of agriculture. Pp. 213–229 in Water Allocation in the Klamath Reclamation Project, 2001: An Assessment of Natural Resource, Economic, Social, and Institutional Issues in the Upper Klamath Basin, W.S. Braunworth, Jr., T. Welch, and R. Hathaway, eds. Corvallis, OR: Oregon State University Extension Service. [Online]. Available: http://eesc.orst.edu/nagcomwebfile/edmat/html/sr/sr1037/sr1037.html [accessed March 10, 2003].

Weddell, B.J. 2000. Relationship Between Flows in the Klamath River and Lower Klamath Lake Prior to 1910. Prepared for the U.S. Department of Interior, Fish and Wildlife Service, Klamath Basin Refuges. Tule Lake, CA. November 28, 2000.

Weidner, C., B. Nixdorf, R. Heinz, B. Wirsing, U. Neumann, and J. Weckesser. 2002. Regulation of cyanobacteria and microcystin dynamics in polymictic shallow lakes. Arch. Hydrobiol. 155(3):383–400.

Weitkamp, L.A., T.C. Wainwright, G.J. Bryant, G.B. Milner, D.J. Teel, R.G. Kope, and R.S. Waples. 1995. Status Review of Coho Salmon from Washington, Oregon, and California. NOAA Technical Memorandum NMFS-NWFSC-24. Seattle, WA: U.S. Dept. of Commerce, National Oceanic and Atmospheric Administration, National Marine Fisheries Service, Northwest Fisheries Science Center. 258pp. [Online]. Available: http://www.nwfsc.noaa.gov/publications/techmemos/tm24/tm24.htm [accessed Oct. 9, 2003].

Welch, E.B., and T. Burke. 2001. Interim Summary Report: Relationship Between Lake Elevation and Water Quality in Upper Klamath Lake, Oregon. Prepared by R2 Resource Consultants, Inc., Redmond, WA, for the Bureau of Indian Affairs. Portland, OR. March 23, 2001.

Wells, H.L. 1881. History of Siskiyou County, California. Oakland, CA: D.J. Stewart and Co. 240 pp. (reprinted in 1971 by Siskiyou Historical Society).

Wells, R.E., and R.W. Simpson. 2001. Northward migration of the Cascadia forearc in the northwestern U.S. and implications for subduction deformation. Earth, Planets Space 53(4):275–283.

Wells, R.E., C.S. Weaver, S. Craig, and R.J. Blakely. 1998. Fore-arc migration in Cascadia and its neotectonic significance. Geology 26(8):759–762.

Welsh, Jr., H.H., G.R. Hodgson, B.C. Harvey, and M.F. Roche. 2001. Distribution of juvenile coho salmon in relation to water temperatures in tributaries of the Mattole River, California. N. Am. J. Fish. Manage. 21(3):464–470.

West, J.R. 1991. A Proposed Strategy to Recover Endemic Spring-Run Chinook Salmon Populations and Their Habitats in the Klamath River Basin. File Report. U.S. Department of Agriculture Forest Service, Pacific Southwest Region. 25 pp. [Online]. Available: http://www.krisweb.com/krisweb_kt/biblio/biblio.htm#Salmon River Sub-basin [accessed Oct. 9, 2003].

West, J.R., O.J. Dix, A.D. Olson, M.V. Anderson, S.A. Fox, and J.H. Power. 1990. Evaluation of Fish Habitat Conditions and Utilization in Salmon, Scott, Shasta and Mid-Klamath Sub-Basin Tributaries 1988–1989. Annual Report for Interagency Agreement 14-16-0001-89508. U.S. Department of Agriculture Forest Service, Klamath National Forest, Yreka, CA. 90 pp.

Wetzel, R.G. 2001. Limnology: Lake and River Ecosystems, 3rd Ed. San Diego, CA: Academic Press.

Wetzel, R.G., and G.E. Likens. 2000. Limnological Analyses, 3rd Ed. New York: Springer.

White, R. 1991. It's Your Misfortune and None of My Own: A History of the American West. Norman, OK: University of Oklahoma Press.

Wilhere, G.F. 2002. Adaptive management in habitat conservation plans. Conserv. Biol. 16(1):20–29.

Williams, G., ed. 1971. Peter Skene Ogden's Snake Country Journals 1827–28 and 1828–29. Publications of the Hudson's Bay Record Society No. 28. London: Hudson's Bay Record Society.

Williamson, C.E., R.S. Stemberger, D.P. Morris, T.M. Frost, and S. G. Paulson. 1996. Ultraviolet radiation in North American lakes: Attenuation estimates from DOC measurements and implications for plankton communities. Limnol. Oceanogr. 41(5):1024–1034.

Wood, T.M. 2001. Sediment Oxygen Demand in Upper Klamath and Agency Lakes, Oregon, 1999. Water-Resources Investigations Report 01-4080. Portland, OR: U.S. Dept. of the Interior, U.S. Geological Survey.

Wood, T.M., G.J. Fuhrer, and J.L. Morace. 1996. Relation between Selected Water-Quality Variables and Lake Level in Upper Klamath and Agency Lakes, Oregon. Water-Resources Investigations Report 96-4079. Portland, OR: U.S. Dept. of the Interior, U.S. Geological Survey.

Wu, J., R.M. Adams, and W.G. Boggess. 2000. Cumulative effects and optimal targeting of conservation efforts: Steelhead trout habitat enhancement in Oregon. Am. J. Agric. Econ. 82(2):400–413.

Appendix A

Statement of Task

The committee will review the government's biological opinions regarding the effects of Klamath Project operations on species in the Klamath River Basin listed under the Endangered Species Act, including coho salmon and shortnose and Lost River suckers. The committee will assess whether the biological opinions are consistent with the available scientific information. It will consider hydrologic and other environmental parameters (including water quality and habitat availability) affecting those species at critical times in their life cycles, the probable consequences to them of not realizing those environmental parameters, and the inter-relationship of these environmental conditions necessary to recover and sustain the listed species.

To complete its charge, the committee will:

1. Review and evaluate the science underlying the Biological Assessments (USBR 2001a,b) and Biological Opinions (USFWS 2001; NMFS 2001).
2. Review and evaluate environmental parameters critical to the survival and recovery of listed species.
3. Identify scientific information relevant to evaluating the effects of project operations that has become available since USFWS and NMFS prepared the biological opinions.
4. Identify gaps in the knowledge and scientific information that are needed to develop comprehensive strategies for recovering listed species and provide an estimate of the time and funding it would require.

A brief interim report will be provided by January 31, 2002. The interim report will focus on the February 2001 biological assessments of the Bureau of Reclamation and the April 2001 biological opinions of the U.S. Fish and Wildlife Service and National Marine Fisheries Service regarding the effects of operations of the Bureau of Reclamation's Klamath Project on listed species. The committee will provide a preliminary assessment of the scientific information used by the Bureau of Reclamation, the Fish and Wildlife Service, and the National Marine Fisheries Service, as cited in those documents, and will consider to what degree the analysis of effects in the biological opinions of the Fish and Wildlife Service and National Marine Fisheries Service is consistent with that scientific information. The committee will identify any relevant scientific information it is aware of that has become available since the Fish and Wildlife Service and National Marine Fisheries Service prepared the biological opinions. The committee will also consider any other relevant scientific information of which it is aware.

The final report will thoroughly address the scientific aspects related to the continued survival of coho salmon and shortnose and Lost River suckers in the Klamath River Basin. The committee will identify gaps in the knowledge and scientific information that are needed and provide approximate estimates of the time and funding needed to fill those gaps, if such estimates are possible. The committee will also provide an assessment of scientific considerations relevant to strategies for promoting the recovery of listed species in the Klamath Basin.

Appendix B

Committee on Endangered and Threatened Fishes in the Klamath River Basin

WILLIAM M. LEWIS JR. (Chair) is professor and director of the Center for Limnology, Cooperative Institute for Research in Environmental Sciences (CIRES) at the University of Colorado. Dr. Lewis earned his PhD from Indiana University (1973) with emphasis on limnology. His research interests, as reflected by over 160 journal articles and books, include productivity and other metabolic aspects of aquatic ecosystems, aquatic food webs, composition of biotic communities, nutrient cycling, and the quality of inland waters. The geographic extent of Dr. Lewis's work encompasses not only the montane and plains areas of Colorado but also Latin America and southeast Asia, where he has conducted extensive studies of tropical aquatic systems. Dr. Lewis has served on many National Research Council committees. He was a member of the National Research Council's Water Science and Technology Board. His current research projects include the use of stable isotopes to define carbon flux in the Orinoco River floodplain, biogeochemistry of the waters of the Orinoco River, metabolic adaptations in planktonic algae, and nutrient regulation in montane waters of the central Rockies.

RICHARD M. ADAMS is professor of agricultural and resource economics at the Oregon State University. Dr. Adams has served as editor of the *American Journal of Agricultural Economics* and associate editor of *Water Resources Research* and the *Journal of Environmental Economics and Management*. He is a member of various government committees dealing with climate change, water resources, and other environmental issues, including service on three National Research Council panels addressing water-resource issues. Dr. Adams's current research interests include the economic effects of air and water pollution, global climate change, and the valuation of nonmarket commodities. He is a Distinguished Fellow of the American Agricultural Economics Association and has published over 150 books, book chapters, and refereed journal articles. Dr. Adams earned his PhD at the University of California Davis (1975).

ELLIS B. COWLING is University Distinguished Professor At-Large of North Carolina State University. He is a forest biologist who became a world leader in air-pollution research and policy. He is director of the Southern Oxidants Study, in which he leads a team of nearly 300 scientists and engineers in a research and assessment program on ozone and particulate-matter pollution in the southeastern states. From 1975 to 1983, he helped to establish the National Atmospheric Deposition Program—the first permanent precipitation-chemistry monitoring program in the United States. Since 1992, he has taught a graduate course and lectured widely on the role of scientists and engineers in public decision-making. Dr. Cowling was elected to membership in the National Academy of Sciences in 1973 and has served on several National Research Council committees and boards.

GENE S. HELFMAN is professor of ecology and a faculty member in the Conservation Ecology and Sustainable Development program at the University of Georgia. He received a BA in zoology from the University of California (1967), an MS in zoology from the University of Hawaii (1973), and a PhD in ecology and systematics from Cornell University (1978). He is on the editorial boards of *COPEIA* and *Environmental Biology of Fishes*. His current projects focus on the conservation of fishes, the effects of land use on fishes, invasive species, homogenization of fish faunas, and behavioral and ecological interactions and their impact on fish conservation.

CHARLES D. D. HOWARD has been an independent consulting engineer since 1969 in water-resources systems analysis. He has provided advice on operations and planning to water and power utilities; provincial, state, and federal governments in Canada and the United States; the United Nations Development Program; and the World Bank. Mr. Howard is the author of many engineering reports and articles in technical journals. In 1998, he received the Julian Hinds Award of the American Society of Civil Engineers. He has participated in a number of National Research Council committees and boards, including the Water Science and Technology Board, 1996–1999. Mr. Howard earned a BS (1960) and an MS (1962) from the University of Alberta and an MS (1966) from the Massachusetts Institute of Technology.

ROBERT J. HUGGETT is professor of zoology and vice president for research and graduate studies at Michigan State University and professor emeritus of marine science at the College of William and Mary. His aquatic-biochemistry research has involved the fate and effects of hazardous substances in aquatic systems with a focus on hydrophobic chemicals and their partitioning in sediment and pore water. From 1994 to 1997, Dr. Huggett was the assistant administrator for research and development for the U.S. Environmental Protection Agency. Dr. Huggett earned his PhD at the College of William and Mary (1977).

NANCY E. LANGSTON is associate professor of environmental studies/forest ecology and management at the University of Wisconsin-Madison. Dr. Langston earned an MPhil at Oxford (1986) and a PhD from the University of Washington (1994). Her research emphasis is on the historical and ecological processes that shape landscape change in western ecosystems. Recent projects include analysis of forest change in fire-adapted ecosystems and restoration alternatives in ponderosa pine forests. Her current projects examine riparian change in the inland West—focusing on the interplay between ranching, irrigation, and wildlife refuge management—and analyses of the potential for adaptive management. She serves on the Board of Directors of the Forest History Society and on the editorial board for *Environmental History*. Her first book, *Forest Dreams, Forest Nightmares: the Paradox of Old Growth in the Inland West*, won the Charles Weyerhaeuser Prize for best book in conservation history (1997), and her forthcoming book is *Where Land and Water Meet: A Western Landscape Transformed.*

JEFFREY F. MOUNT is professor of geology at the University of California, Davis. Dr. Mount's research emphasis is on the geomorphic response of lowland river systems to changes in land use and land cover and the links between hydrogeomorphology and riverine ecology. Current projects include analysis of geomorphology of flood plains, flood-plain response to nonstructural flood-management measures, development of new flood-plain restoration methods, role of hydrologic and sedimentologic residence time in riverine ecosystems, and development of coupled hydrogeomorphic and ecosystem models for environmental monitoring. He earned his PhD from the University of California, Santa Cruz (1980).

PETER B. MOYLE is a professor in the Department of Wildlife, Fish and Conservation Biology at the University of California, Davis. Dr. Moyle earned a BA in zoology from the University of Minnesota (1964), an MS in conservation biology from Cornell University (1966), and a PhD in zoology from the University of Minnesota (1969). His research interests include conservation of aquatic species, habitats, and ecosystems, including salmon;

ecology of fishes of the Sacramento-San Joaquin estuary; ecology of California stream fishes; the impact of introduced aquatic organisms; and the use of flood plains by fish. Dr. Moyle is the author or coauthor of over 170 publications (mostly related to fish in California). His most recent book, *Inland Fishes of California*, was published in 2002.

TAMMY J. NEWCOMB is the Lake Huron Basin coordinator for the Michigan Department of Natural Resources Fisheries Division. In that position, she coordinates ecosystem and watershed management for the Lake Huron drainages and the Lake Huron sport, tribal, and commercial fisheries. Dr. Newcomb is also an adjunct faculty member of the Virginia Polytechnic Institute and State University with a research focus on salmonid population dynamics, watershed and stream habitat management, and stream-temperature modeling. Dr. Newcomb earned her PhD at Michigan State University (1998).

MICHAEL L. PACE is assistant director of the Institute of Ecosystem Studies in Milbrook, New York. Dr. Pace earned his PhD in ecology from the University of Georgia (1981). He has served as chair of the Scientific Advisory Board of the National Center for Ecological Analysis and Synthesis from 2000 to 2001. His research interests focus on aquatic ecosystems. Based on projects conducted in lakes, estuarine mesocosms, and the Hudson River, Dr. Pace's work illustrates that particular species can modify trophic interactions and have enormous influence on the structure and function of ecosystems.

J. B. RUHL teaches law at Florida State University. Professor Ruhl is recognized as a leading authority on endangered-species law and one of the country's most prolific environmental-law scholars. He is coauthor of *The Law of Biodiversity and Ecosystem Management* (Foundation Press 2002). He teaches classes in property, land-use regulation, endangered species, and environmental-business transactions, and he serves as faculty adviser to *The Journal of Land Use & Environmental Law*. He earned his BA (1979) and JD (1982) from the University of Virginia and his LLM from the George Washington University (1986). Professor Ruhl has served as a visiting professor at The George Washington University Law School and a professor at Southern Illinois University School of Law. A member of the American Law Institute and former executive editor of *Natural Resources and the Environment*, he is a former partner in the Austin, Texas, office of Fulbright & Jaworski and is currently of counsel to Smith, Roberts, Elliott & Glen of Austin, Texas.

Index

A

B

C

F

G

H

M

T

U

V

W